Polymer–Graphene Nanocomposites

RSC Nanoscience & Nanotechnology

Series Editors:
Professor Paul O'Brien, *University of Manchester, UK*
Professor Sir Harry Kroto FRS, *University of Sussex, UK*
Professor Ralph Nuzzo, *University of Illinois at Urbana-Champaign, USA*

Titles in the Series:
1: Nanotubes and Nanowires
2: Fullerenes: Principles and Applications
3: Nanocharacterisation
4: Atom Resolved Surface Reactions: Nanocatalysis
5: Biomimetic Nanoceramics in Clinical Use: From Materials to Applications
6: Nanofluidics: Nanoscience and Nanotechnology
7: Bionanodesign: Following Nature's Touch
8: Nano-Society: Pushing the Boundaries of Technology
9: Polymer-based Nanostructures: Medical Applications
10: Metallic and Molecular Interactions in Nanometer Layers, Pores and Particles: New Findings at the Yoctolitre Level
11: Nanocasting: A Versatile Strategy for Creating Nanostructured Porous Materials
12: Titanate and Titania Nanotubes: Synthesis, Properties and Applications
13: Raman Spectroscopy, Fullerenes and Nanotechnology
14: Nanotechnologies in Food
15: Unravelling Single Cell Genomics: Micro and Nanotools
16: Polymer Nanocomposites by Emulsion and Suspension
17: Phage Nanobiotechnology
18: Nanotubes and Nanowires: 2nd Edition
19: Nanostructured Catalysts: Transition Metal Oxides
20: Fullerenes: Principles and Applications, 2nd Edition
21: Biological Interactions with Surface Charge Biomaterials
22: Nanoporous Gold: From an Ancient Technology to a High-Tech Material
23: Nanoparticles in Anti-Microbial Materials: Use and Characterisation
24: Manipulation of Nanoscale Materials: An Introduction to Nanoarchitectonics
25: Towards Efficient Designing of Safe Nanomaterials: Innovative Merge of Computational Approaches and Experimental Techniques
26: Polymer–Graphene Nanocomposites

How to obtain future titles on publication:
A standing order plan is available for this series. A standing order will bring delivery of each new volume immediately on publication.

For further information please contact:
Book Sales Department, Royal Society of Chemistry, Thomas Graham House, Science Park, Milton Road, Cambridge, CB4 0WF, UK
Telephone: +44 (0)1223 420066, Fax: +44 (0)1223 420247
Email: booksales@rsc.org
Visit our website at http://www.rsc.org/Shop/Books/

Polymer–Graphene Nanocomposites

Edited by

Vikas Mittal
*The Petroleum Institute, Chemical Engineering Program, Abu Dhabi,
United Arab Emirates
Email: vmittal@pi.ac.ae*

RSCPublishing

RSC Nanoscience & Nanotechnology No. 26

ISBN: 978-1-84973-567-4
ISSN: 1757-7136

A catalogue record for this book is available from the British Library

Published by The Royal Society of Chemistry,
Thomas Graham House, Science Park, Milton Road,
Cambridge CB4 0WF, UK

Registered Charity Number 207890

For further information see our web site at www.rsc.org

Printed in the United Kingdom by Henry Ling Limited, Dorchester, DT1 1HD, UK

Preface

After clay and nanotubes, graphene has become an immensely important filler for generating polymer nanocomposites. Owing to the possibility of significant improvement of electrical, barrier and mechanical properties of the polymer matrices by the inclusion of very small amounts of graphene, the generated composites are of special attraction. Graphene can also be surface functionalized in order to generate compatibility with a number of polymer matrices and also be used comfortably in *in situ* polymerization as well as controlled polymerization technologies. This book aims to present insights into graphene functionalization and to describe various polymer systems and polymerization methods for generating graphene-filled nanocomposites. Thus, the goal is to present a wide range of possibilities for generating high-potential graphene nanocomposites.

Chapter 1 provides a detailed overview of graphene synthesis and functionalization processes. It also flags up some key challenges such as scaling up production of graphene or graphene-based materials. Chapter 2 demonstrates the means of gelation of graphene oxide. The atom-thick two-dimensional structure of GO sheets makes them gellable at a low critical concentration. The gelation process is observed to be controlled by the balance of bonding and repulsive forces between GO sheets.

In Chapter 3, the basics of the latex-based process for the production of nanocomposites, with examples of its first application to produce CNT–polymer, are given. This is followed by a short overview of the strategies developed to interface graphene and polymers using aqueous-based methodologies. Chapter 4 describes the use of miniemulsion polymerization to generate nanocomposites; this polymerization method allows the formation of polymer lattices containing the graphene nanosheets, which can be exfoliated during the miniemulsion process. Miniemulsions also promote the intercala-

RSC Nanoscience & Nanotechnology No. 26
Polymer–Graphene Nanocomposites
Edited by Vikas Mittal
© The Royal Society of Chemistry 2012
Published by the Royal Society of Chemistry, www.rsc.org

tion of monomers into the modified graphene nanosheets, leading to the formation of exfoliated structures.

Chapter 5 also reviews many kinds of polymer–graphene nanocomposites prepared by *in situ* polymerization, which are advantageous in the functionalization and dispersion of graphene. The obtained nanocomposites exhibit improved properties compared to conventional composites or pure polymers. In Chapter 6, nanocomposites generated by solution-mixing of the graphene oxide filler with chlorinated polyethylene compatibilizers followed by melt-mixing of the solution-mixed master batches with high-density polyethylene are reported. Compatibilizers with two different chlorination contents were used in different amounts in order to analyse their effect on the morphology and properties of resulting nanocomposites. Chapter 7 highlights recent advances in the preparation and applications of pH-sensitive graphene–polymer nanocomposites. Composites can be prepared using either covalent or non-covalent approaches. Chapter 8 reviews recent developments in the preparation, properties and potential applications of dispersible graphene oxide–polymer nanocomposites. In Chapter 9, nanocomposites based on conducting polymers (mainly polyaniline and graphene) prepared by interfacial polymerization are presented. The resulting material was directly obtained as a transparent and free-standing film. In Chapter 11, graphene nanosheet-driven polymer crystallization was reviewed from the perspective of intrachain conformational ordering and under the combined effect of shear flow. Graphene nanosheets exhibited strong nucleating ability in accelerating the crystallization rate of isotactic polypropylene.

Finally, I am indebted to the Royal Society of Chemistry for publishing the book. I dedicate this book to my family and especially to my wife for her continuous support and suggestions for improving the manuscript.

Vikas Mittal
Abu Dhabi

Contents

RSC Nanoscience & Nanotechnology No. 26
Polymer–Graphene Nanocomposites
Edited by Vikas Mittal
© The Royal Society of Chemistry 2012
Published by the Royal Society of Chemistry, www.rsc.org

CHAPTER 1

Graphene Functionalization: A Review

MO SONG* AND DONGYU CAI

Department of Materials, Loughborough University, Loughborough LE11 3TU, United Kingdom
*E-mail: M.Song@lboro.ac.uk

1.1 Introduction

Graphene is a carbon sheet one atom thick, consisting of a two-dimensional honeycomb lattice, which is considered to be the thinnest material in the world.[1,2] It is generally viewed as the basic unit for other carbon materials. Figure 1.1 shows that graphene can become 3D graphite, 1D carbon nanotube (CNT) or 0D fullerenes by the operations of stacking, rolling and wrapping, respectively.[3] Before the discovery of graphene, there was a long-standing argument that strictly 2D crystals are thermodynamically unstable and should not exist. In 2004, this theory was undermined by two physicists (Geim and Novoselov) from Manchester University, who successfully isolated free-standing graphene from 3D graphite using Scotch tape.[1] However, the existence of graphene can compromise with the theory to some extent. One point of view is that strong sp^2 C–C bonds can resist the dislocation of a carbon atom or other crystal defects under thermal fluctuation even at elevated temperature although graphene is a stable material.[4] Another interesting observation was that suspended graphene sheets were not perfectly flat, and out-of-plane deformation in the surface reached 1 nm.[2] The roughening surfaces, with microscopic corrugations in the third dimension, provided the possibility of thermodynamic stability for 2D graphene. This new 2D model

RSC Nanoscience & Nanotechnology No. 26
Polymer–Graphene Nanocomposites
Edited by Vikas Mittal
© The Royal Society of Chemistry 2012
Published by the Royal Society of Chemistry, www.rsc.org

has prompted physicists to discover many exceptional physical phenomena which are not found in other materials; we can only present a brief introduction here.

Graphene has some exceptional physical properties. It is the strongest material ever measured, with a Young's modulus of 1 TPa and tensile stress of 130 GPa, 100 times that of steel.[5] Its thermal conductivity of graphene is high, up to 5000 W m^{-1} K^{-1},[6,7] and its theoretical specific surface area is close to 2630 m^2/g.[8] Its excellent electrical properties are the major reason why graphene has diverted attention from other 2D materials; ambipolar electric properties[1] and quantum Hall effects[9] have been demonstrated in graphene. The carrier mobility historically hits a high value of 200 000cm^2/V S as the electrons transporting across graphene behave like massless relativistic particles.[10] Graphene has been described as a wonder material and may become an exciting multidisciplinary platform attracting physicists, chemists and engineers together to revolutionize the technologies currently used in our society.

Although people have started to dream of seeing graphene-based products, concern remains about future application of graphene, *e.g.* the possibility of low cost and mass production and the assembly of graphene into complex systems with desirable properties. The most difficult handling problem is that

Figure 1.1   Mother of all graphitic forms. Graphene is a 2D building material for carbon materials of all other dimensionalities. It can be wrapped up into 0D buckyballs, rolled into 1D nanotubes or stacked into 3D graphite.[3]

suspended graphene, although it can exist, is unstable and tends to stick together. The functionalization of graphene is considered to be the main route to make suspended graphene stable in a complex environment by introducing a third party into the graphene surface via chemical or physical approaches. We believe that it will play a key role when moving graphene from the laboratory to real-world applications. Progress on functionalizing graphene is making rapid progress, helped by quick learning from experience with CNTs, as both these carbon materials have a similar chemical structure. This chapter attempts to review up-to-date progress in this field and hopefully will benefit those readers who are either already involved with research into graphene or are eager to become involved.

1.2 Fabrication of Graphene

1.2.1 Mechanical Cleavage

The 'Scotch tape' method, as it is sometimes humorously described in the media, is key to the world of graphene. Geim and his coworkers1 used oxygen plasma to etch square mesas (5 μm deep and 20 μm–2 mm width) on top of highly oriented pyrolytic graphite (HOPG), and then pressed the treated HOPG against a layer of wet photoresist spun on the n^+-doped Si substrate topped with a SiO_2 layer. With good adhesion achieved by baking, Scotch tape was used to repeat the action of peeling the mesas to remove thick graphite flakes. Afterwards, the substrate with thin flakes was dipped in acetone and then washed in plenty of water and propanol with the assistance of ultrasound. As a result, the thin flakes (<10 nm thick) attached strongly to the SiO_2 layer owing to Van der Waals and capillary forces. Finally, graphene multilayers (<1.5nm) were selectively detected under atom force microscopy. Later, Geim's group[11] reported another mechanical approach, described as 'rubbing' to successfully extract single-layer graphene from graphite in one step. The of rubbing process was simply described as being similar to drawing with chalk on a blackboard.

Awareness of the problem of low yields, Coleman's group from Dublin attempted to improve the efficiency of mechanical exfoliation in the liquid phase. It was demonstrated by Hernandez *et al.*[12] that graphite could be split into graphene in a well-selected organic solvent by low-power ultrasound, and the concentration of graphene dispersion could reach 0.01 mg/ml with a yield of 1wt%. Repeating the process for recycled graphite sediments potentially allowed improvement of the yield up to 12wt%. The stabilization of graphene was dependent on the surface energy of organic solvents according to the following equation:

$$\frac{\Delta H_{mix}}{V_{mix}} \approx \frac{2}{T_{flake}} (\delta_G - \delta_{sol})^2 \varnothing \qquad (1)$$

where $\delta_i = \sqrt{E^i_{sur}}$ is the square root of the surface energy of phase *i*. The surface energy of graphite is defined as the energy per unit area required to peel two graphene sheets apart. T_{flake} is the thickness of a graphene flake and φ is the graphene volume fraction. The exfoliation takes place when enthalpy of mixing (ΔH_{mix}) is very small, which ensures the negative free energy of mixing (ΔG_{mix}) according to classical thermodynamic theory:

$$\Delta G_{mix} = \Delta H_{mix} - T\Delta S_{mix} \tag{2}$$

where ΔS_{mix} is the entropy of mixing per unit volume. These equations tell us that solvents with surface energy matching that of graphene, such as *N*-methylpyrrolidone, are the best candidates for stabilizing graphene sheets sliced down by ultrasound, ensuring that the graphene–graphene bonding force is balanced by solvent–graphene interaction. Hernandez *et al.*[13] also measured the dispersibility of graphene in 40 solvents and identified the criteria for selecting good solvents to disperse graphene according to basic solubility theory rather than just making rough judgements using surface energy. They found that suitable solvents should have a Hildebrand solubility parameter ~ 23 MPa$^{1/2}$, and Hansen solubility parameters ~ 18 MPa$^{1/2}$, ~ 9.3 MPa$^{1/2}$ and ~ 9.3 MPa$^{1/2}$ for dispersive, polar and hydrogen-bonding components, respectively.

Following this discovery, Coleman's group carried out a series of experiments with the aim of improving the yield. Khan *et al.*[14] reported that the concentration of graphene solution could be increased to 1.2 mg/ml with up to 4wt% monolayers after treating a graphite/NMP mixture continuously with low-power sonication for ~ 460 h. The reduction in mean length (L) and width (W) of graphene flake by sonication ($L \propto t^{-1/2}$ and $W \propto t^{-1/2}$) was the reason for the solvent accommodating more graphene. The length could only be reduced to 1 μm at maximum. A quantitative relation as shown in equation (3) was proposed to relate the concentration (C_G) to sonication time (*t*) as the concentration was defined by dividing mass of average flake by average solvent volume per flake viewed as a sphere with diameter equal to the flake length:

$$C_G \approx \frac{\rho_G LW\tau}{4\pi(\frac{L}{2})^8/3} \approx \frac{6\rho_G}{\pi} \frac{W\tau}{L^2} \propto t^{1/2} \tag{3}$$

where ρ_G is graphitic density and τ is flake thickness. This process was significantly refined to achieve concentrations as high as 63 mg/ml, mainly by redispersing a powder of exfoliated graphene, a few layers thick, formed by filtering the graphene dispersion produced by ultrasonication. Surfactant was used to enhance the yield of graphene in aqueous solutions, a technique adapted from work with CNTs. The dispersion of graphene in sodium cholate aqueous solution was reported to have a concentration of ~ 0.3 mg/ml.[15] The structure of surfactant basically consists of a long tail and an ionic head. Its function here is to charge the surface of graphene flake with ions and generate Coulomb repulsion to prevent the bonding of graphene. The long tail physically attached to graphene creates steric hindrance to prevent aggregation.

It was also reported that a superacid such as chlorosulphonic acid could exfoliate graphite spontaneously without any chemical or physical treatments to increase graphene concentration up to ~2 mg/ml.[16] The mechanism for stabilizing graphene was due to the repulsion between graphene layers induced by acid protonation. Acid strength was key to dispersion quality. The concerns in liquid phase exfoliation are not just the issue of concentration, but include the toxicity and high cost of these organic liquids. Water is an environmentally friendly liquid, but the factors of its wetting ability, surfactants and incompatibility with organic systems will be a bar for future applications. Mechanical exfoliation is simple, but has evident limitations in mass production of graphene for commercial use. This has become a huge challenge for the future of graphene, and it has stimulated massive efforts from chemists to explore a variety of chemical routes to meet the demand for high yields of graphene.

1.2.2 Reduction of Graphene Oxide

1.2.2.1 Chemical Reduction

Apparently, scalability is the catastrophic disadvantage of mechanical exfoliation. Chemists believed that an alternative tool to make graphene in large volume would come from the discovery by Boehm *et al.* in 1962, who found that extremely thin lamellae with a few carbon layers could be produced by reducing a dispersion of graphite oxide (GO) by chemical or thermal means.[17] Chemical reduction took place in the dispersion of GO in dilute alkali with hydrazine, hydrogen sulfide or iron (II) salts, but chemical reduction was not able to repair the damaged carbon structure completely. Here, it is essential to discuss oxidization of graphite using strong chemical oxidants as it led researchers so close to the perfect graphene monolayer nearly half a century ago. The oxidation of graphite should date back to 1859 when B. C. Brodie used the oxidants of potassium chlorate and nitric acid.[18] Nearly 40 years later, L. Staudenmaier[19] improved Bordie's method by changing single feeding of potassium chlorite to a multiple feeding method, and also adding concentrated sulfuric acid to increase the acidity. Nowadays, the most popular method is the one invented by Hummers and Offeman (Hummers method),[20] in which potassium permanganate and concentrated sulfuric acid are used. The defects in the carbon layers of graphite are commonly viewed as the sites for the formation of functional groups after breakdown of π-conjugated structure by oxidants. It is difficult to discern the mechanism of oxidization precisely, because of the intrinsic complexity of graphite defects. For the same reason, it is very challenging to identify the chemical structure of GO as a key to understanding the mechanism of reduction. Figure 1.2 presents a summary of several models so far proposed for the structure of GO. The evolution of these

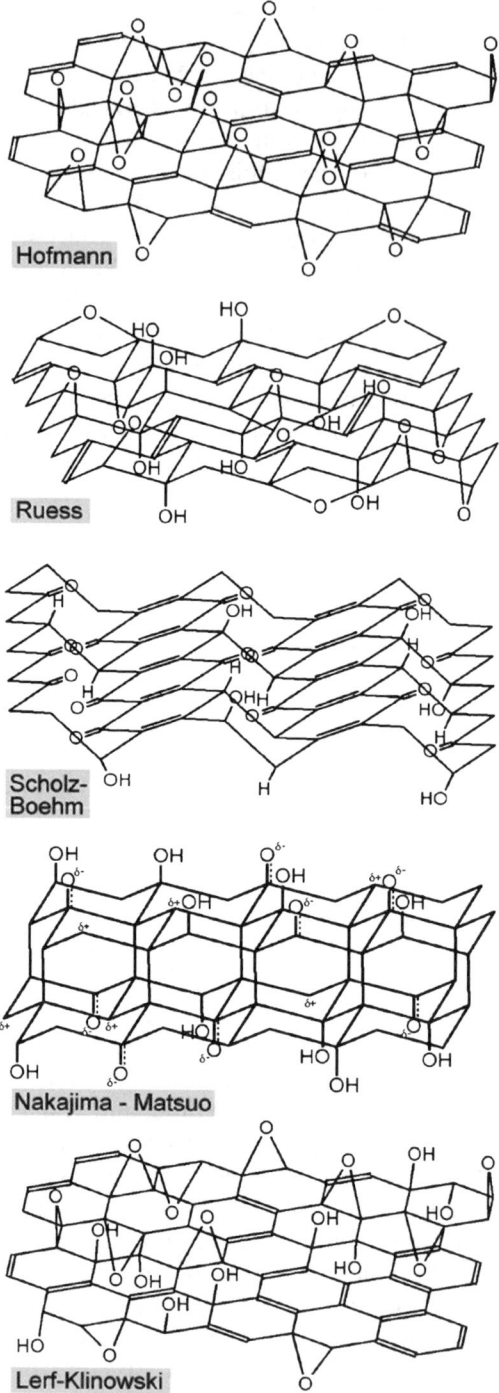

Figure 1.2   Summary of the models proposed for the structure of GO.[24]

models has taken more than half a century since 1939 when Hofmann and Holst proposed the first model that consisted of epoxy groups across a planar carbon structure.[21] In 1946, Ruess modified this model by introducing a hydroxyl group to underpin the observation of hydrogen atoms in GO.[22] The hydroxyl groups repeatedly stood in the fourth position of the aromatic ring and epoxy groups occupied the first and third positions. After another two decades, Scholz and Boehm created a new model that introduced quinoidal species instead of epoxy and ether groups into the cleaved sites of the aromatic backbone.[23] Szabó *et al.*[24] attempted to combine the Ruess and Scholz–Boehm models. Furthermore, the formation of phenolic groups was used to explain the acidity of GO. The fluorination of GO inspired two models analogous to structure of poly(carbon monofluoride) $(CF)_n$ and poly(dicarbon monofluoride) $(C_2F)_n$. Nakajima and Matsuo proposed a stage 2 type $(C_2F)_n$ model containing quasi-pentacovalent carbon in which carbonyl groups hold partial negative charges.[25,26] This model was proposed as negatively charged carbonyl groups are not stable and tend to react with an adjacent charged carbonyl group to form 1,3-ethers. Currently, the most accepted model is that proposed by Lerf, Klinowski and their coworkers.[27,28] The key features of this model include: (1) the basal plane structure of GO contains unoxidized benzene rings and aliphatic six-membered rings with hydroxyl and epoxy (1,2-ether) groups (no 1,3-ethers are detected), and the size of these two regions is dependent on the degree of oxidation; (2) The location of hydroxyl and epoxy groups is very close, and the distribution of functional groups is random; (3) carboxyl groups are formed at the edges of carbon sheets. As presented in Figure 1.3, Ajayan and his coworkers[29] recently confirmed the existence of five- and six-membered ring lactols at the edge of GO in addition to hydroxyl and epoxy groups.

After 2004, Ruoff's group pioneered the preparation of monolayer graphene *via* chemical reduction of GO. The first paper,[30] published in 2006, described the basic route to monolayer graphene in aqueous solution including (1) graphite powder was strongly oxidized to form GO; (2) ultrasonication treatment was further taken to split GO into single layers in aqueous solution; (3) exfoliated GO then went into the reaction with hydrazine to form graphene. Another popular paper from Ruoff's group discussed the possible mechanism of chemical reduction based on the Lerf–Klinowski model of GO.[31] The essence of reduction is the transformation of sp^3-carbon to sp^2-carbon. Figure 1.4 shows that opening epoxide with hydrazine results in further reaction to form an aminoaziridine moiety, and the removal of diimide follows by thermal elimination. In additional, it was believed that the Wharton reaction was one pathway to remove oxygen from quinones next to an epoxide. Several issues should be clarified here. First, the purpose of oxidation is to introduce oxygen-containing functional groups into the basal structure of graphene sheets. The oxygen groups are introduced to reduce the Van der Waals force between the carbon layers within graphite, as well as to improve the stabilization of graphene in water and organic solvents. Secondly, the key difference, compared to Boehm *et al.*'s work, is that ultrasonication is used to

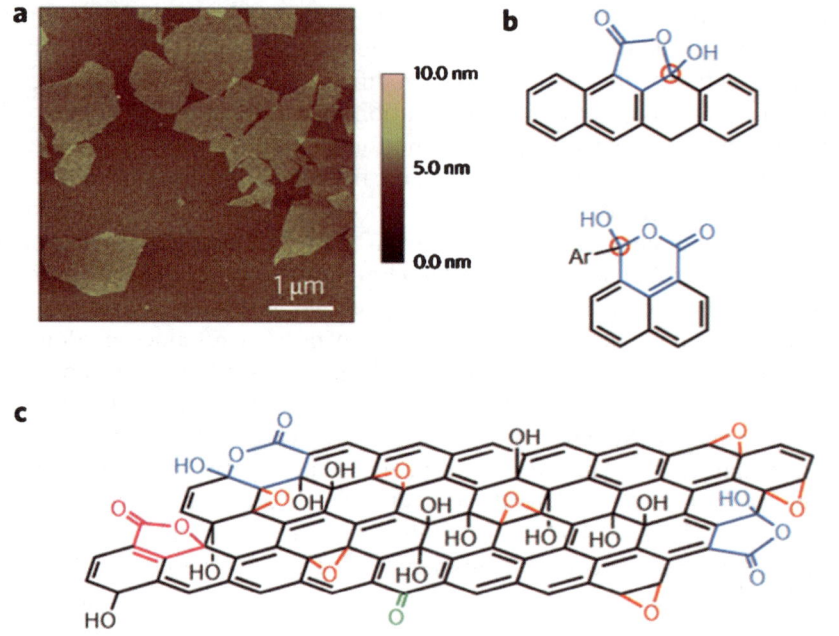

Figure 1.3   Atomic force microscopy (AFM) image and structural model of the graphite oxide (GO) sheets. (a) An AFM image of GO sheets on a silicon substrate. (b) Structure of the five- and six-membered lactol rings. (c) New structural model of GO, taking into account the five- and six-membered lactol rings (blue), ester of a tertiary alcohol (purple), hydroxyl (black), epoxy (red) and ketone (green) functionalities. The model here only shows the chemical connectivity, and not the steric orientation, of these functionalities.[29]

enforce the exfoliation of GO into monolayers for subsequent chemical reduction. Thirdly, this type of graphene appears in publications with a more specific name, reduced graphene oxide (rGO). rGO is not perfect graphene with lower electricity, because small amount of functional groups remain in the basal structure. The stabilization of rGO is the fourth consideration to be taken into account after the functional components are reduced. A polymeric surfactant such as poly(sodium 4-styrensulfonate) needs to be used to stabilize GO in aqueous solution. Finally, Ruoff's work demonstrated a box of tools for testifying the success of reduction. It can readily be observed that the reduction causes a change in colour from yellow (GO) to black (rGO) in water. X-ray photoelectron spectroscopy (XPS) is used to conduct elemental analysis to

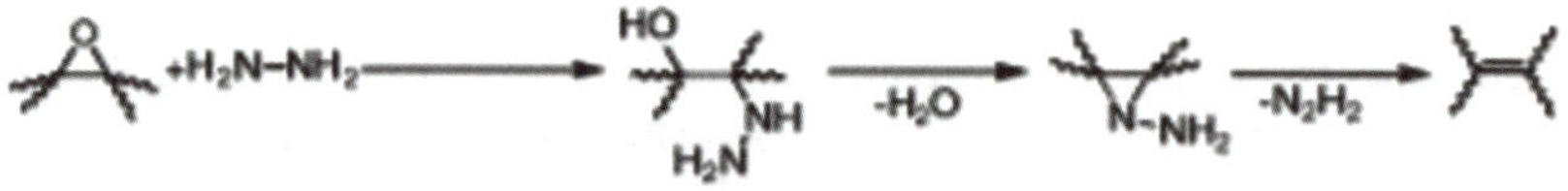

Figure 1.4   A proposed reaction pathway for epoxide reduction with hydrazine.[31]

quantitatively assess the degree of reduction. Other useful tools include measurement of electrical conductivity, Raman spectroscopy, thermogravimetric analysis and solid-state NMR.

The chemical reduction technique still requires improvement aiming at a facile method of producing of high-quality rGO. Owing to the high toxicity of hydrazine and dimethylhydrazine, a great deal of work is going on to explore new reductants, as summarized in **Table 1.1**. Ascorbic acid (vitamin C) was claimed to be an environmentally friendly reductant, which restored the electrical conductivity of rGO up to $\sim 7.7 \times 10^3$ S/m. It seems that when hydroiodic acid or the mixture of hydriodic and acetic acid is used as the reductant the maximum electrical conductivity of rGO is $\sim 3.0 \times 10^4$ S/m. Gao et al.[29] reported an effective reduction protocol as which involves the use of two reductants including sodium borohydride for the deoxygenation of GO followed by concentrated sulfuric acid for dehydration. Thermal annealing is finally used to further reduce the remaining functional groups to obtain highly conductive rGO (2.02×10^4 S/m) with less than 0.5wt% of sulfur and nitrogen

Table 1.1 Summary of reductants for chemical reduction of graphene oxide

Reductants	Reducing temperature (°C)	Reducing time (h)	Electrical conductivity (S/m)	Reference
Hydrazine	100	24	$\sim 2 \times 10^2$	31
Hydroquinone	25	20	–	39
Alkali	50–90	a few minutes	–	40
Sodium borohydride	25	2	$\sim 4.5 \times 10^1$	41
Ascorbic acid (vitamin C)	95	24	$\sim 7.7 \times 10^3$	42
Hydroiodic acid	100	1	$\sim 3 \times 10^4$	43
Hydriodic acid (with acetic acid)	40	40	$\sim 3.0 \times 10^4$	44
*Sulfur-containing compounds	95	3	–	45
Pyrogallol	95	1	$\sim 4.9 \times 10^2$	42
Benzylamine	90	1.5	–	34
Hydroxyl amine	90	1	$\sim 1.1 \times 10^2$	46
Aluminium powder (with hydrochloric acid)	25	0.5	$\sim 2.1 \times 10^3$	47
Iron powder (with hydrochloric acid)	25	6	$\sim 2.3 \times 10^3$	48
Amino acid (L-cysteine)	25	12–72	–	49
Sodium hydrosulfite	60	0.25	$\sim 1.4 \times 10^3$	50
Alcohols	100	24	$\sim 2.2 \times 10^3$	51
Dimethylformamide	153	1	$\sim 1.4 \times 10^3$	35

*Sulfur-containing compounds include $NaHSO_3$, Na_2S, $Na_2S_2O_3$, $SOCl_2$ and SO_2.

impurities. The possible reduction mechanisms proposed for individual cases can be found in the publications listed in **Table 1.1**. However, precise understanding remains an open question for chemists.

With many types of reductants available, another challenge faced by chemists is that the irreversible aggregation of rGO is catastrophic for large-scale production of truly individual graphene sheets. Li *et al.*[32] argued that irreducible carboxylic groups remaining in rGO could act as the charged seeds to create electrostatic repulsion for stabilizing rGO without the need for any surfactant in aqueous solution. The researchers found that this idea worked out once the pH value was simply adjusted to ~10 by adding ammonia to maximize charge density on the surface of unreduced GO. The experimental details were disclosed to warn that overuse of ammonia and hydrazine could result in the destabilization of rGO due to the ionic species formed by the dissociation of excess chemicals, and the stabilization mechanism would fail as the concentration of rGO exceeded 0.5 mg/l. In addition to alkali acting as stabilizer, Si and Samulski[33] reported that *p*-phenyl-SO$_3$H groups chemically introduced into GO could prevent the aggregation of rGO after reduction. In addition, reductants with aromatic rings,[34] polar organic solvents,[35,36] ionic liquids,[37] amphiphilic surfactants[30] and conjugated polymers[38] are also popularly used as external stabilizers for rGO.

1.2.2.2 Other Reducing Methods

In addition to chemical reduction, other routes to reduce GO have also been developed due to interesting chemistry of GO. Schniepp *et al.*[52] reported that direct thermal annealing over 1000 °C could lead to the decomposition of the functional groups in GO to form carbon dioxide, with the electrical conductivity of rGO reaching at least 1×10^3 S/m. The release of carbon dioxide generated vacancies and topological defects on the carbon grid. The mechanism was different from that of chemical reduction, with the concept of repairing the honeycomb structure for GO. Cote *et al.*[53] demonstrated a photothermal reduction of GO by the light generated from a camera flash. It was suggested that a single camera flash at a distance of less than 2 mm could provide enough energy to heat up GO to over 100 °C to switch on the deoxygenating reaction. Electrochemical reduction of GO has also been studied as a chemical-free method. Figure 1.5 illustrates the process for cathodic reduction of GO by electrons, resulting in the deposition of rGO onto the graphite cathode. In the possible reaction mechanism discussed by Zhou *et al.*,[54] it was considered that the addition of hydrogen ions facilitated the reduction due to a protonation effect:

$$Graphene\ oxide + aH^+ + be^- \rightarrow reduced\ graphene + cH_2O$$

Figure 1.6 presents another possible electrochemical reaction proposed by An *et al.*[55] to explain the reduction mechanism when negatively charged GO platelets contact the anode. The electron leaves the platelets to cause oxidation

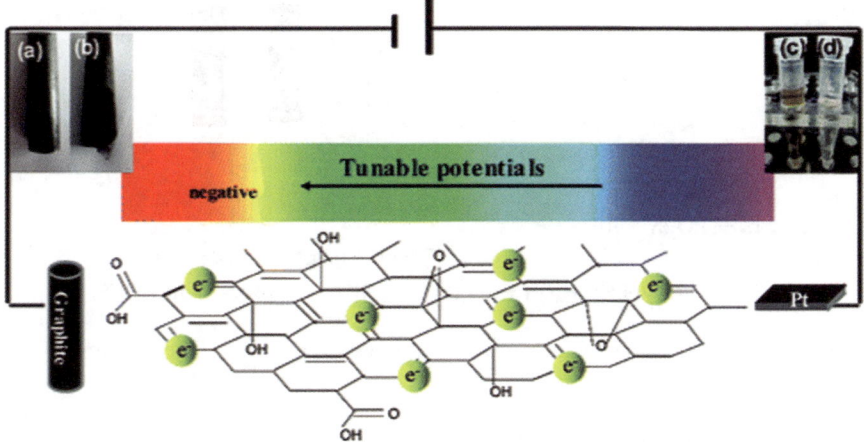

Figure 1.5 Experimental set-up and the optical images of the graphite electrode and the GO suspension before (a,c) and after (b,d) electrochemical reduction.[57]

of carboxylic groups, then carbon dioxide is released following a Koble reaction to form an unpaired electron looking for another unpaired one within the GO framework. The reduction is completed once the pairs of electrons meet to form covalent bonds. The success of both electrochemical strategies is directly indicated by measured electrical conductivity of ~ 8500 and ~ 1043 S/m, respectively. The electrochemical reduction can be flexibly applied to deposit rGO on different substrates and electrodes. As shown in Figure 1.7, William *et al.*[56] explored a photocatalytic route to implement electron transfer reaction to reduce GO, in which titanium dioxide (TiO_2) acted as the source to supply electrons under UV irradiation.

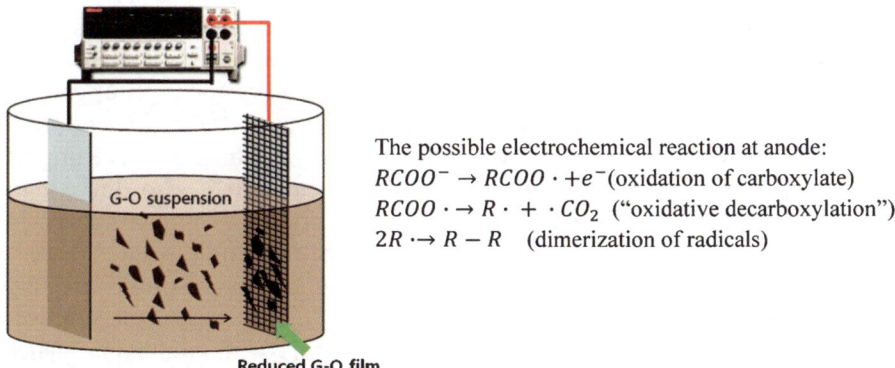

The possible electrochemical reaction at anode:
$RCOO^- \rightarrow RCOO \cdot + e^-$ (oxidation of carboxylate)
$RCOO \cdot \rightarrow R \cdot + \cdot CO_2$ ("oxidative decarboxylation")
$2R \cdot \rightarrow R - R$ (dimerization of radicals)

Figure 1.6 Schematic diagram of the electrophoretic deposition (EPD) process.[55]

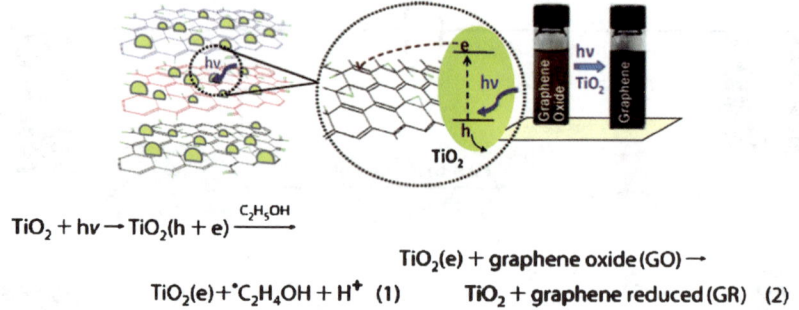

$$TiO_2 + hv \rightarrow TiO_2(h + e) \xrightarrow{C_2H_5OH}$$

$$TiO_2(e) + {}^{\bullet}C_2H_4OH + H^+ \quad (1)$$

$$TiO_2(e) + graphene\ oxide\ (GO) \rightarrow$$

$$TiO_2 + graphene\ reduced\ (GR) \quad (2)$$

Figure 1.7 TiO$_2$–graphene composite and its response under UV excitation[56]

1.2.3 Chemical Vapour Deposition

In contrast to the 'top down' methods discussed above, chemical vapour deposition (CVD) is a 'bottom-up' method capable of producing large-area graphene film on targeted substrates. As with chemical reduction, researchers are inspired by revisiting the discovery reported in the 1960s in which CVD was used to form layers of graphite on Ni and Pt surfaces. With controlling synthesizing parameters, CVD has been recently reported to grow monolayer films of graphene from hydrocarbon sources on various transition metal substrates including Ru,[58] Ir,[59] Co,[60] Ni,[61] Pt[60] and Pd.[60] The choice of metal substrates depends on cost, grain size, etchability and acceptance by the semiconductor industry. For this reason, inexpensive polycrystalline Ni[62–64] and Cu[65] substrates are generating much more interests than other candidates. The criteria for assessing the quality of graphene film include thickness, area, continuousness, evenness and conductivity. CVD growth of graphene refers to thermal decomposition of hydrocarbon precursors catalysed by transition metals at high temperature, which generally undergo the nucleation of graphene islands on the transition metals followed by growth of graphene film with different lattice orientations. The deposition mechanism is closely related to the solubility of carbon in the transition metal. For a transition metal with high carbon solubility such as Ni, CVD growth of graphene film is ruled by surface segregation of carbon at first followed by the precipitation of carbon upon cooling from metastable carbon–metal solid solution.[63] The precipitation of extra carbon to the surface is thought to be the reason for thickening of graphene film over entire polycrystalline Ni substrates. It is also difficult to control the thickness uniformly because the diffusion of carbon is uneven in the of grains and boundary regions. Cu is a kind of transition metal with very low carbon solubility. It has been proved by using a carbon isotope labelling technique that the formation of graphene film is due to catalytic decomposition of gaseous hydrocarbon sources on the Cu surface with minimal influence of surface segregation.[66] This makes it possible for commercial Cu foil to take up uniform monolayer graphene film by CVD over a large area (1 cm × 1 cm), as first discovered by Ruoff and his coworkers.[65] In their follow-up study, more details of the growth of graphene were found by studying the effect of deposition parameters such as temperature (T), methane flow (J_{Me}) and partial pressure (P_{Me}) on the growth rate, domain size and surface coverage of graphene on Cu foils.[67] The scanning electron microscopy (SEM) images shown in Figure 1.8 clearly disclose that high T with low J_{Me} and P_{Me} yield a low density of graphene nuclei, and thus a large domain size. However, the graphene coverage rate was found to decrease with decreasing P_{Me}, and even graphene nuclei could not be formed when J_{Me} and P_{Me} was decreased beyond a critical value. The conclusion was that the surface coverage was dependent on T, J_{Me} and P_{Me} because they determined the saturation level of decomposed methane on the Cu surface. An undersaturated surface resulted in the failure of nucleating graphene. In the case of a saturated surface holding a limited amount of decomposed methane, graphene nuclei grew to a certain island size without any chance to extend. Only a supersaturated surface provided sufficient carbon species to form a continuous film fully covering the Cu substrate. It is believed that a weak interaction between graphene and Cu contributes to the minimal structural disruption formed in graphene film across boundary regions. As summarized by Mattevi *et al.*,[68] CVD conditions including growth pressure and temperature, pre-annealing, H_2/CH_4 flow ratio, growth time and cooling rate have been investigated for the growth of graphene on Cu.

In addition to CVD methods catalysed by metal substrates, epitaxial graphene also can be grown on insulating SiC (0001) substrates by annealing at

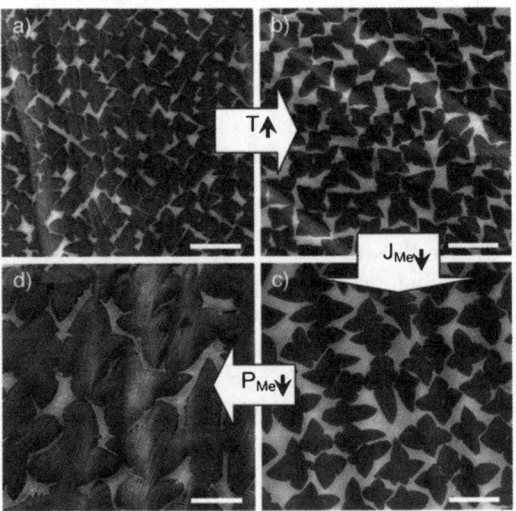

Figure 1.8 SEM images of partially grown graphene under different growth conditions: T (°C)/J_{Md} (sccm)/P_{Me} (mTorr): (a) 985/35/460, (b) 1035/35/460, (c) 1035/7/460, (d) 1035/7/160. Scale bars are 10 μm.[67]

high temperature of 1000–1600 °C, which results in the sublimation of silicon atoms and graphitization of remaining C atoms.[69,70] Instead of studying metal substrates, Tour and his co-workers[71] turned their focus to solid carbon sources, and found that a single uniform layer of graphene could formed on a Cu substrate as spin-coated poly(methyl methacrylate) (PMMA) thin film (~100 nm) was treated at 800–1000°C for 10 min with reductive gas flow (H$_2$/Ar) and under low-pressure conditions. C atoms that extruded from the decomposing PMMA during growth were removed by flowing hydrogen gas to avoid the formation of multilayer graphene. Interestingly, fluorene (C$_{13}$H$_{10}$) and sucrose (table sugar, C$_{12}$H$_{22}$O$_{11}$) also could be used as solid carbon sources to grow monolayer graphene on a Cu catalyst under the same growth conditions as was used for the PMMA-derived graphene.

1.2.4 Synthesis of Graphene Nanoribbons (GNRs)

Precisely controlling the structure of graphene is another challenge, since the graphene obtained by 'top down' methods has a random and irregular shape. The term graphene nanoribbons (GNRs) describes a type of quasi-one-dimensionally structured graphene with an atomically smooth edge. Graphene monolayer is a zero-gap semiconductor. Opening an energy gap is important for the application of graphene in field-effect transistors. Quantum-confined semi-characteristics with bandgap are observed in GNRs as electron transport is confined to a quasi-one-dimensional structure with narrow width.[72,73] Currently, several strategies have been reported to make GNRs.

In a lithographic technique,[73,74] GNR structure was patterned by e-beam lithography followed by oxygen plasma. E-beam resist, hydrogen silsesquiox-ane, was used to protect the GNRs from high-energy damage. Thus, the width of GNRs could be flexibly controlled in the range of 10–100 nm. Tapasztó *et al.*[75] reported a more precise type of lithography conducted in a scanning tunnelling microscope (STM) which was capable of downsizing the width of GNRs to a few nanometres and controlling the crystallographic orientation of the ribbons. The STM tips did the cutting job under a constant bias potential significantly higher than that used for imaging. Similarly, atomic force microscopy (AFM) was another useful tool reported to cut graphene.[76]

Li *et al.*[77] reported chemical unzipping of graphitic materials driven by oxidation. It was believed that cracks were initiated by lining up epoxy groups instead of individual oxygen groups in GO. A sonochemical route was developed to unzip suspended graphene[72] and gas-phase oxidized CNTs[78] in organic solvents. Sonochemistry and ultrahot gas bubbles involved in sonication were responsible for breaking down the graphitic structure. Unzipping of CNTs was also reported to form GNRs by using plasma etching[79] and strong oxidization.[80] The unzipping procedure is schematically illustrated in Figures 1.9 and 1.10, from which it can be seen that scalability is the advantage of oxidization-assisted unzipping. In other research, metallic nanoparticles such as silver,[81] nickel[82] and iron[83] were found to act as knives to

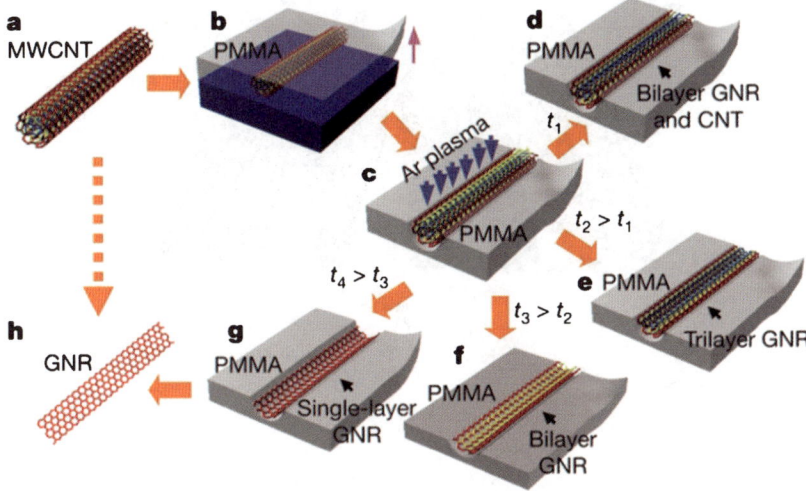

Figure 1.9 Making GNRs from CNTs. (**a**) A pristine multiwalled carbon nanotube (MWCNT) was used as the starting raw material. (**b**) The MWCNT was deposited on a Si substrate and then coated with a PMMA film. (**c**) The PMMA–MWCNT film was peeled from the Si substrate, turned over and then exposed to an Ar plasma. (**d–g**) Several possible products were generated after etching for different times: GNRs with CNT cores were obtained after etching for a short time t_1 (**d**); tri-, bi- and single-layer GNRs were produced after etching for times t_2, t_3 and t_4, respectively ($t_4 > t_3 > t_2 > t_1$; **e–g**). (**h**) The PMMA was removed to release the GNR.[79]

cut graphene according to the hydrogenation mechanism illustrated in Figures 1.11. Graphene and metallic nanoparticles played as carbon sources and catalyst in the reaction, respectively.

A synthetic protocol has also been proposed for bottom-up fabrication of GNRs from aromatic molecules, making it possible to designing the structure of GNRs at the atomic level. Yang *et al.*[84] reported a synthetic route in the liquid state as shown in Figure 1.12. The starting monomer, 1,4-diiodo-2,3,5,6-tetraphenylbenzene (1), first reacted with 4-bromophenylboronic acid (a) via Suzuki–Miyaura coupling to form a hexaphenylbenzene derivative (2). After twofold lithiation with n-butyllithium, the compound (2) reacted with 2-isopropoxy-4,4,5,5-tetramethyl-^{1,3,2}dixaborolane to yield *bis*-boronic ester (3). Then, the compound (3) was polymerized with diiodobenzene to form polyphenylenes (4). Finally, cyclodehydrogenation of polyphenylenes took place with FeCl$_3$ as oxidative reagent at room temperature to form graphene-like polymers (5). Cai and Ruffieux *et al.*[85] provided a thermally activated route (Figure 1.13) to fabricate GNRs on an Au(1,1,1) surface following the steps of dehalogenation, radical addition (C–C coupling) and cyclodehydrogenation. The shape of GNRs could be designed as straight or chevron-type, requiring bianthryl and tetraphenyl-triphenylene precursor monomers, respec-

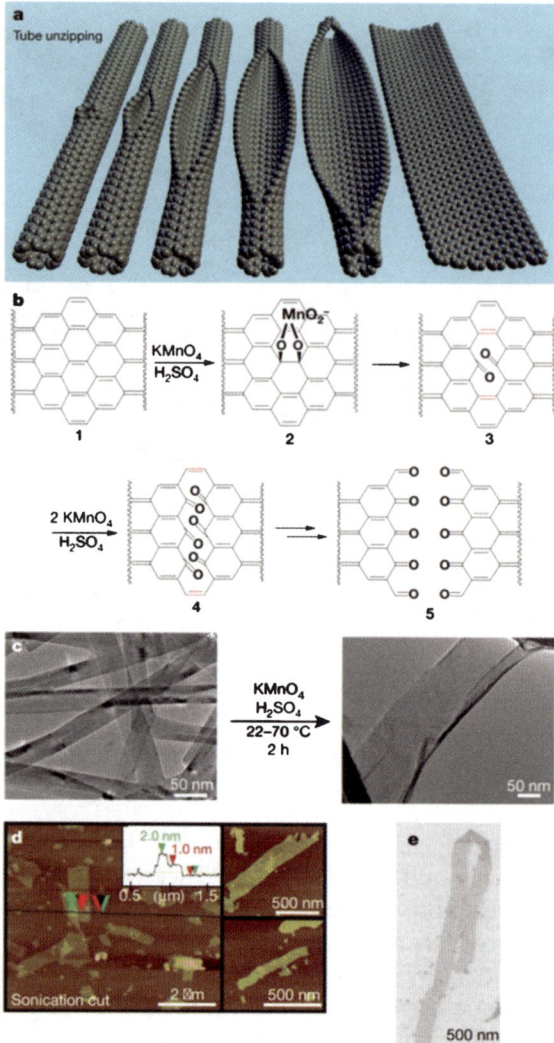

Figure 1.10 Nanoribbon formation and imaging. (**a**) Representation of the gradual unzipping of one wall of a carbon nanotube to form a nanoribbon. Oxygenated sites are not shown. (**b**) The proposed chemical mechanism of nanotube unzipping. The manganate ester in **2** could also be protonated. (**c**) TEM images depicting the transformation of MWCNTs (left) into oxidized nanoribbons (right). The right-hand side of the ribbon is partly folded onto itself. The dark structures are part of the carbon imaging grid. (**d**) AFM images of partly stacked multiple short fragments of nanoribbons that were horizontally cut by tip-ultrasonic treatment of the original oxidation product to facilitate spin-casting onto the mica surface. The height data (inset) indicates that the ribbons are generally single layered. The two small images on the right show some other characteristic nanoribbons. (**e**) SEM image of a folded, 4-μm single-layer nanoribbon on a silicon surface.[80]

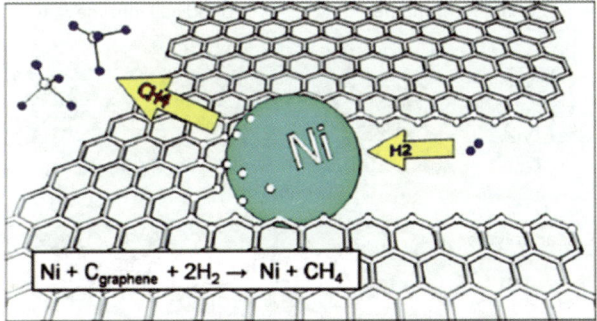

Figure 1.11 Cartoon of a Ni particle etching a graphene sheet (not to scale). Ni nanoparticles absorb carbon from graphene edges which then reacts with H_2 to create methane. (Inset) Summary of the hydrogenation reaction that drives the etching process.[82]

tively. The width of the GNRs was around 1.5 nm according to STM measurement.

1.2.5 Other Methods

Apart from the categories discussed above, some novel methods are explored with the objective of large-scale production of graphene. Wang *et al.*[86]

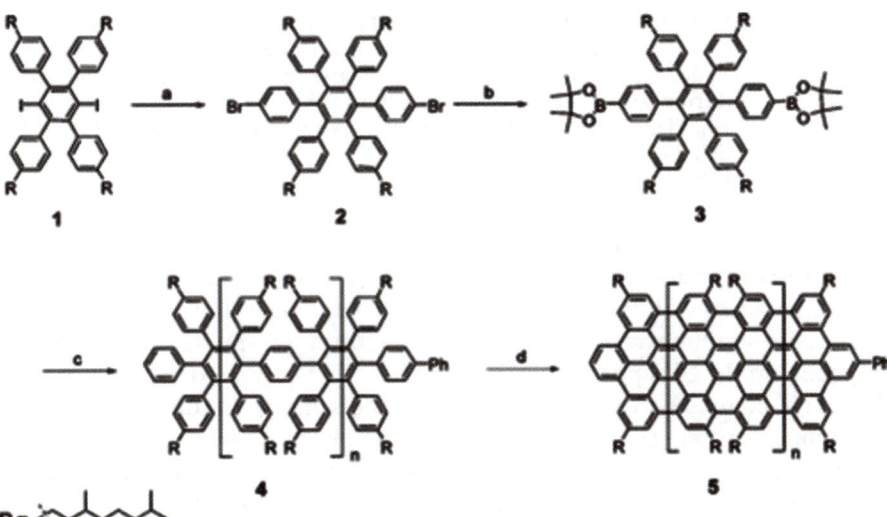

Figure 1.12 Synthesis of graphene nanoribbon. Reagents and conditions: (a) 4-bromophenylboronic acid, Pd (PPh$_3$)$_4$, aliquot 336, K$_2$CO$_3$, toluene, 80°C, 24h, 93%. (b) (i) n-BuLi, THF, -78°C, 1h; (ii)2-isopropoxy-4,4,5,5- tetramethyl ^{1,3,2} dioxaborolane, rt, 2 h, 82%. (c) compound **1**, Pd (PPh$_3$)$_4$, aliquot 336, K$_2$CO$_3$, toluene/H$_2$O reflux, 72 h, 75%. (d) FeCl$_3$, CH$_2$Cl$_2$/CH$_3$NO$_2$, 25°C, 48 h, 65%.[84]

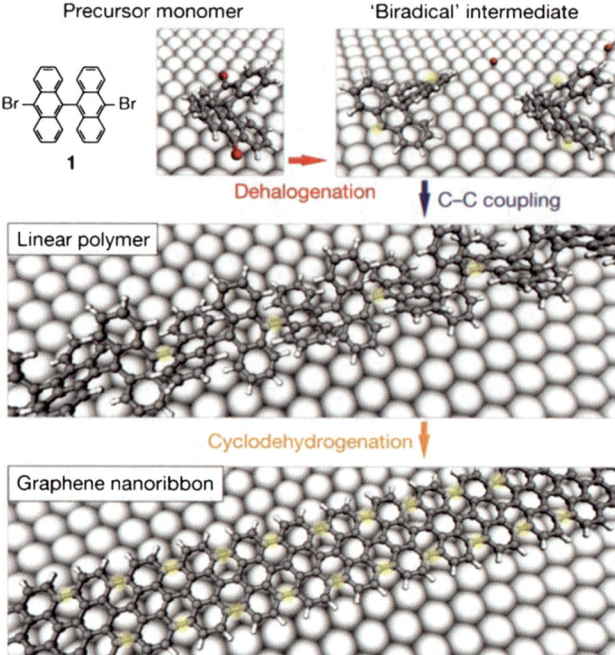

Figure 1.13   Bottom-up fabrication of atomically precise graphene nanoribbons (GNRs). Basic steps for surface-supported GNR synthesis, illustrated with a ball-and-stick model of the example of 10,10′-dibromo-9,9′-bianthryl monomers (**1**). Grey, carbon; white, hydrogen; red, halogens; underlying surface atoms shown by large spheres. Top, dehalogenation during adsorption of the dihalogen functionalized precursor monomers. Middle, formation of linear polymers by covalent interlinking of the dehalogenated intermediates. Bottom, formation of fully aromatic GNRs by cyclodehydrogenation.[85]

reported that graphene nanosheets with 2–10 layers could be produced by arc evaporation of a graphite rod in air. The arc discharge was performed in an electric arc oven with two graphite-rod-based electrodes and a steel chamber. The anode graphite rod was consumed to form several tens of grams of graphene products deposited in the chamber. The proposed mechanism was based on comparative studies on the arc discharge in different gaseous environments such as air, N_2, CO and Ar. It was found that graphene nanosheets, carbon nanohorns and nanospheres were formed in a competitive way, and the yield of nanosheets increased with increasing initial gas pressure as the curvature of the graphene layer was believed to be inhibited by the high pressure. Another mechanism was given by Subrahmanyam *et al.*[87] to explain the growth of graphene with 2–4 layers by arc evaporation of graphite electrodes in a hydrogen atmosphere. The role of hydrogen atoms was to terminate the dangling bonds on the edge of graphene and prevent the formation of a closed structure. Under an atmosphere of diborane[87] and pyridine[87] (or NH$_3$[88]), arc discharge also could lead to the formation of boron- or nitrogen-doped graphene, respectively. A solvothermal reaction was also reported to achieve mass production of graphene at the gram scale.[89] A solvothermal graphene precursor (metal alkoxide encapsulated with ethanol in a clathrate-like structure) was prepared by heating a 1:1 molar ratio of sodium and ethanol in a sealed reactor vessel at 220 °C for 72 h. Single-layer graphene was fabricated by low-temperature flash pyrolysis of the metastable ethanol-rich precursor followed by gentle sonication in ethanol.

1.3 Functionalization of Graphene

First, it is worth looking at the electronic structure of graphene. The basic chemical unit of graphene is hexagonal rings consisting of carbon atoms in the sp^2 hybridization state. The carbon atoms are covalently connected with each other by σ bonds as result of pairing one electron in a 2p orbital. Another electron in a 2p orbital is delocalized across the carbon atoms to form π bonds. Thus, graphene, to some extent, can be viewed as a conjugated macromolecule. Non-covalent functionalization of graphene has the advantage of preventing damage to the conjugated structure. The chosen stabilizers, such as polyvinylpyrrolidone,[38] 1-pyrenebutyrate,[90] and 7,7,8,8- tetracyanoquinodi-methane,[91] generally consist of conjugated structure that can form a strong interaction with the basal plane of graphene *via* π-π stacking. In this section, our focus will be on the covalent functionalization of graphene which involves the transformation from sp^2 to sp^3 hybridization after π bonds are attacked by chemical species.

1.3.1 Functionalization of Graphene with Organic Species

Theoretical organic chemistry provides a box of tools to treat the π bonds in graphene with the purpose of introducing functional groups. First, oxidization of graphite as discussed in section 1.2.2.1 should be the most common approach to generate oxygen-based functional groups onto graphene sheets. It gives the chance to exfoliate graphene sheets in water and organic solvents. It is not difficult to disperse GO in water, due to the hydrophilic nature of the oxygen groups on the surface of carbon sheets.[92,93] However, the solubility of GO in organic solvents is complicated, which initially triggers the research work on functionalizing graphene. Ruoff and his coworkers[92] found that the GO made by the Hummers method cannot be directly exfoliated into dimethylformamide (DMF), which is a strongly polar organic solvent. They used phenyl isocyanate to terminate the hydrophilic oxygen groups and achieved the exfoliation of isocyanate-treated graphene in DMF. We modified the Hummers method and used expandable graphite originally intercalated with sulfuric acid to replace natural graphite as a starting material for oxidization.[94] The resulting GO can be exfoliated into single-layer GO sheets in DMF directly without any chemical treatment, indicating that GO obtained from this experiment is amphiphilic. Paredes *et al.*[95] suggested that organic solvents with higher polarity generally have better dispersing ability. They found that suitable organic solvents to disperse GO include DMF, *N*-methyl-2-pyrrolidone, tetrahydrofuran and ethylene glycol. In contrast, some solvents such as dichloromethane, n-hexane, methanol and *o*-xylene failed to accommodate GO at all. Other solvents including acetone, ethanol, 1-propanol, dimethyl sulfoxide and pyridine could stabilize GO for very short periods, from hours to a few days. Generally, the introduction of non-polar moieties such as alkyl chains to graphene is the way to enhance the solubility of graphene in non-polar organic solvents.[96] Graphene will become amphiphilic after being attached to amphiphilic molecules.[97]

Further chemical treatment of these oxygen seeds, including carboxylic, epoxy and hydroxyl groups, forms a large number of graphene derivatives with broader applications. The amidation[96,98] and esterification[99,100] taking place at carboxylic sites are commonly used to create covalent linkages with organic moieties containing hydroxyl and amide groups. In some cases, the reactivity of carboxylic groups needs to be activated *via* a coupling reaction with thionyl chloride. The carboxylic/hydroxyl groups were also reported to undergo an addition reaction with isocyanate derivatives to form the covalent bonds of an amide/carbamate ester. As the monoisocyanates are used, graphene can be functionalized with aliphatic and aromatic groups connected to isocyanate groups.[92] The use of diisocyanate derivatives can lead to the introduction of organic isocyanate groups onto graphene, providing a further chance to functionalize graphene with organic molecules and polymer chains.[101] The epoxy groups on GO can be converted to hydroxyl groups after a nucleophilic ring-opening reaction with amine groups.[102,103]

Secondly, radical reaction is another route to open π bonds for the attachment of functional groups. It is not difficult to source the ways to produce radicals from well-established previous knowledge, typically featuring the thermal decomposition or photolysis of organic peroxides or azo compounds and redox reaction of hydrogen peroxide and iron. The chemical tails of the radicals will become the functional attachment on to graphene.[104,105] Moreover, Shen *et al.*[105] found that a radical on the surface of graphene could be terminated by atmospheric oxygen or carbon dioxide, which was claimed as a possible new way to make GO. As shown in Figure 1.14, melted benzoyl peroxide (BPO) was intercalated into graphite due to aromatic interactions between graphene and BPO. The decomposition of the intercalated BPO at high temperature could generate volatile gaseous species to violently expand the carbon layer of graphite and meanwhile form some free radicals on the surface of graphene. Thirdly, electron transfer chemistry can be used to understand the reaction between graphene and diazonium salt, resulting in covalent attachment of aryl groups to the basal carbon atoms.[106,107] Sharma *et al.*[106] found that single-layer graphene was almost 10 times as reactive as bi- or multilayers of graphene, and, for the first time, observed that the reactivity of graphene edges was at least 2 times higher

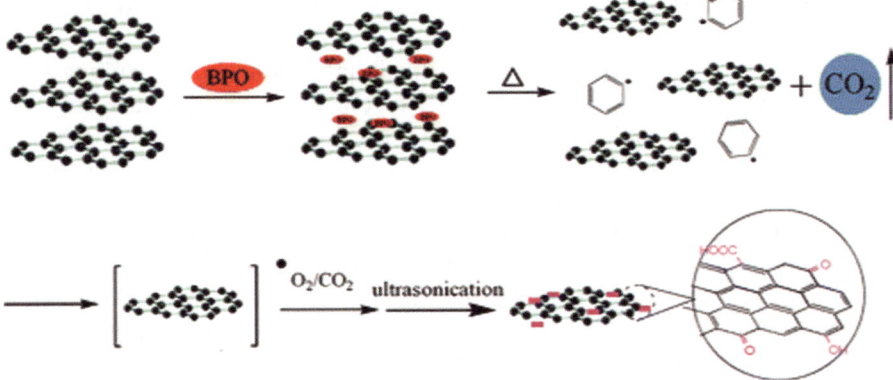

Figure 1.14 Possible mechanism for the preparation of graphene oxide by Ye *et al.* and his co-workers.[105]

than that of the central interior of a single graphene sheet. All these reactivity differences are down to the density of electronic states which determines electron transfer rates in different situations. The final type of reaction discussed here is addition reactions to the C=C bonds in graphene. It was reported that hydrogen atoms produced by plasma treatment of a low-pressure (0.1 mbar) hydrogen–argon mixture (10%H_2) enabled the hydrogenation of graphene to form insulating graphane.[108] Graphane was thermally unstable and the hydrogenation was also found to be reversed at a high temperature of 450 °C. Fluorination of graphene could be carried out by directly exposing graphene to a gas such as xenon difluoride or generating active fluorine atoms by plasma treatment of fluorine gas (CF4 and SF4).[109] Other routes for the addition of graphene include nitrene and 1,3-dipolar cycloaddition of graphene by using azido-phenylalanine[110] and azomethine ylides,[111] respectively. Figure 1.15 shows the functionalization of graphene using nitrene chemistry.

Figure 1.15 General strategy for the preparation of functionalized graphene nanosheet by nitrene chemistry.[110]

It can be seen that functionalized graphene sheets are very stable in the organic solvent.

From the technological aspect, the functionalization above requires pre-exfoliation of GO or graphite, otherwise functional groups can only be attached to the external carbon layers. Consequently, the yield is limited. Englert *et al.*[112] developed a much more efficient method as shown in Figure 1.16 for covalent bulk functionalization of graphene. First, the reductive treatment of graphite with solvated electrons was performed in an inert solvent, 1,2-dimethoxyethane (DME), by using the liquid alloy of sodium and potassium as a very potent electron source. During this process, solvated electrons were formed and absorbed into the layers of graphene until saturation was reached. At the same time, solvated potassium was intercalated into interlayer galleries to balance negative charged graphene layers. The expansion of gallery spacing resulted from the dissolution of potassium cations. After the depletion of the potassium source, charged graphene layers were subsequently exfoliated due to electrostatically repelling forces followed by diazonium functionalization.

It is generally clear that π electron pairs on graphene are vulnerable sites to attack by chemical species. From the theoretical aspect, computational modelling work is also being developed alongside experimental trials to provide a deeper understanding of the formation of the functional attachments at the atomic level. Taking hydrogenation of graphene as an example,[113] it was believed that chemisorption of hydrogen atoms resulted in the transformation from sp^2 to sp^3 hybridization via breaking one of the π bonds, and meanwhile an unpaired electron was created to remain at the neighbouring carbon atoms. Geometrically, an intermediate form of sublattice was formed in graphene by the introduction of the C–H bond, which showed that the values of length and angle of C–C bonds was intermediate between that for sp^2 and sp^3 C–C bonds. Due to the 'delocalized' characteristic of the π-orbitals in graphene, the unpaired electron was smeared in one of the sublattices to make carbon and hydrogen atoms become magnetic. This intermediate state was not stable, and the next hydrogen atom would quickly terminate the unpaired electron and recover energetic equilibrium. In order to minimize geometric frustration, it was also demonstrated that serious atomic distortion was produced with a radius of 5 Å (equal to two periods of the graphene crystal lattice), but was less strong beyond this radius. For energetic stability with minimal additional distortions, the next hydrogen atoms in graphene preferred to be chemisorbed by neighbouring carbon atoms on the other side of the graphene sheet. A computational study on imperfect graphene showed that the defects within graphene were supposed to be the centre of chemical activity, as both the chemisorption and the activation energy of imperfect graphene were calculated to be lower than those of ideal graphene.[114] The defects studied included Stone–Wales defects, bivacancies and nitrogen substitution impurities. Thus, chemisorption on a defect was easier for the first hydrogen atom but an energy barrier existed for the location of the second hydrogen atom as local energy

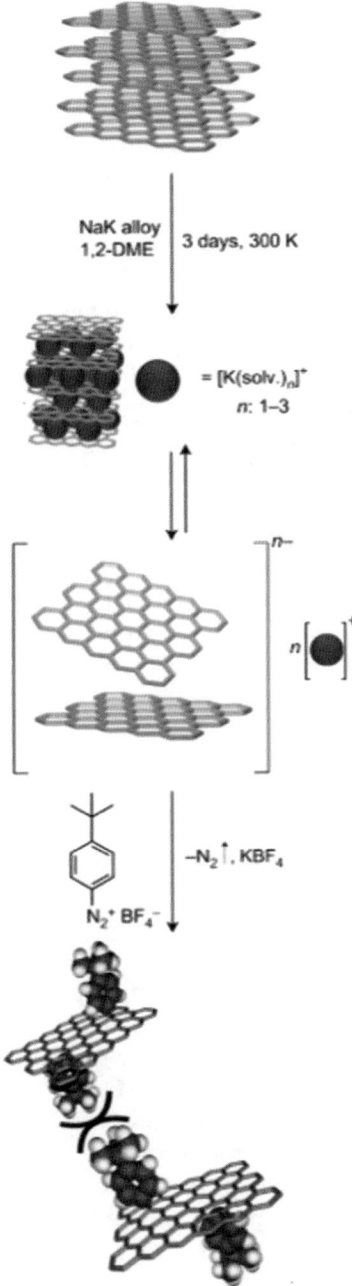

Figure 1.16   Reaction scheme. Representation of the intercalation and exfoliation of graphite with subsequent functionalization of intermediately generated reduced graphene yielding 4-t-butylphenyl functionalized graphene (double bonds in the basal planes have been omitted for clarity).1,2-DME, 1,2-dimethoxyethane.[112]

minima were formed during the first step, and this prevented further functionalization in addition to the defective sites. It was also pointed out that the presence of unpaired electrons due to broken bonds was considered to be the reason for the graphene edge to have higher chemical activity than bulk graphene. Similar computational studies were also used to understand the functionalization of graphene by other chemical species at the atomic level.

1.3.2   Functionalization of Graphene with Macromolecules

Grafting macromolecules onto graphene has also generated a great deal of interest since, in comparison with functional groups, macromolecular chains show improved thermal stability and exhibit stronger steric hindrance to prevent the aggregation of graphene. Furthermore, they allow covalent integration of graphene into more complex organic systems to develop novel composite materials. This section introduces the main strategies as reported for covalently grafting graphene with long-chain attachments. Profiting from experience with CNTs, research in this field has developed rapidly. Some researchers favoured dividing these strategies into two categories: 'grafting from' and 'grafting to' methods.[115] The key difference lies in that 'grafting from' methods are concerned with the growth of macromolecules initiated from the surface of graphene, and 'grafting to' methods are about linking pre-synthesized macromolecules to graphene.

1.3.2.1   'Grafting From' Methods

In terms of 'grafting from', let us start from atom transfer radical polymerization (ATPR), which has shown great advantages in controlling the structure of polymers. The initiating seeds can be kept alive and hardly experience termination by impurities from the environments in comparison with anionic or ionic polymerization. A copolymer can be designed with controlled length of blocked segments. It is interesting to have the terminal alkyl halide converted to diverse functionalities *via* organic chemistry. As shown in Figure 1.17, Lee *et al.* reported[116] a typical ATRP procedure to graft the graphene with polystyrene (PS), PMMA, and poly(butyl acrylate), respectively, in which 2-bromo-2-methylpropionyl bromide (BMPB) was the initiator. The key step was binding BMPB with the surface of GO via the esterification-like reaction of the acyl bromide in BMPB with the hydroxyl groups in GO. Other researchers also studied other ways to connect of the initiator with graphene. Gonçalves *et al.*[117] used ethylene glycol to terminate the carboxylic groups in GO for the enrichment of the hydroxyl groups followed by the similar route to that reported by Lee *et al.*[116] Yang *et al.*[118] attempted to convert the carboxylic groups to amine groups that could work together with the hydroxyl groups for the attachment of the initiator. It was found that poly(2- (dimethylamino)ethyl methacrylate) (PDMAEMA) chains grown onto the surface of graphene could assist the carbon sheets to accept

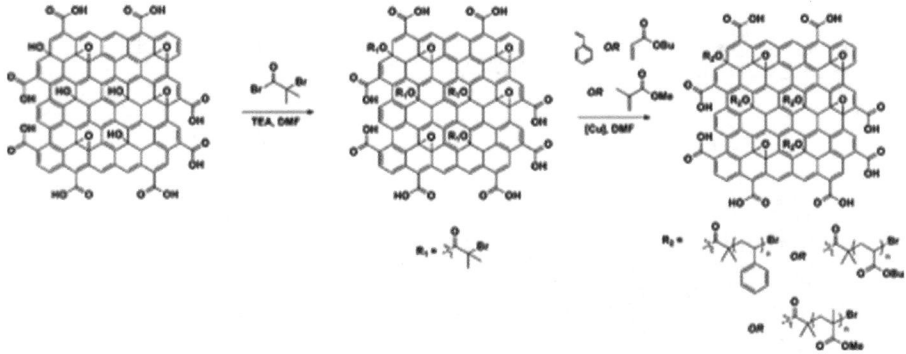

Figure 1.17 Synthesis of surface-functionalized graphene oxide via attachment of an atom transfer radical polymerization (ATRP) initiator (α-bromoisobutyryl bromide) followed by polymerization of styrene, butyl acrylate or methyl methacrylate.[116]

poly(ethylene dimethacrylate-co-methacrylic acid) particles via the interaction of hydrogen bonding. Fang *et al.* argued that more caution was required when decorating rGO with atom transfer radical polymerization (ATRP) due to the aggregation tendency of rGO. The stabilization could be compromised by the degree of reduction, size of graphene sheets and extra help from surfactants.[119] A subsequent study by the same authors disclosed that the rGO could be covalently attached to hydroxyl groups via the diazonium reaction, and subsequently the grafting density could be controlled by varying the concentration of diazonium compounds.[120]

Other 'grafting from' routes previously appearing in research on CNTs were also taken into account for graphene. Shen *et al.*[97] reported a route of *in situ* free radical polymerization for grafting rGO with a PS–polyacrylamide (PAM) copolymer. The sp^2 bond in rGO joined in the copolymerization with styrene and acrylamide initiated by benzoyl peroxide. Huang *et al.*[121] investigated grafting GO with polypropylene chains using *in situ* Ziegler–Natta polymerization as shown in Figure 1.18. The oxygen groups allowed settlement of the Mg/Ti catalyst on to a single GO sheet at the nanoscale. Deng *et al.*[122] proposed a protocol following single electron transfer living radical polymerization (SET-LRP). Figure 1.19 reveals that the process was basically similar to that of ATRP, but was applied with a different catalytic cycle. Cu (I)X disproportionated into Cu (0) and Cu (II)X$_2$. An electron was activated from the outer sphere of Cu (0) and deactivated with Cu (II)X$_2$/N-ligand, which required low energy at room temperature or below in comparison with the ATRP process. Tris (hydroxylmethyl) aminomethane (TRIS) was used to open the epoxy ring on graphene to increase the number of hydroxyl groups. Wang *et al.*[123] grafted GO from the monomers of acrylic acid and *N*-isopropylacrylamide using Ce (IV)-induced redox polymerization initiated by the redox pair of Ce^{4+} and hydroxyl groups on GO. Etmimi *et al.*[124] took advantage of the esterification reaction (Figure 1.20) to link the hydroxyl

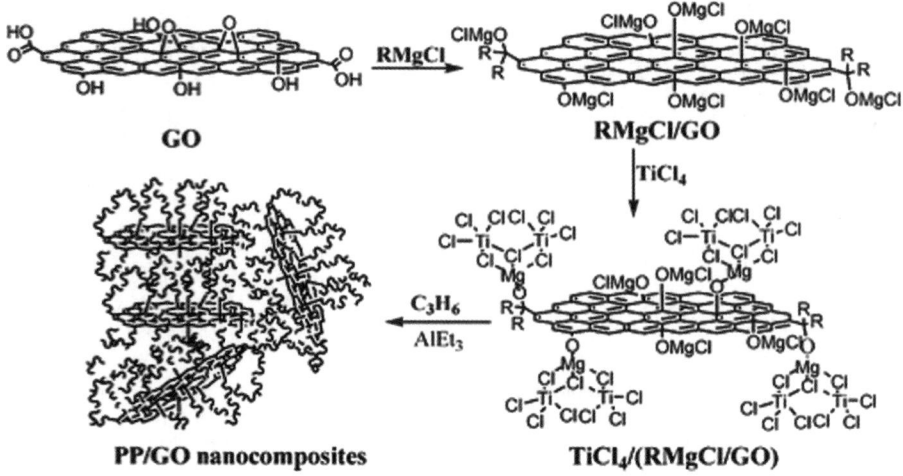

Figure 1.18   Fabrication of PP/GO nanocomposites by *in situ* Ziegler–Natta polymerization.[121]

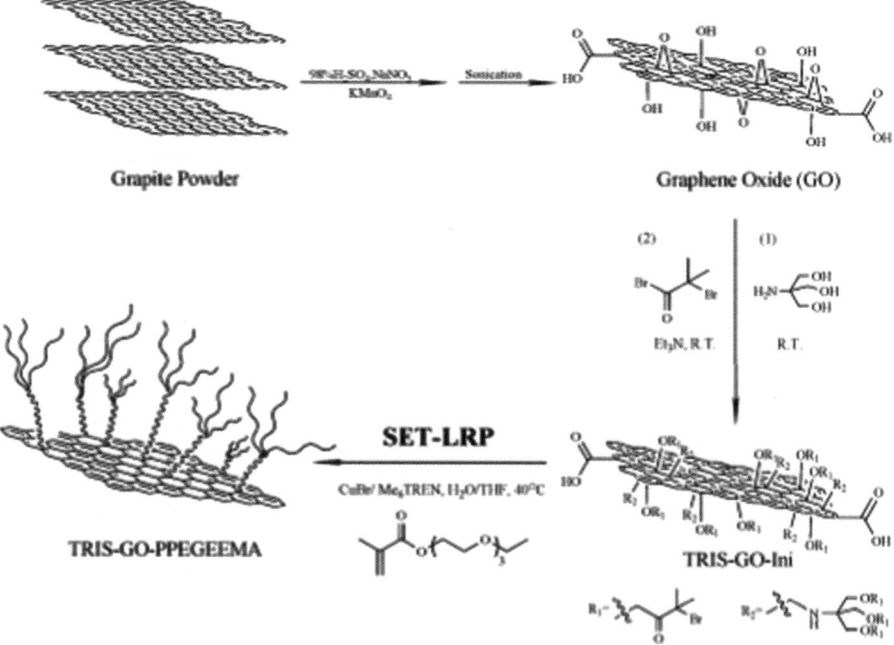

Figure 1.19   *In situ* growth of PPEGEEMA polymer chains *via* single electron transfer living radical polymerization (SET-LRP) from the surface of TRIS-modified graphene oxide sheets.[122]

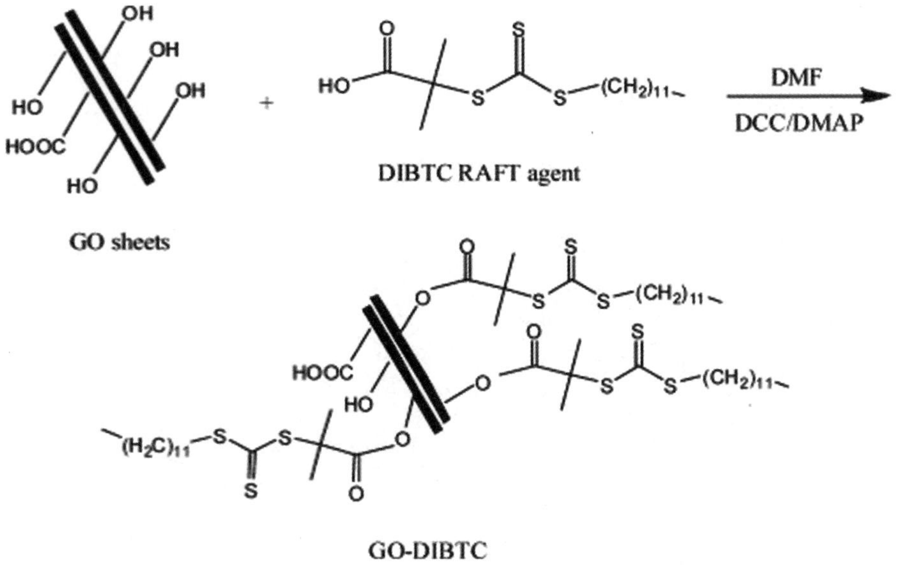

Figure 1.20 The overall synthesis route for the preparation of RAFT-immobilized GO nanosheets (DMF, *N,N*-dimethylformamide; DCC, 1,3-dicyclohexyl carbodiimide; DMAP: 4-dimethylaminopyridine).[124]

groups on GO with dodecyl isobutyric acid trithiocarbonate (DIBTC). With azobisisobutyronitrile (AIBN) as radical initiator, DIBTC acted as a chain transfer agent for reversible addition-fragmentation chain transfer (RAFT) polymerization of PS on the surface of GO. Apart from the routes of living radical polymerization, polyurethane (PU) chains can grow from the hydroxyl and carboxylic groups on GO following the route of polycondensation.[101]

1.3.2.2 'Grafting To' Methods

The essence of 'grafting from' methods is mainly about immobilizing initiators on the surface of graphene available for further polymerization. The principle of 'grafting to' methods seems simpler: it requires either graphene or polymers to have functional groups with the potential to form covalent linkages. In 'grafting from' methods, polymers are in the minority compared to graphene and act as dispersants to enhance the compatibility of graphene with organic solvents or polymeric matrices. In most cases, 'grafting to' methods are more likely to integrate graphene into polymeric matrices to form composites with covalent interfaces. A review by Salavagione *et al.*[115] summarizing the work relating to 'grafting to' methods shown in **Table 1.2**. Esterification/amidation is used to covalently link the carboxylic groups on GO with polymers containing hydroxyl or amine groups such as polyethylene glcol (PEG),[125] poly(vinyl alcohol) (PVA),[99,100] polyvinyl chloride (PVC),[126] poly(ethylenei-

Table 1.2 Reactions employed for grafting polymers to graphene and its derivatives115

Polymer	Form of graphene	Type of reaction	Notable findings	Reference
PEG	GO	Esterification	Drug delivery ability	125
PVA	GO, rGO	Esterification	Stereoselectivity	100
PVA	GO	Esterification		99
PEI	rGO	Amidation	Formation of hybrid carbon films for supercapacitors	127
PVC	GO, rGO	Esterification	Better mechanical properties	126
TPAPAM	GO	Amidation	Non-volatile rewritable memory effect	128
P3HT	GO	Esterification	Enhanced power conversion efficiency	129
PS, PEG	rGO	Nitrene cycloaddition	Improved dispersability, electrical conductivity	110
PAc	Graphene	Nitrene cycloaddition	Better solubility	130
Epoxy	GO	Cross-linking by amine-induced ring opening	Improved mechanical properties	131
PNIPAM	GO	Cross-linking by esterification/nucleophilic substitution	Thermal and pH responsive hydrogels	134
MA-g-PE	GO	Maleic ring-opening	Improved mechanical properties	133
Poly-L-lysine	GO	Nucleophile epoxy ring-opening	Amplified biosensor	132
PNIPAM	GO	ATNRC	Improved dispersability in water and organic solvents	135
PMMA	rGO	Radical grafting	Electrical conductivity	138
PS	GO	Click chemistry	Solubility in organic media	139
PNIPAM	GO	Click chemistry	Drugs delivery	137
PEG	Graphene	Click chemistry	Water-dispersability	140
PGMA, PMMA, PHEA, PMA, PBMA, PHEMA, PS, PAM, PNIPAM	GO	Radical coupling	Higher solubility, low intrinsic viscosity, electrical conductivity	136
NYLON	GO	Condensation	Better mechanical properties	141
PS-PAM	rGO	Radical coupling	Amphiphilic graphene	97

mine) (PEI),[127] triphenylamine-based polyazomethine (TPAPAM)[128] and poly(3-hexylthiophene) (P3HT).[129] Nitrene chemistry was reported to simply generate covalent linkages *via* cycloadditions of azide groups on the end of polymers (PEG, PS and polyacetylene (PAc)) with C=C bonds in graphene.[110,130] The ring-opening reaction of epoxides is commonly used to form covalent bonds between epoxy and graphene prefunctionalized with amine groups.[131] This type of reaction also can be used to open the epoxides on graphene sheets by amine groups on the polymer.[132] Maleic rings also can be opened by the amine-functionalized graphene for attaching maleic acid grafted polyethylene to graphene.[133] Sometimes, a reactive third party is introduced to react with both polymer and functionalized graphene. *e.g.* the covalent bonding can be formed via esterification of epichlorohydrin with the carboxylic groups in poly(*N*-isopropylacrylamide) (PNIPAM) and GO.[134] In another study, grafting PNIPAM to graphene was achieved by atom transfer nitroxide radical coupling (ATNRC) of Br-terminated PNIPAM and 2,2,6,6-tetramethylpiperidine-1-oxyl-modified graphene catalysed by a CuBr/ *N*,*N*,*N'*,*N'*,*N''*-pentamethyldiethylenetriamine system.[135] Click chemistry (1,3-dipolar azide-alkyne cycloaddition) is also powerful in the 'grafting to' method. Figure 1.21 shows a typical click chemistry route for grafting PNIPAM to rGO. PMMA end-capped by a cleavable alkoxyamine (PMMA-ONR$_2$) could become PMMA macroradicals after C-O bonds were broken under mild heating. The addition of the macroradicals to C=C bonds

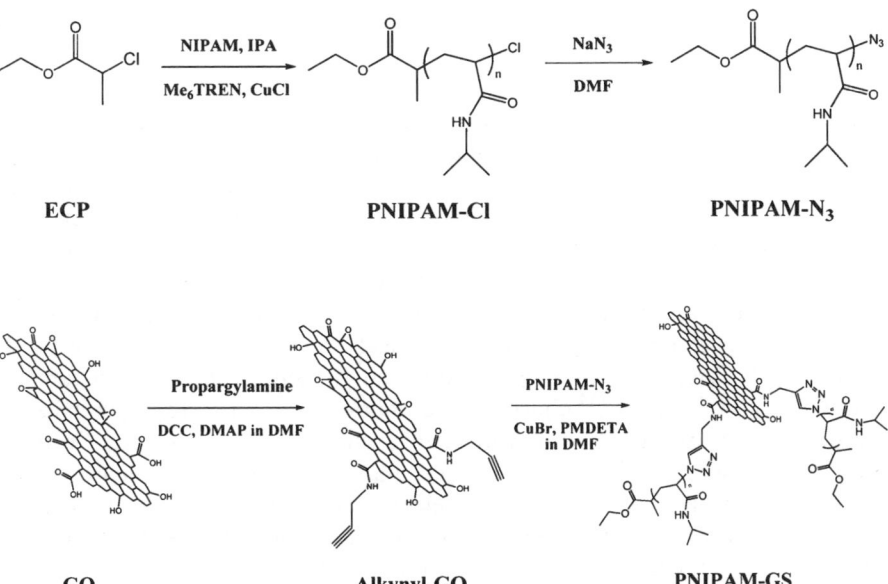

Figure 1.21   Synthetic procedure for the functionalization of reduced graphene oxide with poly(N-isopropylacrylamide) by click chemistry.[137]

in graphene resulted in the grafting of PMMA chains to graphene. Kan *et al.* reported a more universal approach for preparing 2D molecular brushes by grafting macromolecule to graphene *via* radical coupling.[136] The macromolecular radicals were simply prepared by free radical polymerization of monomers including glycidyl methacrylate (GMA), methyl acrylate (MMA), hydroxyethyl acrylate (HEA), methyl acrylate (MA), butyl methacrylate (BMA), hydroxyethyl methacrylate (HEMA), styrene (St), acrylamide (AM) and acrylamides (acrylamide (AM) and *N*-isopropylacrylamide (NIPAAm)). The solubility of graphene in different solvents could be flexibly controlled with these macromolecular brushes. A similar strategy was used to prepare amphiphilic graphene by grafting with amphiliphic polystyrene-polyacrylamide (PS-PAM).[97]

1.3.3 Functionalization of Graphene with Inorganic Nanoparticles (INPs)

INPs have attracted a huge amount of attention due to their remarkable properties. A great deal of research has shown that they can be absorbed onto graphene to act as inorganic functional species. Metal oxide and metal nanoparticles are studied in most cases. This type of functionalization is leading to a group of novel organic–inorganic composites for energy-related applications. In this section we discuss several popular strategies involved with the various interaction forces that drive this absorption process to occur spontaneously.

1.3.3.1 Direct Mixing

Direct mixing is usually used to functionalize graphene with pre-synthesized nanoparticles using organic compounds as binders. Electrostatic interaction is the key driving force when graphene and nanoparticles are oppositely charged. Sun *et al.*[142] prepared a graphene dispersion stabilized by Nafion with fluorobackbones, and found that commercial TiO_2 (P25) was bound to negatively charged graphene due to an electrostatic attractive force. Yang *et al.*[143] reported another strategy driven by mutual electrostatic interaction, in which Co_2O_3 nanoparticles were positively charged by grafting with aminopropyltrimethoxysilane (APS) followed by self-assembly to negatively charged GO. Hong *et al.*[144] charged gold nanoparticles and graphene using 1-pyrenebutyrate (positive) and 4-dimethylaminopyridine (negative), respectively. Then self-assembly of the gold nanoparticles onto functionalized graphene was simulated by the mutual electrostatic attraction. π–π interaction was another driving force specifically for assembling of the nanoparticles attached with benzene rings onto graphene. Feng *et al.*[145] used benzyl mercaptan to introduce benzene rings to the surface of cadmium sulfide (CdS) quantum dots. It was believed that the attached benzene rings interacted with rGO via π–π interaction, resulting in the deposition of quantum dots onto the graphene. In addition, Liu *et al.*[146] took advantage of the adhesive characteristics of polymer to assist graphene to capture nanoparticles. As shown in Figure 1.22, GO was reduced and decorated by bovine serum albumin (BSA), which is an amphiphilic biopolymer. 'Adhesive' graphene could be glued to single or mixed types of nanoparticles. The TEM images in Figure 1.23 show the morphology of the graphene sheet coated with gold nanoparticles, and the density of gold nanoparticles on the surface could be well controlled by this method. Kamat and coworkers[56,147] believed that electron transfer could cause the self-assembly of nanoparticles onto graphene sheets. In their strategy (shown in Figure 1.7), electron-hole pairs were generated by the UV irradiation of TiO_2 nanoparticles dispersed in ethanol. While a fraction of holes were scavenged by ethanol, excess electrons become trapped at the surface defect Ti^+ sites. These electrons could transfer from excited TiO_2 to GO to form rGO, which took up TiO_2 at the same time. Interestingly, it was found that there were some electrons remaining alive on rGO which continued to reduce Ag^+ to form graphene–TiO_2–Ag nanocomposites.[147] It was argued that TiO_2 could directly coordinate with the hydroxyl groups on GO to become the functional attachment. The low solubility of TiO_2 in aqueous solution limited the anchoring efficiency. A two-phase self-assembly process was developed to improve the efficiency.[148] GO was stabilized in deionized water. Meanwhile, a dispersion of TiO_2 nanorods in toluene was prepared by a low-temperature hydrolysis approach with oleic acid as stabilizer. The self-assembly of TiO_2 nanorods onto GO sheets was found at the water–toluene interface after the solution of TiO_2 nanorods and GO were mixed together. There have also been attempts to establish covalent

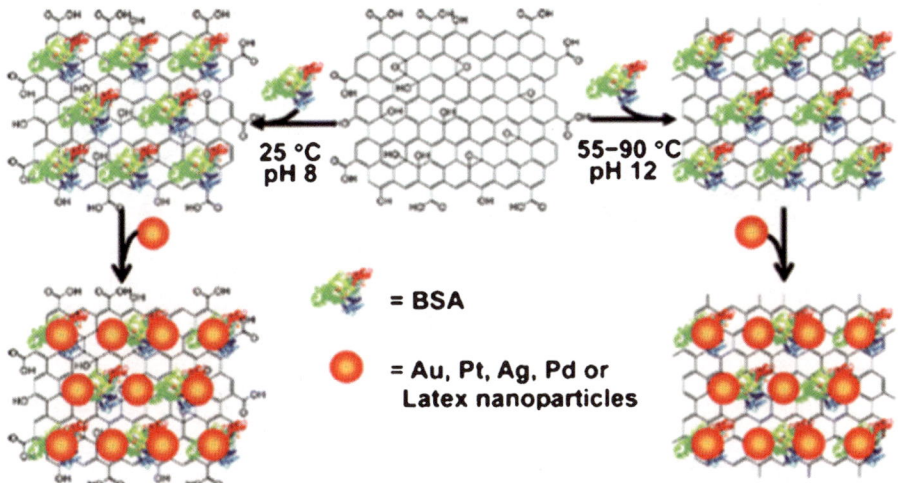

Figure 1.22 Protein-based decoration and reduction of GO, leading to a general nanoplatform for nanoparticle assembly.[146]

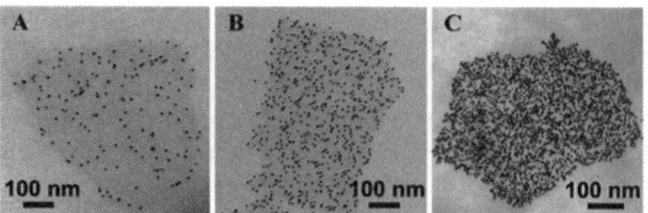

Figure 1.23 TEM images of Au nanoparticle (AuNP) decorated bovine serum
albumin (BSA)–GO with well-controlled densities of AuNPs. AuNP
densities were varied by increasing the concentrations of BSA from
0.5 mg mL^{-1} (A) to 20 mg mL^{-1} (B), during the preparation of BSA–
GO. NaCl was omitted for the samples in (A) and (B). (C) AuNP
density was further increased (in comparison with (B)) by adding 0.1 M
NaCl to the assembly system as in (B).[146]

linkages to connect GO and nanoparticles, such as fullerene molecules[149] and
polyhedral oligomeric silsesquioxane (POSS)[150] with carboxylic acid and
amine groups *via* esterification/amidation, respectively. The schematic route
for the attachment of POSS is illustrated in Figure 1.24.

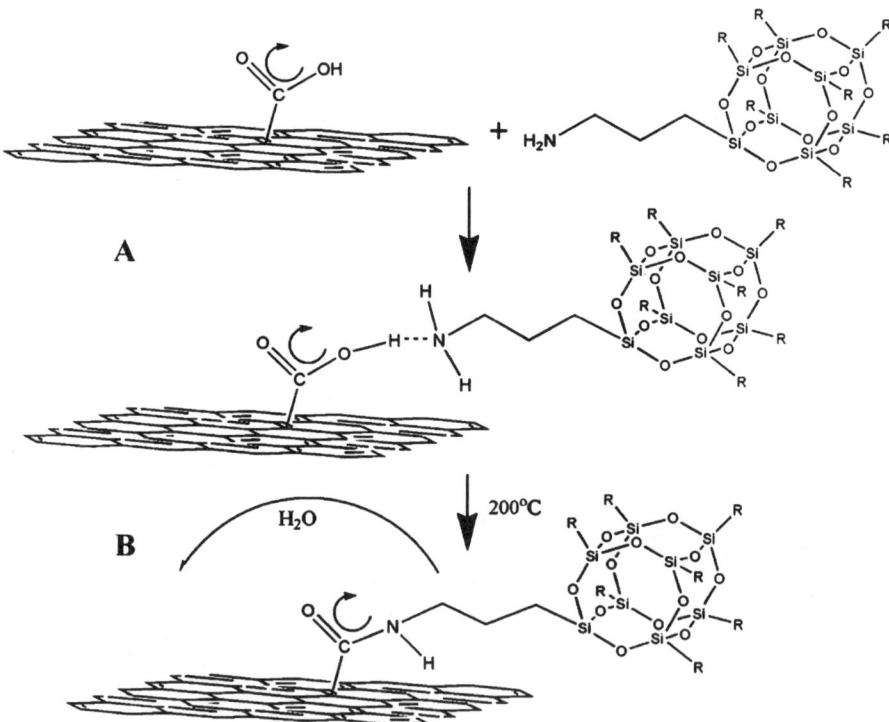

Figure 1.24   Schematic chemical reaction process in POSS and GO.[150]

1.3.3.2 In situ *Synthesis*

This method generally involves the reduction of metal or metal oxide precursors in the presence of GO or rGO. Here, the discussion of this method will be divided into the consideration of metal and metal oxide nanoparticles.

Regarding gold nanoparticles, Muszynski *et al.*[151] chemically reduced HAuCl$_4$ with NaBH$_4$ in graphene–octadecylamine suspension. Octadecylamine was chemically linked to GO for improving the solubility of graphene sheets in THF. It was considered that simple physisorption resulted in coating gold nanoparticles onto graphene spontaneously. Xu *et al.*[152] reported a similar absorption mechanism. A mixture of water and ethylene glycol was used, reducing metallic salts to form noble metal nanoparticles (Au, Pt and Pd) in the presence of GO. After deposition onto GO, it was found that the metal nanoparticles played a catalytic role in the reduction of GO with ethylene glycol to form a graphene nanomat to support the metal nanoparticles. Guo *et al.*[153] attempted to deposit two kinds of metal nanoparticles (Pt and Pd) onto rGO. In this method, H$_2$PdCl$_4$ was reduced by formic acid followed by a second reduction of K$_2$PtCl$_4$ with ascorbic acid. Pt nanoparticles grew onto the surface of Pd to form Pt-on-Pd bimetallic nanodendrites with an average size of 15 nm. They exhibited much higher electrocatalytic activity toward methanol oxidation reaction than the platinum black and commercial E-TEK Pt/C catalysts. Li *et al.*[154] found that the carboxylic groups in GO could be coupled with a reducing agent that was an amino-terminated ionic liquid (IL-NH$_2$), through the interaction between $-$COOH and $-$NH$_2$. It made the carboxylic groups act as the nucleating sites for growth of gold nanoparticles. As a result, the coating density of gold nanoparticles could be flexibly controlled by reducing GO or increasing carboxylic content by using 3,4,9,10-perylene tetracarboxylic acid (PTCA) to functionalize GO *via* π–π interaction. Zhou *et al.*[155] revealed a new deposition method without using an organic reducing agent. GO or rGO was coated onto 3-aminopropyltriethoxylsilane-modified Si/SiO$_X$ substrates, respectively. The oxygen functionalities on graphene immobilized with Ag$^+$ served as nucleation sites, and the growth of Ag nanoparticles occurred owe to the reduction of Ag$^+$ by the electrons supplied by conjugated domains of GO or rGO. Kong *et al.*[156] proposed a similar mechanism for depositing Au nanoparticles on rGO. It was believed that the electron-induced reduction likely resulted from galvanic displacement and redox reaction due to the relative potential difference between rGO and Au$^+$. Further to this discovery, the gold nanoparticles formed could be used as seeds to further initiate the growth of gold nanorods on the surface of rGO from the solution of cetyltrimethylammoniumbromide (CATB), HAuCl$_4$ and ascorbic acid.[157] For comparison, the seeds of pre-synthesized gold nanoparticles also could be transferred onto rGO pre-coated with pyrene ethylene glycol for further growth of gold nanorods. Huang *et al.*[158] created hexagonal close-packed gold square sheets (AuSSs), with an edge length of 200–500 nm and a thickness of ∼2.4 nm, on the surface of graphene by heating a mixture of GO, HAuCl₄ and 1-amino-9-octadecene in a solution of hexane and ethanol. The existence of GO significantly affected the yield of AuSSs. Without GO, AuSSs in a large size distribution (30–500 nm) were obtained, with the coexistence of gold nanowires and nanoparticles. A possible mechanism as illustrated in Figure 1.25 was proposed: mixing 1-amino-9-octadecene molecules with HAuCl₄ in hexane together with GO sheets resulted in partial reduction of Au^{4+} to form Au^+, which was complexed with 1-amino-9-octadecene molecules and absorbed onto GO through long alkylamine chains. With the addition of ethanol, 1-amino-9-octadecene-AuCl could be slowly reduced to form square-like gold seeds on GO sheets, which underwent further growth to form well-defined AuSSs on GO.

In situ deposition of metal oxide nanoparticles is also promoted by the hydrolysis of oxide precursors anchored onto negatively charged oxygen functionalities of GO. For instance, Figure 1.26 shows the schematic route for the preparation of graphene–TiO_2 composites *via* hydrolysis reaction of absorbed titanium(IV) butoxide and water.[159] Additionally, CuO,[160] SnO_2,[161] and MnO_2 nanoneedles[162] were deposited onto graphene sheets following a similar route. In some work, microwave energy was used to initiate the process of hydrolysis.[163] However, it was found that deposition efficiency was significantly limited by low number density of the oxygen functionalities. Wang *et al.*[164] used an ionic surfactant, sodium dodecyl sulfate (SDS), to improve the number of negative charges on the surface graphene. The hydrophobic tails of SDS were absorbed into surface of graphene sheets, and the hydrophilic sulfate heads were strongly bonded with TiO_2 precursors. The amount of TiO_2 coated to graphene was improved compared to the case without SDS. In a hydrothermal process (Figure 1.27), Co_2O_3 nanoparticles were coated graphene by heating $Co(OH)_2$–graphene composites at 450 °C following the reaction of $Co(NO_3)_2$ and ammonia.[165]

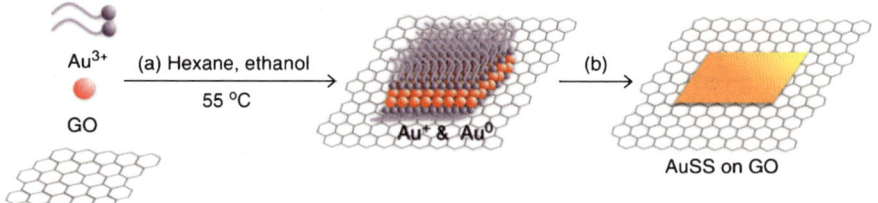

Figure 1.25   Step (a): 1-amino-9-octadecene, Au^{3+} and GO sheets in a mixed solvent of hexane and ethanol are heated at 55 °C. Au^{3+} is partially reduced to Au^+ and complexed with 1-amino-9-octadecene. Step (b): The formed 1-amino-9-octadecene–AuCl complex, adsorbed on the GO surface, finally becomes the Au square sheet (AuSS).[158]

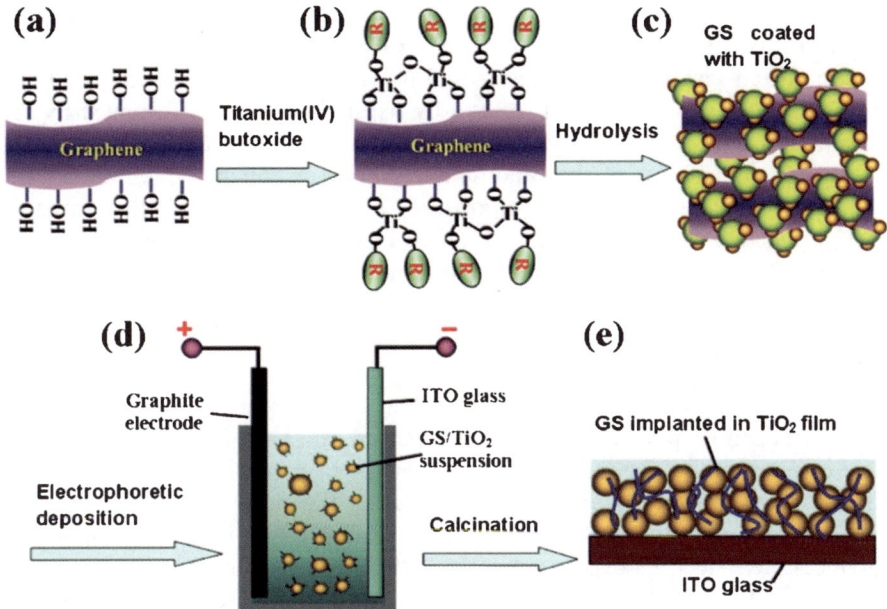

Figure 1.26 Schematic flowchart of *in situ* incorporation of graphene sheet (GS) in nanostructured TiO₂ films: (a) GS prepared by chemical exfoliation with residual oxygen-containing functional groups, such as hydroxyl. (b) Schematic of titanium(IV) butoxide grafted on the reduced GS surfaces by chemisorption. (c) Schematic diagram of GS coated with TiO₂ colloids after hydrolysis. (d) Illustration of the electrophoretic deposition process used to prepare GS/TiO₂ composite films. (e) Schematic representation of the structure of the GS/TiO₂ composite film after calcination.[159]

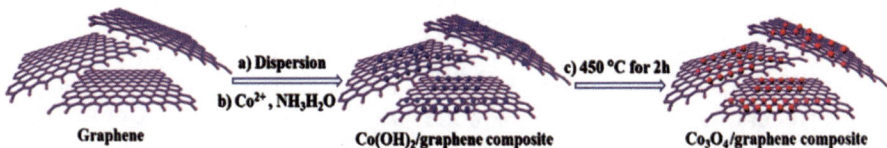

Figure 1.27 Schematic representation of the fabrication process of Co₃O₄/graphene composite: (a) Dispersion of chemically derived graphene in isopropyl alcohol/water (1:1, v/v) solution. (b) Formation of Co(OH)₂–graphene composite in basic solution. (c) Phase transformation from Co(OH)₂–graphene composite to Co₃O₄–graphene composite by calcination.[165]

1.4 Functionalized Graphene–Polymer Nanocomposites (FPNs)

There are high expectations that graphene can tackle several key technological challenges in the industries such as aerospace, automotive, defence, electronics and energy. Functionalization provides one of the key tools to deliver graphene as an engineering material. GO is a great ambassador for functionalized graphene. As discussed in section 2.2, the door to mass production of graphene is first opened by oxidizing graphite, followed by the reduction of GO. Although the quality of rGO is still being argued, researchers have sensed the bonus of the functionalization of graphene which allows the formation of graphene-based composites. Fabrication of graphene–inorganic nanoparticle composites has been reviewed in section 3.4. Here, we attempt to specifically highlight the application of functionalized graphene in polymer nanocomposites.

1.4.1 Fabrication

Functionalized graphene (FG) can be dispersed in a variety of solvents due to the existence of its organic attachments. It is easy to figure out a solvent-assisted method to fabricate FPNs by selecting the right solvents compatible with polymer and FG. GO is a typical FG usually used in FPNs. Numerous polymers have been incorporated with the FG *via* this simple method such as PMMA,[166] PU,[167] PVA,[168] PS,[169] and polycaprolactone (PCL).[170] *In situ* polymerization of pre-polymers or monomers in the presence of FG is a synthetic approach to prepare FPNs. Thermosetting of epoxy in presence of the FG is a typical example of this method.[171] *In situ* emulsion polymerization of styrene monomers in presence of GO nanosheets exfoliated in water was reported to prepare FG/PS nanocomposites.[172] The methods introduced in section 3.2 can be used to make FG join in the polymerization and bond with polymeric matrices. Although exfoliation of the FG in solvents is very successful in the laboratory, its limitations for melting processing of FPNs, a high-throughput approach, are clear to see. Thermal shock of GO can yield single-layer GSs as the oxygen groups are reduced. These exfoliated rGO sheets have been incorporated into polycarbonate (PC) by melt compounding, resulting in the enhancement in the electrical properties of PC.[173] However, minor mechanical improvement might result from the weak interface as the oxygen functionalities were eliminated. This problem may be solved by using GO grafted with macromolecules, which has higher thermal stability than GO. A novel process developed by our group could be the solution to tackle this challenge as well.[174] This technique consists of coating polymer powders such as polyethylene, polypropylene and nylon with exfoliated FG in water or organic solvents. After removal of the liquids, the FG-coated powders are suitable for processing by twin-extruder and injection moulding.

1.4.2 Mechanical Properties

The breaking strength and Young's modulus of graphene reach 42 N m^{-1} and 1 TPa, respectively, making it the strongest material ever measured.[5] However, it is believed that FG can do a better reinforcing job than pristine graphene, as the strength of the interface is central to the mechanical enhancement of PNCs instead of the intrinsic strength of the nanofillers themselves. The organic functionalities are capable of enhancing the compatibility between graphene and polymeric matrices, *e.g.* via hydrogen bonding. In addition, the wrinkled surface of graphene can mechanically interlock with polymer chains to improve the interface. Owing to these interactions, Ramanathan *et al.*[166] found that PMMA chains were restricted to confined polymeric regions on the substantial surfaces of each exfoliated sheet. The percolation of the confined regions resulted in an unprecedented increase in the glass transition temperature (T_g) of PMMA. The elastic modulus and ultimate strength of the PMMA were improved by nearly 80% and 20%, respectively, by the incorporation of 1 wt% FG. The excellent reinforcing effect of GO was also found in a PVA system due to hydrogen bonding between PVA and GO.[168,175] The reinforcement could also be attributed to graphene-nucleated crystalline interface as an increase in the crystallinity of FG-filled PVA was observed.[176] A more effective interface can be engineered when the FG is covalently bonded with polymeric matrices. Very stiff PU nanocomposites were achieved due to the strong covalent interface resulting from the reaction between the hydroxyl groups on FG and the isocyanate group on the end of PU chains.[167] The Young's modulus of the PU has been improved by nearly 9 times with the addition of 4.4 wt% FG. In a nano-scratch test, the scratch depth of the indenter in materials is recorded along with the scratch length at a certain scratch rate, which reflects the protective ability of the surface coatings for the substrates. The scratch depth profiles in Figure 1.28 show that the incorporation of 4.4 wt% FG resulted in a nearly 80% decrease in the scratch depth. The remarkable improvement in scratch resistance pointed to the promising application of these composite materials in surface coating. In a silicone system, it was found that hydroxyl groups on the FG could covalently bonded to the SiH-containing component during the curing of silicone elastomer.[177] The modulus of the silicone foam with 0.25 wt% FG increased by over 200% in comparison with pure silicone foam. Rafiee *et al.*[171] reported that only 0.125 wt% FG resulted in a remarkable increase in the fracture toughness and fracture energy up to $\sim 65\%$ and $\sim 115\%$, respectively. Also, the reduction in the rate of crack propagation in the epoxy reached ~ 25-fold as a result of the addition of 0.125 wt% FG. It was considered that the 2D structure and wrinkled surface enabled graphene to deflect cracks far more effectively than 1D CNTs or low-aspect-ratio nanoparticles. NASA and Princeton researchers[178] disclosed a similar investigation on FG–epoxy nanocomposites, but the toughness of the nanocomposites seemed not to be improved with the addition of the FG up to 0.5 wt%. The addition of ^{18}C-modified FG significantly reduced the toughness of the epoxy. Like other nanofillers, FG-induced toughening is confronting complications. More experimental and theoretical work needs to be carried out to fully understand the toughening mechanism, which will be one of the core issues in the mechanical enhancement of FPNs in the future.

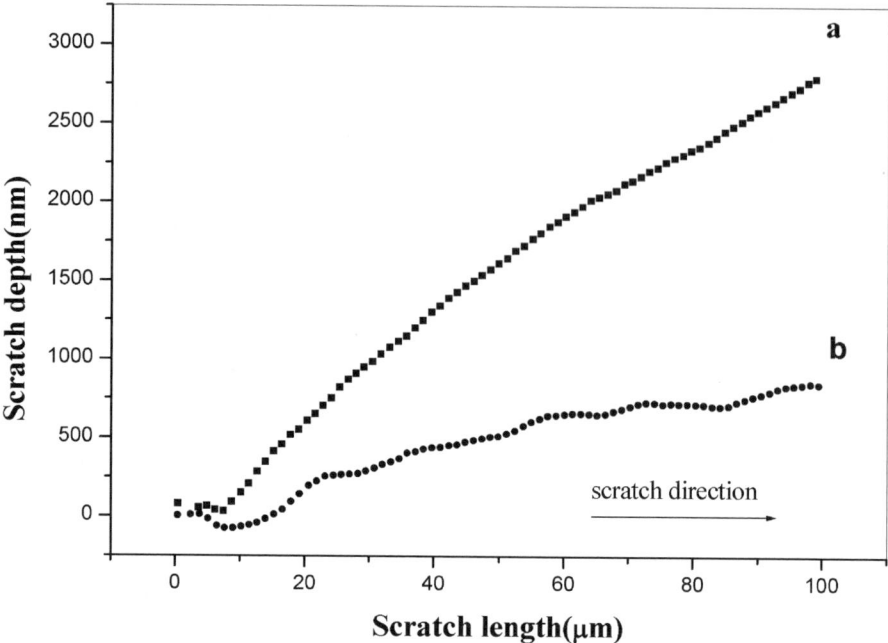

Figure 1.28 Nano-scratch depth profiles for the PU (a) and the 4.4 wt% functionalized graphene–PU nanocomposite (b) at a scratch rate of 3 μm/s. Reproduced from reference 167.

1.4.3 Electrical Properties

Although functionalization of graphene is necessary for the issue of dispersion, the FG with destructive conjugated structure contributes little to the conductivity of FPNs. Reduction of FG pre-mixed in polymeric matrices is the way to achieve conductive FPNs. Reduction methods are discussed in section 2.2.1. Strictly, rGO should be intermediate between pure graphene and FG since a small amount of oxygen groups still remain on carbon sheets after reduction, although it is referred to 'graphene' in some articles. Ruoff and his coworkers[169] first used rGO to improve the conductivity of PS. The solubility of the isocyanate-modified GO in an organic solvent paved the way to good dispersion of the GSs in the PS matrix. Chemical reduction of the GO using hydrazine was carried out in the presence of PS to avoid reaggregation of the rGO with small amount of functionalities. The percolation threshold was found to be as lower as 0.1 vol% and the maximum electrical conductivity of the nanocomposites could reach 0.1 S m^{-1} with the addition of 1 vol% rGO. Low percolation threshold and high value of maximum electrical conductivity are two goals in the development of semiconductive PNCs. It is now clear that the percolation threshold depends on the aspect ratio of nanofillers and the free space for the settlement of nanofillers. A latex technology has been reported to reduce the free space for nanofillers in a PS matrix.[179] The concept was simply achieved by mixing dispersion of surfactant-stabilized graphene and PS latex in water. During film formation of PS latex, the graphene is pushed into the free spaces in the PS latex to form conductive pathways in the PS matrix instead of being randomly dispersed. Using this method, the percolation threshold could be lowered to 0.6 wt%. The maximum electrical conductivity is generally limited by two factors: the intrinsic conductivity of the nanofillers and electron loss at the junctions of the conductive pathway formed by the nanofillers in polymeric matrices.[180] The rGO is conductive but its conductivity is not comparable to the pure graphene. The mechanical cleaving of pristine graphene from graphite in organic solvents has shown its advantage in fabricating quality graphene with high conductivity.[12] However, low production yield will be the limit in real applications. This is unavoidable in PNCs, since polymer chains may easily penetrate into the junctions and increase the electron loss transferring through nanofiller networks.[180] This is why the maximum conductivities of PNCs are commonly 2–4 orders of magnitude lower than the intrinsic conductivities of the nanofillers.

Solvent-free processing of semiconductive FPNs might be more welcomed by industry. Kim *et al.*[181] melt-compounded polyester with the FG prepared by partial pyrolysis of GO and achieved a low percolation threshold of 0.3 vol%, much lower than that required for graphite (3 vol%). Ansari *et al.*[182] also investigated the percolation threshold of poly(vinylidene fluoride) (PVDF) melt-compounded with FG and expanded graphite, respectively. The percolation threshold for expanded graphite was around 5 wt%, 2.5 times that for the FG (2 wt%). However, Steurer *et al.*[183] found the maximum conductivities of some thermoplastics melt-compounded with FG failed to outperform other conductive carbon fillers such as carbon black and expanded graphite (both of which have been commercial products for a long time). Another similar study disclosed that the rGO also underperforms commercial CNTs.[179] For this reason, the FG may lose the battle to carbon black and expanded GNPs for semiconductive PNCs. If FG is to be the main player in future, the low-cost fabrication of highly conductive FG will be the goal.

1.4.4 Thermal Properties

It has been confirmed that nanofillers can act as barriers to prevent the propagation of heat generated from the external environment in polymeric matrices, resulting in improved thermal stability of polymers.[184] A study on the thermal stability of GO by thermal gravimetric analysis (TGA) showed that GO started to lose some mass below 100 °C and the elimination of oxygen functionalities and the sublimation of carbon backbone occurred at 248 °C and 652 °C, respectively.[185] Graphene grafted with polymer chains showed better thermal stability than GO.[119] The onset decomposition temperature of PS–FG was much higher than that of GO. So, further functionalization of GO will extend the application of FG when a polymer needs to be processed at high temperature. With regard to the thermal stability of FPNs, it was reported that

1 wt% GO can improve the thermal degradation temperature of PMMA from $\sim 285\,°C$ to $\sim 342\,°C$.[166] The thermal degradation temperature of silicone foam increased by $57.7\,°C$, form $\sim 450\,°C$ to $\sim 507\,°C$, as 0.25 wt% GO was added.[177]

The high thermal conductivity of graphene has simulated researchers to explore the application of FPNs as thermal management materials. Experience with CNT–polymer nanocomposites has shown that the strong phonon scattering in the interface destroys our hopes of using a small amount of highly conductive CNTs to achieve substantial enhancement in the thermal conductivity of polymers.[186] Some unique composite structures are designed to reduce the interfacial strong phonon scattering.[187,188] Similar results are found in FPNs. The addition of GNPs to very high level is the dominant approach to reduce interfacial thermal resistance and yield a desirable value of thermal conductivity. However, the high loading of graphite is affordable, given the low cost of graphite in comparison with expensive CNTs. Yu et al.[189] exfoliated chemically intercalated graphite by a quick thermal shock at high temperature into slightly oxidized GNPs, which were still thermally and electrically conductive. 24-hour sonication was further applied to yield good dispersion of the GNPs in epoxy matrix. The thermal conductivity of epoxy was improved to be $6.44\ \mathrm{W\ m^{-1}\ K^{-1}}$ when 25 vol% GNPs was added. As shown in Figure 1.29, multilayer GNPs beat CNTs in improving the thermal conductivity of polymers for the following reasons, as suggested by the authors: (1) with their flat surface and large surface area, the GNPs could form stronger interactions with polymeric matrices than CNTs; (2) the rigid GNPs could keep their high aspect ratio in comparison with flexible CNTs. Later, they created a unique microstructure shown in Figure 1.30 to reduce the

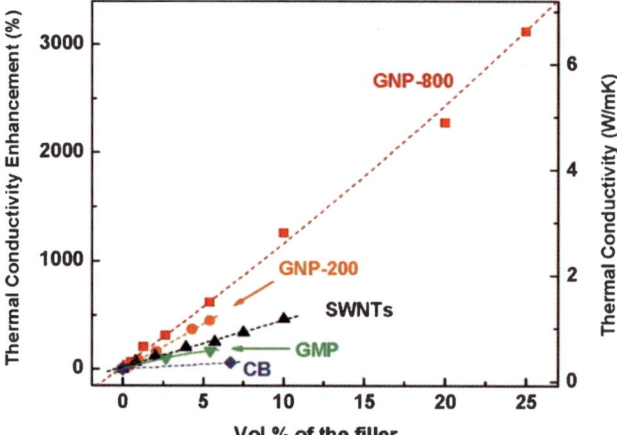

Figure 1.29   Thermal conductivity enhancement of epoxy-based composites at 30 °C. The graphitic fillers used were: graphitic microparticles (GMP), graphene nanoplatelets (GNPs) exfoliated at 200 °C (GNP-200) and 800 °C (GNP-800), carbon black (CB) and purified single-walled carbon nanotubes (SWCNTs).[189]

interfacial phonon scattering by using the combination of single-walled carbon nanotubes (SWCNTs) and GNPs as a hybrid nanofiller for epoxy.[190] It was suggested that flexible SWCNTs could bridge planar GNPs *via* Van der Waals attraction and extend the contacting area of SWCNT–GNP junctions for more phonon transfer. A synergistic effect of the hybrid nanofiller was present in the

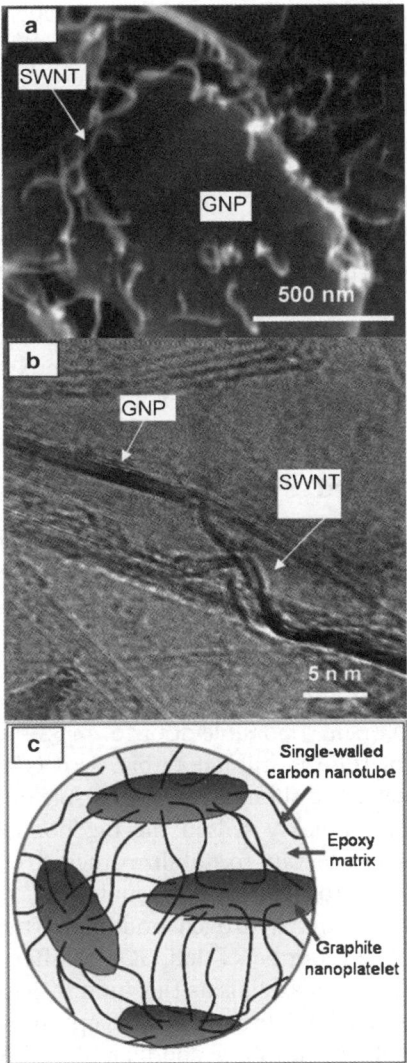

Figure 1.30 (a) SEM and (b) TEM images of the cross-section of graphene nanoplatelets (GNP)-single-walled carbon nanotubes (SWCNT) hybrid filler–epoxy composite. Note that SWCNTs are bridging adjacent GNPs, and SWCNT ends are extended along the nanoplatelet surfaces. (c) Schematic representation of GNP–SWCNT network in polymer matrix.[190]

enhancement of the thermal conductivity of epoxy nanocomposites. The optimum combination was 1:3 and the optimum loading of the hybrid filler should be in the range 10–20 wt%. The synergistic effect disappeared when the addition of the hybrid fillers was over 30 wt%. Ganguli *et al.*[191] found that 20 wt% silane-functionalized thermally expanded graphite enhanced the thermal conductivity of epoxy from 0.2 W m^{-1} K^{-1} to 5.8 W m^{-1} K^{-1}. It was found that silane functionalities could form covalent bonding with epoxy and improve the interfacial heat transfer between the two components by reducing acoustic impedance mismatch in the interfacial area. It is important to investigate whether high loading of GNPs would seriously affect the ductility of polymers. Veca *et al.*[192] thermally expanded graphite by applying alcohol and oxidative acid treatment with the assistance of long time and vigorous sonication, by which the carbon nanosheets was well-dispersed in epoxy matrix with a thickness less than 10 nm. The incorporation of 33 vol% carbon nanosheets could improve the in-plane thermal conductivity of epoxy nanocomposites to 80 W m^{-1} K^{-1}, which was 5–10 times that of the average value for the across-plane direction. This highly anisotropic nature resulted from the 2D structure of GSs. Interestingly, the epoxy nanocomposites still had good ductile properties even with 33 vol% carbon nanosheets.

1.5 Conclusions and Perspective

There are great expectations that graphene will tackle several key technological challenges in the aerospace, automotive, defence, electronics and energy industries. Functionalization provides one of key tools to deliver graphene as an engineering material. GO is a great ambassador for FG, which is obtained by exfoliation of oxidized graphite. The existence of oxygen functionalities facilitates the process of exfoliation as well as long-term stability in the liquid state. Reduction of GO opens a possible door for mass production of graphene. However, the quality of rGO is still debatable, as it is difficult to remove the functionalities completely. We should never be suspicious about the beauty of nature. Researchers have quickly sensed the big bonuses of using GO. For instance, GO can act as a surfactant to stabilize other carbon nanomaterials such as CNTs and fullerene to form all carbon materials.[193] GO shows excellent ability for film formation, and hydrogen bonding is the source of the super strength of GO paper.[194] It is our belief that GO will offer a great opportunity to generate functional thin-film materials as this kind of paper is used as a carrier for other nanoparticles. We have tested this concept to make hybrid CNT–GO films, which exhibited high electrical conductivity.[195] Very recently, it was reported that high-strength graphene fibres could be wet-spun from a highly concentrated aqueous GO solution.[196] The oxygen functionalities not only let us use graphene more easily, but also offer a great chance for graphene to be further functionalized with other chemical species. These diverse chemical species can assist the dispersion of GSs into various polymeric matrices. Moreover, they are particularly useful for improving the interface strength.

We also reviewed the possibility of establishing a hierarchical structure combining graphene and other INPs. The functionalities on the surface of graphene play a key role in bonding two components together. These novel inorganic–inorganic hybrids have shown promising applications in fabricating energy-related devices such as photovoltaic devices, fuel cells and super-capacitors. It has been reported that graphene–TiO_2composites outperformed pure TiO_2as electron acceptor in dye-sensitized solar cells. The reason for improved photoelectrical conversion was that the 2D graphene could work as a bridge to bring about faster electron transport, lower recombination and higher light scattering.[197] In a fuel cell system, it was found that the need for an expensive Pt-based catalyst could be reduced as the electrocatalytic activity of Pt could be enhanced by depositing it onto graphene. Anchoring Pt onto the surface graphene prevented the aggregation of Pt and increased its effective surface area.[198] The reaction of oxygen groups with carbonaceous species could reduce the likelihood of carbon monoxide poisoning the catalyst during methanol oxidation.[199] Ruoff and his coworkers first demonstrated that the specific capacitance of a rGO-based supercapacitor reached 135 F g^{-1} in the aqueous electrolyte.[8] This type of supercapacitor worked as an electrical double layer capacitor (EDLC) in which the capacitance relied on the conductivity and specific surface area of the electrodes. Reduction of specific surface area could be caused by restacking of rGO during film formation. Incorporation of other conductive carbon nanomaterials such as carbon black[200] and CNTs[127] into rGO has been proposed to create nanopores among graphene layers. A specific capacitance of 386 F g^{-1} at a scan rate of 10 mV s^{-1} was obtained in a supercapacitor in which a rGO–CNT hybrid was used as the electrode.[201] Unlike EDLCs, the working mechanism of a pseudo-capacitor mainly involves rapid redox reactions taking place in a metal-oxide-based electrode. When a graphene–MnO_2 composite was used as electrode, the specific capacitance of 310 F g^{-1} at 2 mV s^{-1} was nearly three times higher that obtained from a solely MnO_2 electrode.[163] In the presence of the conductive rGO network, the conductivity of electrode was increased as well as the effective interfacial area between MnO_2 and the electrolyte, both of which promoted the electrochemical reaction of MnO_2.

In addition to the oxygen functionalities of GO, π-bonds in graphene are another active site for direct covalent and non-covalent functionalization. π-bonds can be treated following many conventional organic reactions such as oxidization, radical chemistry and addition. The hydrogenation of graphene is reversible, which is leading researchers to explore the application of graphene in hydrogen storage. As an example, it has also been used by theorists to model functionalization, which provided deeper understanding of the basic principles of functionalization. Non-covalent functionalization *via* π–π stacking or other physical interaction was briefly introduced. It was favoured because the integrity of graphene could be maintained.

Since the discovery of graphene in 2004, research has progressed rapidly. There is no doubt that the commercial use of graphene will be central to the future of the material. The key challenge is scaling-up production of graphene or graphene-based materials. Several breakthrough results, introduced above, have brought hope. However, reproducibility is still an issue as it is difficult to control the uniformity of individual GSs using a 'top down' method, and it may also be affected by the irregular edge of graphene and randomly dispersed functionalities on graphene sheets. State-of-the-art techniques are also needed to achieve well-controlled microstructure of FG and its derivatives. With so much investment from so many countries, we hope the first commercial graphene products will become a reality in the near future.

References

1. K. S. Novoselov, A. K. Geim, S. V. Morozov, J. Y. Zhang, S. V. Dubonos, I. V. Grigorieva, A. A. Firsov, *Science*, 2004, **306**, 666.
2. J. C. Meyer, A. K. Geim, M. I. Katsnelson, K. S. Novoselov, T. J. Booth, S. Roth, *Nature*, 2007, **446**, 60.
3. A. K. Geim, K. S. Novoselov, *Nat. Mater.*, 2007, **6**, 183.
4. N. D. Mermin, *Phys. Rev.*, 1968, **176**, 250.
5. C. Lee, X. Wei, J. W. Kysar, J. Hone, *Science*, 2008, **321**, 385.
6. K. Saito, J. Nakamura, A. Natori, *Phys. Rev. B*, 2007, **76**, 115409.
7. A. A. Balandin, S. Ghosh, W. Bao, I. Calizo, D. Teweldebrahn, F. Miao, C. N. Lau, *Nano Lett.*, 2008, **8**, 902.
8. M. D. Stoller, S. Park, Y. Zhu, J. An, R. S. Ruoff, *Nano Lett.*, 2008, **8**, 3498.
9. K. S. Novoselov, Z. Jiang, Y. Zhang, S. V. Morozov, H. L. Stormer, U. Zeitler, J. C. Maan, G. S. Boebinger, P. Kim, A. K. Geim, *Science*, 2007, **315**, 1379.
10. X. Du, I. Skachko, A. Barker, E. Y. Andrei, *Nat. Nanotechnol.*, 2008, **3**, 491.
11. K. S. Novoselov, D. Jiang, F. Schedin, T. J. Booth, V. V. Khotkevich, S. V. Morozov, A. K. Geim, *Proc Natl. Acad. Sci U S A*, 2005, **102**, 10451.
12. Y. Hernandez, V. Nicolosi, M. Lotya, F. M. Blighe, Z. Sun, S. De, I. T. McGovern, B. Holland, M. Byrne, Y. K. Gun'ko, J. J. Boland, P. Niraj, G. Duesberg, S. Krishnamurthy, R. Goodhue, J. Hutchison, V. Scardaci, A. C. Ferrari, J. N. Coleman, *Nat. Nanotechnol.*, 2008, **3**, 563.
13. Y. Hernandez, M. Lotya, D. Rickard, S. D. Bergin, J. N. Coleman, *Langmuir*, 2010, **26**, 3208.
14. U. Khan, H. Porwal, A. O'Neill, K. Nawaz, P. May, J. N. Coleman, *Langmuir*, 2011, **27**, 9077.
15. M. Lotya, P. J. King, U. Khan, S. De, J. N. Coleman, *ACS Nano*, 2010, **4**, 3155.
16. N. Behabtu, J. R. Lomeda, M. J. Green, A. L. Higginbotham, A. Sinitskii, D. V. Kosynkin, D. Tsentalovich, A. N. G. Parra-Vasquez, J. Schmidt, E. Kesselman, Y. Cohen, Y. Talmon, J. M. Tour, M. Pasquali, *Nat. Nanotechnol.*, 2010, **5**, 406.
17. D. R. Dreyer, R. S. Ruoff, C. W. Bielawski, *Angew. Chem. Int. Ed.*, 2010, **49**, 9336.

18. B. C. Brodie, Philos. Trans. R. Soc. London, 149, 1859, 249.
19. L. Staudenmaier, *Ber. Dtsch. Chem. Ges.*, 1898, **31**, 1481.
20. W. S. Hummers, R. E. Offeman, *J. Am. Chem. Soc.*, 1958, **80**, 1339.
21. U. Hofmann, R. Holst, *Ber. Dtsch. Chem. Ges.*, 1939, **72**, 754.
22. G. Ruess, *Monatsh. Chem.*, 1946, **76**, 381.
23. W. Scholz, H. P. Boehm, *Anorg. Allg.Chem.*, 1969, **369**, 327.
24. T. Szabó, O. Berkesi, P. Forgó, K. Josepovits, Y. Sanakis, D. Petridis, I. Dékány, *Chem. Mater.*, 2006, **18**, 2740.
25. T. Nakajima, A. Mabuchi, R. Hagiwara, *Carbon*, 1988, **26**, 357.
26. T. Nakajima, Y. Matsuo, *Carbon*, 1994, **32**, 469.
27. A. Lerf, H. He, M. Forster, J. Klinowski, *J. Phys. Chem. B*, 1998, **102**, 4477.
28. H. He, J. Klinowski, M. Forster, A. Lerf, *Chem. Phys. Lett.*, 1998, **287**, 53.
29. W. Gao, L. B. Alemany, L. Ci, P. M. Ajayan, *Nat. Chem.*, 2009, **1**, 403.
30. S. Stankovich, R. D. Piner, X. Chen, N. Wu, S. T. Nguyen, R. S. Ruoff, *J. Mater. Chem.*, 2006, **16**, 155.
31. S. Stankovich, D. A. Dikin, R. D. Piner, K. A. Kohlhaas, A. Kleinhammes, Y. Jia, Y. Wu, S. T. Nguyen, R. S. Ruoff, *Carbon*, 2007, **45**, 1558.
32. D. Li, M. B. Müller, S. Gilje, R. B. Kaner, G. G. Wallace, *Nat. Nanotechnol.*, 2008, **3**, 101.
33. Y. Si, E. T. Samulski, *Nano Lett.*, 2008, **8**, 1679.
34. S. Liu, J. Tian, L. Wang, X. Sun, *Carbon*, 2011, **49**, 3158.
35. K. Ai, Y. Liu, L. Lu, X. Cheng, L. Huo, *J. Chem. Mater.*, 2011, **21**, 3365.
36. S. Dubin, S. Gilje, K. Wang, V. C. Tung, K. Cha, A. S. Hall, J. Farrar, R. Varshneya, Y. Yang, R. B. Kaner, *ACS Nano*, 2010, **4**, 3845.
37. Y. Liang, D. Wu, X. Feng, K. Müllen, *Adv. Mater.*, 2009, **21**, 1679.
38. Y. Zhang, W. Hu, B. Li, C. Peng, C. Fan, Q. Huang, *Nanotechnology*, 2011, **22**, 345601.
39. G. Wang, J. Yang, J. Park, X. Gou, B. Wang, H. Liu, J. Yao, *J. Phys. Chem. C*, 2008, **112**, 8192.
40. X. Fan, W. Peng, Y. Li, X. Li, S. Z. Wang G., F. Zhang, *Adv. Mater.*, 2008, **20**, 4490.
41. H. J. Shin, K. K. Kim, A. Benayad, S. M. Yoon, H. K. Park, I. S. Jung, M. H. Jin, H. K. Jeong, J. M. Kim, J. Y. Choi, Y. H. Lee, *Adv. Funct. Mater.*, 2009, **19**, 1987.
42. M. J. Fernández-Merino, L. Guardia, J. I. Paredes, S. Villar-Rodil, P. Solís-Fernández, A. Martínez-Alonso, J. M. D. Tascón, *J. Phys. Chem. C*, 2010, **114**, 6426.
43. S. Pei, J. Zhao, J. Du, W. Ren, H. M. Cheng, *Carbon*, 2010, **48**, 4466.
44. I. K. Moon, J. Lee, R. S. Ruoff, H. Lee, *Nat. Commun.*, 2010, **1**, 73.
45. W. Chen, L. Yan, P. R. Bangal, *J. Phys. Chem. C*, 2010, **114**, 19885.
46. X. Zhou, J. Zhang, H. Wu, H. Yang, J. Zhang, S. Guo, *J. Phys. Chem. C*, 2011, **115**, 11957.
47. Z. Fan, K. Wang, T. Wei, J. Yan, L. Song, B. Shao, *Carbon*, 2010, **48**, 1686.

48. Z. Fan, W. Kai, J. Yan, T. Wei, L. Zhi, J. Feng, Y. Ren, L. Song, F. Wei, *ACS Nano*, 2011, **5**, 191.

49. D. Chen, L. Li, L. Guo, *Nanotechnology*, 2011, **22**, 325601.

50. T. Zhou, F. Chen, K. Liu, H. Deng, Q. Zhang, J. Feng, Q. Fu, *Nanotechnology*, 2011, **22**, 045704.

51. D. R. Dreyer, S. Murali, Y. Zhu, R. S. Ruoff, C. W. Bielawski, *J. Chem. Mater.*, 2011, **21**, 3443.

52. H. C. Schniepp, J. L. Li, M. J. McAllister, H. Sai, M. Herrera-Alonso, D. H. Adamson, R. K. Prud'homme, R. Car, D. A. Saville, I. A. Aksay, *J. Phys. Chem. B*, 2006, **110**, 8535.

53. L. J. Cote, R. Cruz-Silva, J. Huang, *J. Am. Chem. Soc.*, 2009, **131**, 11027.

54. M. Zhou, Y. Wang, Y. Zhai, J. Zhai, W. Ren, F. Wang, S. Dong, *Chem. Eur. J.*, 2009, **15**, 6116.

55. S. J. An, Y. Zhu, S. H. Lee, M. D. Stoller, T. Emilsson, S. Park, A. Velamakanni, J. An, R. S. Ruoff, *J. Phys. Chem. Lett.*, 2010, **1**, 1259.

56. G. Williams, B. Seger, P. V. Kamat, *ACS Nano*, 2008, **2**, 1487.

57. H. L. Guo, X. F. Wang, Q. Y. Qian, F. B. Wang, X. H. Xia, *ACS Nano*, 2009, **3**, 2653.

58. P. W. Sutter, J. I. Flege, E. A. Sutter, *Nat. Mater.*, 2008, **7**, 406.

59. J. Coraux, A. T. N'Diaye, C. Busse, T. Michely, *Nano Lett.*, 2008, **8**, 565.

60. J. C. Hamilton, J. M. Blakely, *Surf. Sci.*, 1980, **91**, 199.

61. M. Eizenberg, J. M. Blakely, *Surf. Sci.*, 1979, **82**, 228.

62. K. S. Kim, Y. Zhao, H. Jiang, S. Y. Lee, J. M. Kim, K. S. Kim, J. H. Ahn, P. Kim, J. Y. Choi, B. H. Hong, *Nature*, 2009, **457**, 706.

63. A. N. Obraztsov, E. A. Obraztsova, A. V. Tyurnina, A. A. Zolotukhin, *Carbon*, 2007, **45**, 2017.

64. A. Rina, X. Jia, J. Ho, D. Nezich, H. Son, V. Bulovic, M. S. Dresselhaus, J. Kong, *Nano Lett.*, 2009, **9**, 30.

65. X. Li, W. Cai, J. An, S. Kim, J. Nah, D. Yang, R. Piner, A. Velamakanni, I. Jung, E. Tutuc, S. K. Banerjee, L. Colombo, R. S. Ruoff, *Science*, 2009, **324**, 1312.

66. W. Cai, R. D. Piner, F. J. Stadermann, S. Park, M. A. Shaibat, Y. Ishii, D. Yang, A. Velamakanni, J. S. An, M. Stoller, J. An, D. Chen, R. S. Ruoff, *Science*, 2008, **321**, 1815.

67. X. Li, C. W. Magnuson, A. Venugopal, J. An, J. W. Suk, B. Han, M. Borysiak, W. Cai, A. Velamakanni, Y. Zhu, L. Fu, E. M. Vogel, E. Voelkl, L. Colombo, R. S. Ruoff, *Nano Lett.*, 2010, **10**, 4328.

68. C. Mattevi, H. Kim, M. Chhowalla, *J. Mater. Chem.*, 2011, **21**, 3324.

69. W. A. Heer, C. Bergera, X. Wu, P. N. Firsta, E. H. Conrada, X. Lia, T. Lia, M. Sprinklea, J. Hassa, M. L. Sadowski, M. Potemski, G. Martinez, *Solid State Commun.*, 2007, **143**, 92.

70. Y. M. Lin, C. Dimitrakopoulos, K. A. Jenkins, D. B. Farmer, H. Chiu Y., *Science*, 2007, **327**, 662.

71. Z. Sun, Z. Yan, J. Yao, E. Beitler, Y. Zhu, J. M. Tour, *Nature*, 2010, **468**, 549.

72. X. Li, X. Wang, L. Zhang, S. Lee, H. Dai, *Science*, 2008, **319**, 1229.

73. M. Y. Han, B. Özyilmaz, Y. Zhang, P. Kim, *Phys. Rev. Lett.*, 2007, **98**, 206805.
74. Z. Chen, Y. M. Lin, M. J. Rooks, P. Avouris, *Physica E*, 2007, **40**, 228.
75. L. Tapasztó, G. Dobrik, P. Lambin, L. P. Biró, *Nat. Nanotechnol.*, 2008, **3**, 397.
76. S. Masubuchi, M. Ono, K. Yoshida, K. Hirakawa, T. Machida, *Appl. Phys. Lett.*, 2009, **94**, 082107.
77. J. L. Li, K. N. Kudin, M. J. McAllister, R. K. Prud'homme, I. A. Aksay, R. Car, *Phys. Rev. Lett.*, 2006, **96**, 176101.
78. L. Jiao, X. Wang, G. Diankov, H. Wang, H. Dai, *Nat. Nanotechnol.*, 2010, **5**, 321.
79. L. Jiao, L. Zhang, X. Wang, G. Diankov, H. Dai, *Nature*, 2009, **458**, 877.
80. D. V. Kosynkin, A. L. Higginbotham, A. Sinitskii, J. R. Lomeda, A. Dimiev, B. K. Price, J. M. Tour, *Nature*, 2009, **458**, 872.
81. N. Severin, S. Kirstein, I. M. Sokolov, J. P. Rade, *Nano Lett.*, 2009, **9**, 457.
82. L. C. Campos, V. R. Manfrinato, J. D. Sanchez-Yamagishi, J. Kong, P. Jarillo-Herrero, *Nano Letters*, 2009, **9**, 2600.
83. S. S. Datta, D. R. Strachan, S. M. Khamis, A. T. C. Johnson, *Nano Lett.*, 2008, **8**, 1912.
84. X. Yang, X. Dou, A. Rouhanipour, L. Zhi, H. J. Räder, K. Müllen, *J. Am. Chem. Soc.*, 2008, **130**, 4216.
85. J. Cai, P. Ruffieux, R. Jaafar, M. Bieri, T. Braun, S. Blankenburg, M. Muoth, A. P. Seitsonen, M. Saleh, X. Feng, K. Müllen, R. Fasel, *Nature*, 2010, **466**, 470.
86. Z. Wang, N. Li, Z. Shi, Z. Gu, *Nanotechnology*, 2010, **21**, 175602.
87. K. S. Subrahmanyam, L. S. Panchakarla, A. Govindaraj, C. N. R. Rao, *J. Phys. Chem. C*, 2009, **113**, 4257.
88. N. Li, Z. Wang, K. Zhao, Z. Shi, Z. Gu, S. Xu, *Carbon*, 2010, **48**, 255.
89. M. Choucair, P. Thordarson, J. A. Stride, *Nat. Nanotechnol.*, 2009, **4**, 30.
90. Y. Xu, H. Bai, G. Lu, C. Li, G. Shi, *J. Am. Chem. Soc.*, **130**, 5856.
91. R. Hao, W. Qian, L. Zhang, Y. Hou, *Chem. Comm.*, 2008, 6576.
92. S. Stankovich, R. D. Piner, S. T. Nguyen, R. S. Ruoff, , *Carbon*, **44**, 3342.
93. G. I. Titelman, V. Gelman, S. Bron, R. L. Khalfin, Y. Cohen, H. Bianco-Peled, *Carbon*, 2005, **43**, 641.
94. D. Cai, M. Song, *J. Mater. Chem.*, 2007, **17**, 3678.
95. J. I. Paredes, S. Rodil-Villar, A. Martínez-Alonso, J. M. D. Tascón, *Langmuir*, 2008, **24**, 10560.
96. S. Niyogi, E. Bekyarova, M. E. Itkis, J. L. McWilliams, M. A. Hamon, R. C. Haddon, *J. Am. Chem. Soc.*, 2006, **128**, 7720.
97. J. Shen, Y. Hu, C. Li, C. Qin, M. Ye, *Small*, 2009, **5**, 82.
98. Y. F. Xu, Z. B. Liu, X. L. Zhang, Y. Wang, J. G. Tian, Y. Huang, Y. F. Ma, X. Y. Zhang, Y. S. Chen, *Adv. Mater.*, 2009, **21**, 1275.
99. L. M. Veca, F. Lu, M. J. Meziani, L. Cao, P. Zhang, G. Qi, L. Qu, M. Shrestha, Y. P. Sun, *Chem. Commun.*, 2009, 2565.

100. H. J. Salavagione, M. A. Gomez, G. Martinez,, *Macromolecules*, 2009, **42**, 6331.
101. X. Wang, Y. Hu, L. Song, H. Yang, W. Xing, H. Lu, *J. Mater. Chem.*, 2011, **21**, 4222.
102. H. Yang, C. Shan, F. Li, D. Han, Q. Zhang, L. Niu, *Chem. Commun.*, 2009, 3880.
103. S. Wang, P. J. Chia, L. L. Chua, L. H. Zhao, R. Q. Png, S. Sivaramakrishnan, M. Zhou, G. S. R. Goh, R. H. Friend, A. T. S. Wee, P. K. H. Ho, , *Adv. Mater.*, **20**, 3440.
104. H. Liu, S. Ryu, Z. Chen, M. L. Steigerwald, C. Nuckolls, L. E. Brus, *J. Am. Chem. Soc.*, 2009, **131**, 17099.
105. J. Shen, Y. Hu, M. Shi, X. Lu, C. Qin, C. Li, M. Ye, *Chem. Mater.*, 2009, **21**, 3514.
106. R. Sharma, J. H. Baik, C. J. Perera, M. S. Strano, *Nano Lett.*, 2010, **10**, 398.
107. R. Salvio, S. Krabbenborg, W. J. M. Naber, A. H. Velders, D. N. Reinhoudt, W. G. Wiel van der, *Chem. Eur. J.*, 2009, **15**, 8235.
108. D. C. Elias, R. R. Nair, T. M. G. Mohiuddin, S. V. Morozov, P. Blake, M. P. Halsall, A. C. Ferrari, D. W. Boukhvalov, M. I. Katsnelson, A. K. Geim, K. S. Novoselov, *Science*, 2009, **323**, 610.
109. S. H. Cheng, K. Zou, F. Okino, H. R. Gutierrez, A. Gupta, N. Shen, P. C. Eklund, J. O. Sofo, J. Zhu, *Phys. Rev. B*, 2010, **81**, 205435.
110. H. He, C. Gao, *Chem. Mater.*, 2010, **22**, 5054.
111. M. Quintana, K. Spyrou, M. Grzelczak, W. R. Browne, P. Rudolf, M. Prato, *ACS Nano*, 2010, **4**, 3527.
112. J. M. Englert, C. Dotzer, G. Yang, M. Schmid, C. Papp, J. M. Gottfried, H. P. Steinürck, E. Spiecker, F. Hauke, A. Hirsch, *Nat. Chem.*, 2011, **3**, 279.
113. D. W. Boukhvalov, M. I. Katsnelson, A. I. Lichtenstein, *Phys. Rev. B*, 2008, **77**, 035427.
114. D. W. Boukhvalov, M. I. Katsnelson, *Nano Lett.*, 2008, **8**, 4373.
115. H. J. Salavagione, G. Martínez, G. Ellis, *Macromol. Rapid Commun.*, 2011, **32**, 1771.
116. S. H. Lee, D. R. Dreyer, J. An, A. Velamakanni, R. D. Piner, S. Park, Y. Zhu, S. O. Kim, C. W. Bielawski, R. S. Ruoff, *Macromol. Rapid Commun.*, 2010, **31**, 281.
117. G. Gonçalves, P. A. A. P. Marques, A. Barros-Timmons, I. Bdkin, M. K. Singh, N. Emami, J. Grácio, *J. Mater. Chem.*, 2010, **20**, 9927.
118. Y. Yang, J. Wang, J. Zhang, J. Liu, X. Yang, H. Zhao, *Langmuir*, 2009, **25**, 11808.
119. M. Fang, K. Wang, H. Lu, Y. Yang, S. Nutt, *J. Mater. Chem.*, 2009, **19**, 7098.
120. M. Fang, K. Wang, H. Lu, Y. Yang, S. Nutt, *J. Mater. Chem.*, 2010, **20**, 1982.
121. Y. Huang, Y. Qin, Y. Zhou, H. Niu, Z. Z. Yu, J. Y. Dong, *Chem. Mater.*, 2010, **22**, 4096.
122. Y. Deng, Y. Li, J. Dai, M. Lang, X. Huang, *J. Polym. Sci. Part A: Polym. Chem.*, 2011, **49**, 4747.

123. B. Wang, D. Yang, J. Z. Zhang, C. Xi, J. Hu, *J. Phys. Chem. C*, 2011, **115**, 24636.
124. H. M. Etmimi, M. P. Tonge, R. D. Sanderson, *J. Polym. Sci. Part A: Polym. Chem.*, 2011, **49**, 1621.
125. Z. Liu, J. T. Robinson, X. Sun, H. Dai, *J. Am. Chem. Soc.*, 2008, **130**, 10876.
126. H. J. Salavagione, G. Martínez, *Macromolecules*, 2011, **44**, 2685.
127. D. Yu, L. Dai, *J. Phys. Chem. Lett.*, 2010, **1**, 467.
128. X. D. Zhuang, Y. Chen, G. Liu, P. P. Li, C. X. Zhu, E. T. Kang, K. G. Noeh, B. Zhang, J. H. Zhu, Y. X. Li, *Adv. Mater.*, 2010, **22**, 1731.
129. D. Yu, Y. Yang, M. Durstock, J. B. Baek, L. Da, *ACS Nano*, 2010, **4**, 5633.
130. X. Xu, Q. Luo, W. Lv, Y. Dong, Y. Lin, Q. Yang, A. Shen, D. Pang, J. Hu, J. Qin, Z. Li, *Macromol. Chem. Phys.*, 2011, **212**, 768.
131. M. Fang, Z. Zhang, J. Li, H. Zhang, H. Lu, Y. Yang, *J. Mater. Chem.*, 2010, **20**, 9635.
132. C. Shan, H. Yang, D. Han, Q. Zhang, A. Ivaska, L. Niu, *Langmuir*, 2009, **25**, 12030.
133. Y. Lin, J. Jin, M. Song, *J. Mater.Chem.*, 2011, **21**, 3455.
134. S. Sun, P. Wu, *J. Mater. Chem.*, 2011, **21**, 4095.
135. Y. Deng, Y. Li, J. Dai, M. Lang, X. Huang, *J. Polym. Sci. , Part A: Polym. Chem.*, 2011, **49**, 1582.
136. L. Kan, Z. Xu, C. Gao, *Macromolecules*, 2011, **44**, 444.
137. Y. Pan, H. Bao, N. G. Sahoo, T. Wu, L. Li, *Adv. Funct. Mater.*, 2011, **21**, 2754.
138. D. Vuluga, J. M. Thomassin, I. Molenberg, I. Huynen, B. Gilbert, C. Jerome, M. Alexandre, C. Detrembleur, *Chem. Commun.*, 2011, **47**, 2544.
139. S. Sun, Y. Cao, J. Feng, P. Wu, *J. Mater. Chem.*, 2010, **20**, 5605.
140. Z. Jin, T. P. McNicholas, C. J. Shih, Q. H. Wang, G. L. C. Paulus, A. J. Hilmer, S. Shimizu, M. S. Strano, *Chem. Mater.*, 2011, **23**, 3362.
141. Z. Xu, C. Gao, *Macromolecules*, 2010, **43**, 6716.
142. S. Sun, L. Gao, Y. Liu, *Appl. Phys. Lett.*, 2010, **96**, 083113.
143. S. Yang, X. Feng, S. Ivanovici, K. Müllen, *Angew.Chem. Int. Ed.*, 2010, **49**, 8408.
144. W. Hong, H. Bai, Y. Xu, Z. Yao, Z. Gu, G. Shi, *J. Phys. Chem. C*, 2010, **114**, 1822.
145. M. Feng, R. Sun, H. Zhan, Y. Chen, *Nanotechnology*, 2010, **21**, 075601.
146. J. Liu, S. Fu, B. Yuan, Y. Li, Z. Deng, *J. Am. Chem. Soc.,* 2010, **2010**, 7279.
147. I. V. Lightcap, T. H. Kosel, P. V. Kamat, *Nano Lett.*, 2010, **10**, 577.
148. J. Liu, H. Bai, Y. Wang, Z. Liu, X. Zhang, D. D. Sun, *Adv. Funct. Mater.*, 2010, **20**, 4175.
149. Y. Zhang, L. Ren, S. Wang, A. Marathe, J. Chaudhuri, G. Li, *J. Mater. Chem.*, 2011, **21**, 5386.
150. J. Jin, X. Wang, M. Song, *J. Nanosci.Nanotechnol.*, 2011, **11**, 7715.
151. R. Muszynski, B. Seger, P. V. Kamat, *J. Phys. Chem. C*, 2008, **112**, 5263.

152. C. Xu, X. Wang, J. Zhu, *J. Phys. Chem. C*, 2008, **112**, 19841.
153. S. Guo, S. Dong, E. Wang, *ACS Nano*, 2010, **4**, 547.
154. F. Li, H. Yang, C. Shan, Q. Zhang, D. Han, A. Ivaska, L. Niu, *J. Mater. Chem.*, 2009, **19**, 4022.
155. X. Zhou, X. Huang, X. Qi, S. Wu, C. Xue, F. Y. Boey, Q. Yan, P. Chen, H. Zhang, *J. Phys. Chem. C*, 2009, **113**, 10842.
156. B. S. Kong, J. Geng, H. T. Jung, *Chem. Commun.*, 2174..
157. Y. K. Kim, H. K. Na, Y. W. Lee, H. Jang, S. W. Han, D. H. Min, , *Chem. Commun.*, 2009, **46**, 3185.
158. X. Huang, S. Li, Y. Huang, S. Wu, X. Zhou, S. Li, C. L. Gan, F. Boey, C. A. Mirkin, H. Zhang, *Nat. Commun.*, 2011, **2**, 292.
159. Y. B. Tang, C. S. Lee, J. Xu, Z. T. Liu, Z. H. Chen, Z. He, Y. L. Cao, G. Yuan, H. Song, L. Chen, L. Luo, H. M. Chen, W. J. Zhang, I. Bello, S. T. Lee, *ACS Nano*, 2010, **4**, 3482.
160. J. Zhu, G. Zeng, F. Nie, X. Xu, S. Chen, Q. Han, X. Wang, *Nanoscale*, 2010, **2**, 988.
161. D. Wang, R. Kou, D. Choi, Z. Yang, Z. Nie, J. Li, L. V. Saraf, *ACS Nano*, 2010, **4**, 1587.
162. S. Chen, J. Zhu, X. Wu, Q. Han, X. Wang, *ACS Nano*, 2010, **4**, 2822.
163. J. Yan, Z. Fan, T. Wei, W. Qian, M. Zhang, F. Wei, *Carbon*, 2010, **48**, 3825.
164. D. Wang, D. Choi, J. Li, Z. Yang, Z. Nie, R. Kou, D. Hu, C. Wang, L. V. Saraf, J. Zhang, I. A. Aksay, J. Liu, *ACS Nano*, 2009, **3**, 907.
165. Z. S. Wu, W. Ren, L. Wen, L. Gao, J. Zhao, Z. Chen, G. Zhou, F. Li, *ACS Nano*, 2010, **4**, 3187.
166. T. Ramanathan, A. A. Abdala, S. Stankovich, D. A. Dikin, M. Herrera-Alonso, R. D. Piner, D. H. Adamson, H. C. Schniepp, X. Chen, R. S. Ruoff, S. T. Nguyen, I. A. Aksay, R. K. Prud'homme, L. C. Brinson, *Nat. Nanotechnol.*, 2008, **3**, 327.
167. D. Cai, K. Yusoh, M. Song, *Nanotechnology*, 2009, **20**, 85712.
168. Y. Xu, W. Hong, H. Bai, C. Li, G. Shi, *Carbon*, 2009, **47**, 3538.
169. S. Stankovich, D. A. Dikin, G. H. B. Dommett, K. M. Kohlhaas, E. J. Zimney, E. A. Stach, R. D. Piner, S. T. Nguyen, R. S. Ruoff, *Nature*, 2006, **442**, 282.
170. D. Cai, M. Song, *Nanotechnology*, 2009, **20**, 315708.
171. M. A. Rafiee, J. Rafiee, I. Srivastava, Z. Wang, H. Song, Z. Z. Yu, N. Koratkar, *Small*, 2009, **6**, 179.
172. H. Hu, X. Wang, J. Wang, L. Wan, F. Liu, H. Zheng, R. Chen, C. Xu, *Chem. Phys. Lett.*, 2010, **484**, 247.
173. H. Kim, C. W. Macosko, *Polymer*, 2009, **50**, 3797.
174. *Br. Pat.*, 0717937.7, 2007.
175. J. Liang, Y. Huang, L. Zhang, Y. Wang, Y. Ma, T. Guo, Y. Chen, *Adv. Func. Mater.*, 2009, **19**, 1.
176. B. Das, K. E. Prasad, U. Ramamurty, C. N. R. Rao, *Nanotechnology*, 2009, **20**, 125705.

177. R. Verdejo, F. Barroso-Bujans, M. A. Rodriguez-Perez, J. A. Saja, M. A. Lopez-Manchado, *J. Mater.Chem.*, 2008, **18**, 2221.
178. S. G. Miller, M. A. Meador, I. Aksay, R. K. Prud'homme, http://www.grc.nasa.gov/WWW/RT/2006/RX/RX23P-miller.html.
179. E. Tkalya, M. Ghislandi, A. Alekseev, C. Koning, J. Loos, *J. Mater.Chem*, 2010, **20**, 3035.
180. X. Sun, M. Song, *Macromol. Theory Simul.*, 2009, **18**, 155.
181. H. Kim, C. W. Macosko, *Macromolecules*, 2008, **41**, 3317.
182. S. Ansari, E. P. Giannelis, *J. Polym. Sci. : Part B: Polym. Phys.*, 2009, **47**, 888.
183. P. Steurer, R. Wissert, R. Thomann, R. Mülhaupt, *Macromol. Rapid Commun.*, 2009, **30**, 316.
184. T. Kashiwagi, F. Du, J. F. Douglas, K. I. Winey, R. H. Harris, J. R. Shields, *Nature Mater.*, 2005, **4**, 928.
185. H. K. Jeong, Y. P. Lee, M. H. Jin, E. S. Kim, J. J. Bae, Y. H. Lee, *Chem. Phys. Lett.*, 2009, **470**, 255.
186. F. H. Gojny, M. II. G. Wichmann, B. Fiedler, I. A. Kinloch, W. Bauhofer, A. H. Windle, K. Schulte, *Polymer*, 2006, **47**, 2036.
187. H. Huang, C. Liu, Y. Wu, S. Fan, *Adv. Mater.*, 2005, **17**, 1652.
188. R. Haggenmueller, C. Guthy, J. R. Lukes, J. E. Fischer, K. I. Winey, *Macromelcules*, 2007, **40**, 2417.
189. A. Yu, P. Ramesh, M. E. Itkis, E. Bekyarova, R. C. Haddon, *J. Phys. Chem. C*, 2007, **111**, 7565.
190. A. Yu, P. Ramesh, X. Sun, E. Bekyarova, M. E. Itkis, R. C. Haddon, *Adv. Mater.*, 2008, **20**, 4740.
191. S. Ganguli, A. K. Roy, D. P. Anderson, *Carbon*, 2008, **46**, 806.
192. L. M. Veca, M. J. Meziani, W. Wang, X. Wang, F. Lu, P. Zhang, Y. Lin, R. Fee, J. W. Connell, Y. P. Sun, *Adv. Mater.*, 2009, **21**, 2088.
193. V. C. Tung, J. H. Huang, I. D. Tevis, F. Kim, J. Kim, C. W. Chu, S. I. Stupp, J. Huang, *J. Am. Chem. Soc.*, 2011, **133**, 4940.
194. D. A. Dikin, S. Stankovich, E. J. Zimney, R. D. Piner, G. H. B. Dommett, G. Evmenenko, S. Nguyen, R. S. Ruoff, *Nature*, 2007, **448**, 457.
195. D. Cai, M. Song, C. Xu, *Adv. Mater.*, 2008, **20**, 1706.
196. Z. Xu, C. Gao, *Nat. Commun.*, 2011, **2**, 571.
197. N. Yang, J. Zhai, D. Wang, Y. Chen, L. Jiang, *ACS Nano*, 2010, **4**, 887.
198. E. Yoo, T. Okata, T. Akita, M. Kohyama, J. Nakamura, I. Honma, *Nano Lett.*, 2009, **9**, 2255.
199. Y. Li, W. Gao, L. Ci, C. Wang, P. M. Ajayan, *Carbon*, 2010, **48**, 1124.
200. J. Yan, T. Wei, B. Shao, F. Ma, Z. Fan, M. Zhang, C. Zheng, Y. Shang, W. Qian, F. Wei, *Carbon*, 2010, **48**, 1731.
201. Z. Fan, J. Yan, L. Zhi, Q. Zhang, T. Wei, J. Feng, M. Zhang, W. Qian, F. Wei, *Adv. Mater.*, 2010, **22**, 3723.

CHAPTER 2

Gelation of Graphene Oxide

GAOQUAN SHI

Department of Chemistry, Tsinghua University, Beijing 100084, People's Republic of China
E-mail: gshi@tsinghua.edu.cn

2.1   Introduction

Graphene oxide (GO) has attracted a great deal of attention in recent years, mainly because it is a unique precursor of chemically derived graphene.[1] GO is usually synthesized by exfoliation of graphite oxide (multilayered GO) prepared by the oxidation of natural graphite powder in acidic media.[2–5] The typical method of synthesizing graphite oxide was developed by Hummers and co-workers.[6] A GO sheet obtained from the graphite oxide synthesized by Hummers' method usually has a lateral dimension in the range of submicrometres to several tens of micrometres and a thickness of ~ 1 nm.[7,8] This thickness is much greater than that of a clean graphene sheet (~ 0.3 nm) because of the presence of various oxygenated groups (*e.g.* hydroxyl, epoxide and carboxyl).[9–11] The hydroxyl and epoxide groups of GO are mainly on its basal plane and carboxyl groups are predominately at its edges.[12] As a result, the basal plane of a GO sheet is more hydrophobic than its edge. It behaves like a two-dimensional (2D) amphiphilic macromolecule.[13] The atom-thick 2D structure of GO sheets provides them with various new supramolecular behaviours compared with conventional low-dimensional counterparts such as carbon dots and carbon nanotubes (CNTs).[14] For example, GO sheets can be assembled into various macrostructures such as Langmuir–Blodgett (LB) and paper-like films.[15–17] In particular, the three-dimensional (3D) assembly of GO into gels can partly preserve its inherent structure and properties.[18] These

RSC Nanoscience & Nanotechnology No. 26
Polymer–Graphene Nanocomposites
Edited by Vikas Mittal
© The Royal Society of Chemistry 2012
Published by the Royal Society of Chemistry, www.rsc.org

GO or reduced GO (rGO)-based gels have potential applications in energy-related systems, sensor, catalysis and biotechnology.[18–20] In this chapter, the gelation behaviours of GO and rGO are summarized and discussed.

2.2 GO-Based Gels

2.2.1 Acid-Induced Gelation

A GO sheet can be regarded as a single layer graphite brings various hydrophilic oxygenated functional groups (Figure 2.1).[21] Thus, GO sheets can be dispersed in water to form a stable colloidal dispersion. It was also believed that the electrostatic repulsion between GO sheets, resulting from their ionized carboxyl groups, prevented their aggregation in aqueous medium.[22,23] Therefore, acidification of a GO dispersion would weaken the electrostatic repulsion. The zeta potential of a GO dispersion was measured to be increased with the decrease in its pH value, indicating the protonation of carboxyl groups.[21] GO sheets are unstable in a strong acidic aqueous medium because of insufficient mutual repulsion. As the concentration of GO (C_{GO}) was low (<2 mg mL^{-1}), a GO flocculant was formed. The edge-to-area ratio of a GO sheet increases with the decrease of its lateral dimension. Thus, in aqueous media with the same pH value, the smaller GO sheets should have higher solubility than their larger counterparts because of higher densities of ionized – COOH groups. Acidifying a GO dispersion to proper pH values can selectively precipitate large GO sheets. Actually, GO sheets prepared by Hummers' method can be separated into two portions, with large or small lateral dimensions, from their aqueous dispersion.[7] This method is based on the selective precipitation of GO sheets with lateral dimensions mostly ($>90\%$) larger than 40 µm^2 at a pH value of 4.0.

On the other hand, if C_{GO} was sufficiently high (>2 mg mL^{-1}), a hydrogel instead of an amorphous precipitate was formed upon acidification.[21] This concentration is called the critical gelation concentration (CGC) of GO. The CGC value depends on the lateral dimension and oxidation extent of GO

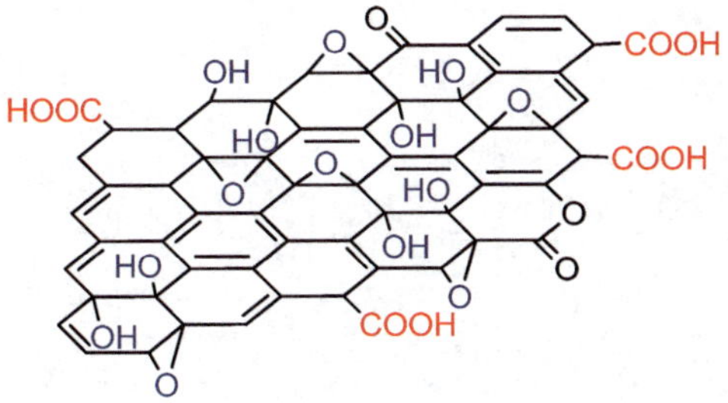

Figure 2.1 Chemical structure of GO. Hydrophilic groups are in red (carboxyl groups) and blue (hydroxyl and epoxy groups). Reproduced with permission from reference 21, © 2011 ACS.

sheets. Larger GO sheets with fewer oxygenated groups have lower CGC. At concentrations higher than the CGC, GO sheets in their dispersion can come into contact with each other to form a dynamical 3D network. This assumption has been confirmed by the fact that the volume of a GO dispersion can be maintained after freeze drying.[21] Furthermore, lyophilized GO dispersion shows a 3D network composed of GO sheets (Figure 2.2). As the bonding forces between GO sheets were sufficiently enhanced by acidification, a sol–gel transition would occur. The morphology of the GO hydrogel (GOH) prepared by acidification is identical to that shown in Figure 2.2. Its pore sizes are in the range of submicrometres to a few micrometres. Its pore walls consist of GO sheets and they are cross-linked through hydrogen bonding between their oxygenated groups and π-stacking between their conjugated domains.

The lateral dimensions of GO sheets have strong effects on their gelation behaviour. As the lateral dimensions of most GO sheets are as large as several micrometres, a stable hydrogel can easily be formed in the GO dispersion with C_{GO} greater than CGC (>2 mg mL^{-1}) by acidification to a pH value around 0.6.[21] However, it is difficult to make a stable hydrogel from GO sheets with lateral dimensions less than 1 μm even at a much higher concentration (*e.g.* 9 mg mL^{-1}). In this case, acidification of GO resulted in precipitation. These phenomena can be explained as follows. GO sheets are randomly orientated in the hydrogel, while they adopt parallel arrangement in their sediment. The latter staking mode is energetically favourable because of the larger contact area between GO sheets. However, the mobility of large GO sheets in solution is greatly limited. As a result, it is difficult to adjust their orientation to be

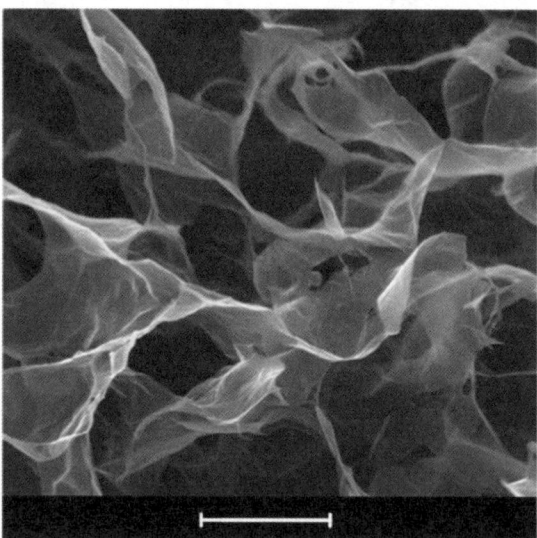

Figure 2.2 Scanning electron microscopy (SEM) image of freeze-dried GO dispersion with a concentration of 5 mg mL^{-1}, scale bar = 10 μm. Reproduced with permission from reference 21, © 2011 ACS.

parallel with each other. Compared with gelation, precipitation is a kinetically slower process. Moreover, the large conjugated basal planes make GO sheets stiff, forming a stable network. However, if the sizes of GO sheets are reduced to than 1 μm, they can change their conformations and positions in solution more easily. Consequently, acidified GO sheets trend to aggregate in an energetically favourable layer-by-layer manner to form a precipitate. The stability of a hydrogel is decided by the relative strength of repulsion and binding forces between GO sheets. For example, over-acidifying the dispersion of large GO sheets would also lead to the formation of irregular aggregates because the repulsion force would be too weak.[21]

2.2.2 Cross-linker-Induced Gelation

The addition of cross-linkers such as metal ions, polymers or small organic molecules to GO dispersions can promote the gelation of GO sheets because of increasing their bonding forces. Multivalent anions including Mg^{2+}, Ca^{2+}, Cu^{2+}, Pb^{2+}, Cr^{3+} and Fe^{3+} are effective cross-linkers.[21] The anions partially neutralize the negative charges of GO sheets and bonding them together into networks through electrostatic interaction. This mechanism has been confirmed by adding a chelating agent, ethylene diamine tetraacetic acid (EDTA), to the GO/metal hydrogel to remove the metal ions. Upon addition of EDTA, the hydrogels decomposed quickly due to the loss of metal ions. This phenomenon also confirmed that GO sheets are in contact with each other in their concentrated solution and small metal ions can interact with individual GO sheets to form 3D networks. If a reductive multivalent ion such as Fe^{2+} was used, GO sheets were reduced and a conductive graphene hydrogel was obtained.[24] Simultaneously, Fe^{2+} ions were transformed to α-FeOOH or Fe_3O_4 nanoparticles at different pH values. The resulting composites have potential applications in removing heavy ions and oils or industrial water purification. On the other hand, it was found that monovalent ions such as Li^+, Na^+, K^+, Ag^+ could not induce the gelation of GO.[21] This is possibly due to that these ions cannot provide a sufficiently strong electrostatic force to bind neighbouring GO sheets.

Small quaternary ammonium salts, such as cetyltrimethyl ammonium bromide (CTAB) and tetramethylammonium chloride (TMAC), can also induce the formation of GO hydrogels.[21] This is mainly because electrostatic interaction (ESI) is a long-range force and it is usually stronger than a hydrogen bond. Moreover, CTAB was found to have a lower CGC value (0.3 mg mL^{-1}) than TMAC (1.9 mg mL^{-1}), indicating that the long hydrophobic chain on CTAB increased its interaction with GO sheets.

The addition of various polymers can result in the formation of GO–polymer composite hydrogels. Poly(vinyl alcohol) (PVA) was first used to blend with GO dispersion to form a composite hydrogel.[25] In this case, hydrogen bonding between PVA chains and GO sheets was recognized as the driving force. In this hydrogel, GO sheets form a network and PVA acts as a physical cross-linking agent. This is different from the case of classical nanocomposite hydrogels, in which nanomaterials usually provide cross-linking sites for the polymer gelators.[26] The GO–PVA composite hydrogels show 3D networks with a morphology similar to that of pure GOH (Figure 2.3A), indicating that the polymer additive did not change the conformation of GO sheets. X-diffraction studies indicate that PVA chains are coated on the surfaces of GO sheets and increased the d-space of GO sheets (Figure 2.3B). The XRD pattern of pure freeze-dried GO shows a diffraction peak at $2\theta = 11.4°$, corresponding to an interplanar spacing of 7.76 Å.[5] This result reflects that GO sheets were aggregated after lyophilization. Pure PVA exhibits a characteristic diffraction peak at $2\theta = 19.5°$, resulting from its (101) crystal planes.[27] The XRD patterns of GO–PVA composites are different from those of pure GO and PVA, mainly due to the strong interaction between these two components. With the increase of PVA–GO weight ratio ($r_{P/G}$), the peak in region II is weakened and broadened, and its position shifted to lower angles. This is because the coating of PVA on GO sheets induced an enlargement of GO interplanar spacing. When $r_{P/G}$ was increased to be greater than 1:10, a new peak is appeared in region 1, which further shifted to smaller angles with the increase of PVA content, and finally to $2\theta = 5.26°$ (16.8 Å) as $r_{P/G} = 1:1$. This new peak is attributed to PVA-spaced GO interplanar spacing. When $r_{P/G} < 1:2$, only peak I was observed and peak II disappeared, indicating the absence of direct GO–GO packing mode in these composites. Moreover, another new peak around 18° is observed (peak III), corresponding to a spacing of 4.9 Å, which can be assigned to the spacing between PVA chains and GO sheets or the distance between PVA chains. The GO–PVA composite hydrogel is pH sensitive; it is gellable in acidic media but undergoes gel–sol transition under alkaline conditions.[25] This is also due to the interaction force

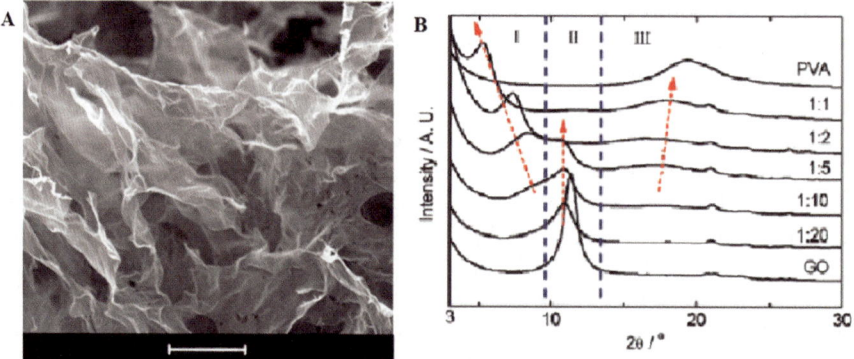

Figure 2.3   (A) SEM images of lyophilized GO–PVA blend with $r_{P/G} = 1:20$. (B) XRD patterns of lyophilized PVA, GO and GO–PVA blends with various $r_{P/G}$. Reproduced with permission from reference 25, © 2010 RSC.

between GO sheets being modulated by the pH value. Thus, it can be used for pH-controlled selective drug release.

A GO–PVA composite hydrogel has also been prepared by treating the GO–PVA aqueous mixture through a repeated freeze/thaw procedure, consisting of a freezing step (12 h at $-22\ ^{\circ}$C) followed by a thawing step (3 h at 25 $^{\circ}$C).[28] In this case, the mechanical property of PVA hydrogel was greatly improved by adding a small amount of GO nanofiller. Compared to pure PVA hydrogels, a 132% increase in tensile strength and a 36% improvement of compressive strength were achieved with the addition of 0.8 wt% of GO.

Similar to PVA, poly(ethylene oxide) (PEO) and poly(vinyl pyrrolidone) (PVP) are also hydrogen bond acceptors, whereas hydroxypropylcellulose is both a hydrogen bond donor and an acceptor.[21] Thus, they can form hydrogen bonds with adjacent GO sheets, providing an additional bonding force for the gelation of GO. The XRD results of GO–PEO composite hydrogel are similar to that of GO–PVA, indicating that the PEO chains are coated on GO sheets. In comparison, all the XRD patterns of GO–PVP composites show a peak associated with GO sheets at the angle range of $2\theta = 10.5^{\circ} - 12.5^{\circ}$ (Figure 2.4). The position of this peak is close to that of pure GO sheets at $2\theta = 11.5^{\circ}$ (the interplanar distance of GO sheets was calculated to be 1.54 nm). However, its intensity decreases remarkably, and its full width at half-maximum (fwhm) value increases simultaneously with the increase of PVP content. These results indicate that the introduction of PVP chains partly prevented the stacking of GO sheets. The XRD peak associated with PVP intercalated GO sheets was not observed in the experimental angle scale, possibly due to their large d-space.

ESI is another driving force for supramolecular self-assembly, and the assembly of GO sheets induced by ESI has also been realized in many systems.

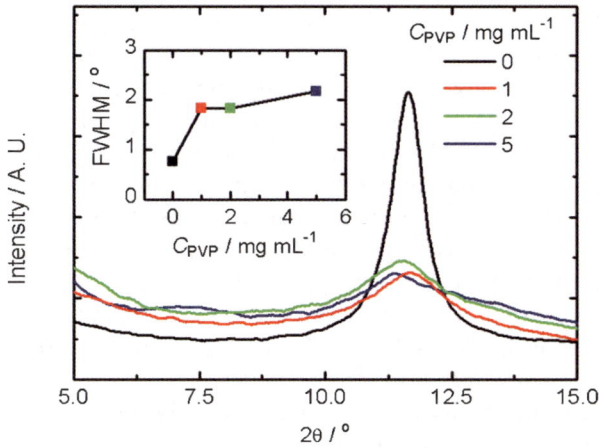

Figure 2.4 X-ray diffraction patterns of lyophilized GO and GO–PVP hydrogels with concentration of GO $= 5$ mg mL^{-1}. Reproduced with permission from reference 21, © 2011 ACS.

For example, poly(dimethyldiallylammonium chloride) (PDDA) is an effective cross-linker for forming a hydrogel with GO.[21] In this system, ESI between the quaternary ammonium groups of PDDA and the carboxyl groups of GO sheets is the dominating driving force for gelation. The CGC of PDDA was measured to be as low as 0.1 mg mL^{-1}. This value is much lower than that of PVA, implying that ESI is more effective than hydrogen bonding for GO gelation.

Graphene oxide–conducting polymer (GO/CP) composite hydrogels can be prepared by *in situ* chemical polymerization of corresponding aromatic monomers in aqueous dispersions of GO sheets.[29] Thin conducting polymer films were coated on GO sheets because of the π–π stacking and ESI interactions between positively charged conjugated polymer chains and negatively charged GO sheets. During the polymerization process, GO sheets acted as a template of growing polymer layers and the polymer chains acted as a cross-linker of GO networks. GO–polypyrrole (PPy), GO–poly(3,4-ethylenedioxythiophene) (PEDOT) and GO–polyaniline (PANi) hydrogels were obtained by this technique. Among them, GO–PPy composite hydrogels were tested to have low critical hydrogel concentrations (<1%, by weight), high storage moduli (>10 kPa) and electrical conductivity, and strong electrochemical activity. A gas sensor based on a typical GO–PPy hydrogel showed high sensitivity towards ammonia gas.

Chemically grafting polymer chains onto GO sheets is also an effective route to GO composite hydrogel. Recently, Wu and co-workers prepared a GO–poly(*N*-isopropylacrylamide) (PNIPAM) hydrogel network by covalently bonding GO sheets and PNIPAM-co-AA microgels directly in water (Figure 2.5).[30] This composite hydrogel exhibited dual thermal and pH responses with good reversibility because of the presence of PNIPAM component and residual carboxyl groups. A infrared (IR)-responsive hydrogel has been prepared by grafting PNIPAM chains to GO sheets functionalized with glycidyl methacrylate.[31] The superior thermal conductivity and photo-thermal conversion capability of GO greatly enhanced the performance of the IR-responsive PNIPAM hydrogel. This material has potential applications in actuators, such as in microelectromechanical systems, microfluidic devices and labs on chips. GO–polyacrylamide (GO–PAM) hydrogels were also prepared by *in situ* polymerization.[32] The PAM chains increased the dispersibility of GO sheets in the composite hydrogels and consequently significantly improved their mechanical properties. The compressive strength of the GO–PAM hydrogel loaded with 1 wt% GO increased sixfold in comparison to that of pure PAM hydrogel. The GO–PAM-based hydrogels were responsive to external stimuli such as pH and electric fields.

Small biomolecules including peptides and carbohydrates, and biomacro-molecules such as protein and DNA, have also been explored as cross-linkers of GO-based gels. For example, a series of amphiphilic molecules each possessing a polar carbohydrate headgroup attached to a hydrophobic pyrene group have been used as the gelators for GO.[33] The gelation process depends

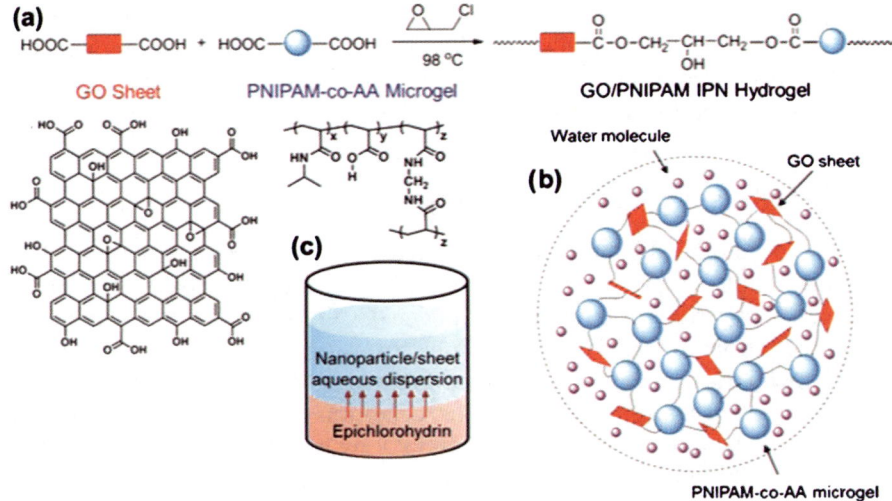

Figure 2.5 (a) Formation of GO–PNIPAM interpenetrating (IPN) hydrogel networks via the reaction between epichlorohydrin (ECH) and carboxyl groups in GO sheets and PNIPAM-co-AA microgels. (b) Structural sketch of GO–PNIPAM IPN hydrogel. (c) The sealed reaction tube was incubated at 98 °C while ECH permeated into the aqueous phase to induce the cross-linking reaction. Reproduced with permission from reference 30, © 2011 RSC.

on the molecular structure of the gelator. The driving forces for the gelation process were studied to be π–π stacking and hydrogen bonding interactions. These GO-based composite hydrogels can be applied to remove dye from aqueous solution.

A pyrene-containing peptide was also used as the cross-linker of GO hydrogel.[34] Rheological studies reveal that the storage modulus (G') of a graphene-containing hybrid organogel is seven times more rigid than that of native organogel. This suggests that inclusion of graphene into an organogel makes the gel a more elastic and solid-like soft material. A haemoglobin–GO composite hygrogel can be prepared by shaking an aqueous mixture containing 4.4 mg mL^{-1} haemoglobin and 10 mg mL^{-1} GO.[35] The electrostatic attraction between positively charged haemoglobin molecules and negatively charged GO sheets is responsible for the formation of this composite hydrogel.

Double stranded DNA (ds-DNA) chains are negatively charged and their base groups are bonded in pairs and located inside the strains. Therefore, ds-DNA strains have weak interaction with GO sheets. On the other hand, ds-DNA strains can be unwound to single strained chains (ss-DNA) by heating.[36] The base groups of ss-DNA chains are exposed and can interact with GO sheets through π–π stacking interaction. Actually, it was found that in an aqueous mixture of 3 mg mL^{-1} GO and 5 mg mL^{-1} ds-DNA gelation occurred after heating at 90 °C for 5 min. The hydrogel possesses high mechanical strength and environmental stability because of the GO framework and the strong interactions between both components. Furthermore, the high specific surface area of the hydrogel also provides it with a large dye-loading capacity. The hydrogel also exhibited a self-healing property because of reversible transformation between ds- and ss-DNA chains.

2.3   Reduced GO-Based Gels

GO can be chemically or thermally reduced to conductive graphene. The resulting graphene materials are called reduced graphene oxide (rGO) or chemically converted (or modified) graphene (CCG, or CMG).[37,38] rGO sheets are more hydrophobic and have weaker electrostatic repulsion and stronger π–π stacking interaction than those of their GO precursors.[38,39] Thus, they can be self-assembled to hydrogels with 3D interpenetrating networks even without a cross-linker. Unlike the self-assembled GO hydrogels formed by physical interactions, the rGO-based hydrogels have conductive frameworks.

2.3.1   Hydrothermal or Solvothermal Reduction

Typical rGO hydrogel (rGOH) can be easily prepared by heating 2 mg mL^{-1} homogeneous graphene oxide aqueous dispersion sealed in a Teflon-lined autoclave at 180 °C for 12 h.[40] The as-prepared rGOH containing about 2.6 wt% of hydrothermally reduced GO (or graphene) and 97.4 wt% water is mechanically strong. The rGOH has a well-defined and interconnected 3D porous network as imaged by scanning electron microscopy (SEM) of its freeze-dried samples (Figure 2.6). The pore sizes of the network are in the range of submicrometres to several micrometres and the pore walls consist of thin layers of stacked graphene sheets. The partial overlapping or coalescing of flexible graphene sheets resulted in the formation of physical cross-linking sites of the rGOHs framework. The rGOH has an electrical conductivity of ~5 mS cm^{-1}. Furthermore, it is thermally stable in the temperature range 25–100 °C and its storage modulus (450−490 kPa) is about 1−3 orders of magnitude higher than those of conventional self-assembled hydrogels. The conductive property and porous structure mean this rGOH can be used for the electrodes of supercapacitors.[40] The rGOH prepared *via* hydrothermal reduction of GO dispersions can be further reduced with hydrazine (Hz) or hydroiodic acid (HI) to improve their conductivities.[41] The conductivities of these chemically reduced graphene hydrogels are an order of magnitude higher than that of the original rGOH. A supercapacitor based on Hz-reduced rGOH exhibited a high specific capacitance of 220 F g^{-1} at 1 A g^{-1}, and 74% of this capacitance was maintained as the discharging current density was increased up to 100 A g^{-1}. Furthermore, it showed high power density and long cycle life. The 3D rGOH matrix was also used as a scaffold for proliferation of a MG63 cell line.[42] The experimental results indicate that guided filopodia protrusions of MG63 on the hydrogel were observed on the third day of cell

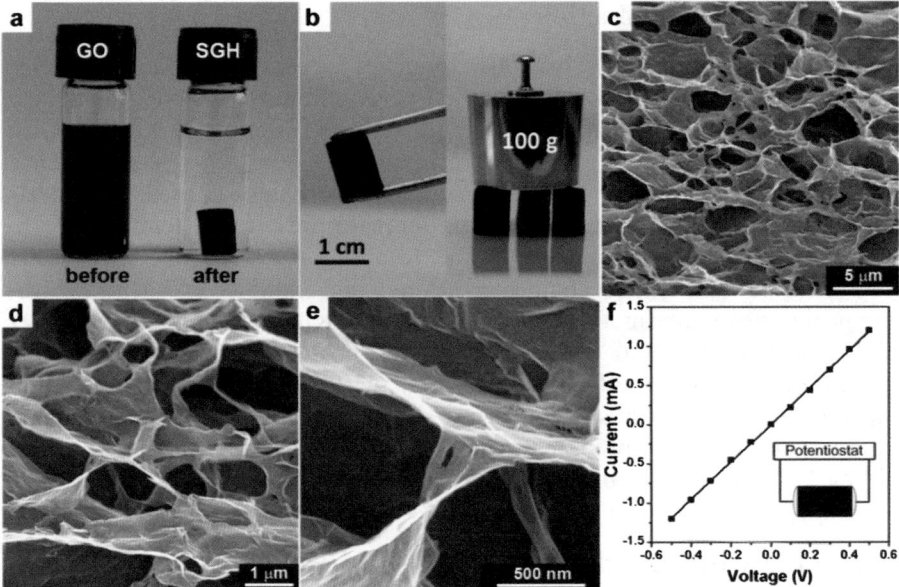

Figure 2.6   (a) Photographs of a 2 mg mL^{-1} homogeneous GO aqueous dispersion before and after hydrothermal reduction at 180 °C for 12 h; (b) photographs of a strong SGH allowing easy handling and supporting weight; (c–e) SEM images with different magnifications of the SGH interior microstructures; (f) room temperature I–V curve of the SGH exhibiting ohmic characteristics; the inset shows the two-probe method used for the conductivity measurements. Reproduced with permission from reference 40, © 2011 ACS.

culture, demonstrating compatibility of the graphene hydrogel structure for bioapplications.

An organogel of rGO can be prepared by changing the aqueous medium to a thermally stable organic solvent.[43] GO sheets were tested for dispersibility in propylene carbonate (PC). A 2 mg mL^{-1} GO dispersion was found to be stable and no precipitation was observed after ageing for 1 month. The GO sheets dispersed in PC are monolayers without aggregation. After solvothermal reduction of 2 mg mL^{-1} GO dispersion in PC, a columniform of rGO organogel (rGO-OG) was produced. The formation of this rGO-OG is also attributed to the regional overlapping and coalescing of flexible reduced GO nanosheets through π–π interaction. The interconnected graphene nanosheets provide rGO-OG with a high electrical conductivity of 2 S m^{-1}, much higher than that of rGOH (5 mS cm^{-1}). This is mainly because the solvothermal reduction of GO in PC is more effective than in an aqueous medium. The as-formed rGO-OG can be used directly for the electrodes of supercapacitors with PC electrolyte. The supercapacitors can be operated in a wide voltage range of 0–3 V and possess high specific capacity, energy and power densities.

2.3.2 Chemical Reduction

GO can be reduced by various reducing agents such as hydrazine monohydrate[44], sodium borohydride[45], hydroquinone[46] or strongly alkaline solutions[47]. However, when a strong reducing agent is applied, rGO sheets in aggregates are formed as powdery precipitates. On the other hand, when a weak organic reducing agent such as sodium ascorbate was used, GO sheets were reduced gradually at relative low temperatures (room temperature to 90 °C) to form a stable rGOH.[48] The resulting rGOH is also electrically conductive (1 S m^{-1}), mechanically strong, and exhibits excellent electrochemical performance in a supercapacitor. The rGOHs can be prepared under mild and environmental friendly conditions, which offers many potential applications in biotechnology, such as cell scaffold, enzyme immobilization and biocatalysis. Composite hydrogels of rGO and CNTs or noble metal nanoparticles were also prepared by using vitamin C as the reducing agent and adding CNTs or metal salt to the GO dispersion.[49]

2.3.3 Electrochemical Reduction

The shapes of the graphene-based gels described above depend on those of the containers used for carrying out the gelation reactions. A cylindrical gel is usually obtained when a columniform bottle is used because of the uniform assembling of graphene sheets from different directions to form a more compact structure. Therefore, it is difficult to shape them into thin films with desired structures. Recently, we developed an electrochemical technique for forming thin films of rGOH on Au foils.[50] The rGOH films were deposited on Au electrodes by electrolysing 3 mg mL^{-1} GO aqueous suspension containing 0.1 M lithium perchlorate (LiClO$_4$) at an applied potential of -1.2 V for 10 s. At negative potential, GO sheets were reduced to conductive rGO. The resulting electrochemically reduced GO sheets are more hydrophobic and have weaker electrostatic repulsion and stronger π–π stacking interaction than of their GO precursors. Thus, they self-assembled to form 3D interpenetrating networks and deposited onto the substrate electrode, driven by the electric field.[40] The as-formed rGOH film consists of two parts: a basal layer of approximately 200 nm thick and a porous 3D interpenetrating layer with a thickness of about 20 μm. The pore sizes of the network are in the range of several micrometres to more than 10 μm, and the pore walls are nearly vertical to the surface of the Au electrode. A double-layer capacitor based on these electrochemically reduced GO films (ErGO-DLC) showed extraordinary high rate performance. At 120 Hz the ErGO-DLC exhibited a phase angle of $-84°$, a specific capacitance of 283 mF cm^{-2} and a resistor–capacitor (RC) time constant of 1.35 ms, making it capable of replacing a commercial aluminium electrolyte supercapacitor (AEC) for a 120 Hz filtering application. Furthermore, the thickness of the ErGO films can be modified by controlling the electrochemical deposition time. Thus, the performance of the ErGO-DLCs can also be modulated.

2.4 Conclusion

The atom-thick two-dimensional structure of GO sheets makes them gellable at a low critical concentration. The gelation process is controlled by the balance of bonding and repulsion forces between GO sheets. Acidification, reduction or addition of cross-linkers can promote the gelation of GO because of weakening the repulsion force. GO-based hydrogels have large specific surface areas and improved mechanical properties. They have potential applications in sensors, actuators, electrocatalysis and drug delivery. Reduction of GO can produce conductive graphene gels with high electrical conductivity, good thermal and environmental stability and excellent mechanical properties. They have been investigated for the application in supercapacitors, cell growth and absorption of heavy metal ions or oils. The structures and properties of GO or rGO can be easily modified, and various binders can be used to promote the gelation processes. Therefore, it is reasonable to expect that the family of graphene-based gels will expand dramatically in the near future.

Acknowledgements

This work was supported by the National Basic Research Program of China (973 Program, 2012CB933402) and the Natural Science Foundation of China (51161120361, 91027028).

References

1. K. P. Lou, Q. L. Bao, G. Eda, M. Chhowalla, *Nat. Chem.*, 2010, **2**, 1015.
2. S. Stankovich, R. D. Piner, X. Q. Chen, N. Q. Wu, S. T. Nguyen, R. S. Ruoff, *J. Mater. Chem.*, 2006, **16**, 155
3. S. Stankovich, D. A. Dikin, G. H. B. Dommett, K. M. Kohlhaas, E. J. Zimney, E. A. Stach, R. D. Piner, S. T. Nguyen, R. S. Ruoff, *Nature*, 2006, **442**, 282.
4. Y. X. Xu, H. Bai, G. W. Lu, C. Li, G. Q. Shi, *J. Am. Chem. Soc.*, 2008, **130**, 5856.
5. Y. X. Xu, L. Zhao, H. Bai, W. J. Hong, C. Li,G. Q. Shi, *J. Am. Chem. Soc.*, 2009, **131**, 13490.
6. W. S. Hummers, R. E. Offeman, *J. Am. Chem. Soc.*, 1958, **80**, 1339.
7. X. L. Wang, H. Bai, G. Q. Shi, *J. Am. Chem. Soc.*, 2011, **133**, 6338.
8. L. J. Cote, F. Kim, J. X. Huang, *J. Am. Chem. Soc.*, 2009, **131**, 1043.
9. S. Park, R. S. Ruoff. *Nature Nanotechnol.* 2009, **4**, 217.
10. X. L. Li, G. Y. Zhang, X. D. Bai, X. M. Sun, X. R. Wang, E. Wang, H. J. Dai, *Nature Nanotechnol.*, 2008, **3**, 538.
11. L. Zhang, J. J. Liang, Y. Huang, Y. F. Ma, Y. Wang, Y. S. Chen, *Carbon*, 2009, **47**, 3365.

12. T. Szabo, O. Berkesi, P. Forgo, K. Josepovits, Y. Sanakis, D. Petridis, I. Dekany, *Chem. Mater.*, 2006, **18**, 2740.
13. J. Kim, L. J. Cote, F. Kim, W. Yuan, K. R. Shull, J. X. Huang, *J. Am. Chem. Soc.*, 2010, **132**, 8180.
14. Y. X. Xu, G. Q. Shi, *J. Mater. Chem.*, 2011, **21**, 3311.
15. N. V. Medhekar, A. Ramasubramaniam, R. S. Ruoff, V. B. Shenoy, *ACS Nano* 2010, **4**, 2300.
16. D. A. Dikin, S. Stankovich, E. J. Zimney, R. D. Piner, G. H. B. Dommett, G. Evmenenko, S. T. Nguyen, R. S. Ruoff, *Nature*, 2007, **448**, 457.
17. O. C. Compton, S. T. Nguyen, *Small*, 2010, **6**, 711.
18. J. L. Vickery, A. J. Patil, S. Mann, *Adv. Mater.* 2009, **21**, 2180.
19. Q. Wu, Y. Q. Sun, H. Bai, G. Q. Shi, *Phys. Chem. Chem. Phys.*, 2011, **13**, 11193; B. Adhikari, A. Biswas, A. Banerjee, *Langmuir* 2012, **28**, 1460.
20. B. Adhikari, A. Biswas, A. Banerjee, *Langmuir* 2012, **28**, 1460.
21. H. Bai, C. Li, X. L. Wang, G. Q. Shi, *J. Phys. Chem. C*, 2011, **115**, 5545.
22. W. Gao, L. B. Alemany, L. Ci, P. M. Ajayan, *Nat. Chem.* 2009, **1**, 403.
23. D. Li, M. B. Muller, S. Gilje, R. B. Kaner, G. G. Wallace, *Nat. Nanotechnol.* 2008, **3**, 101.
24. H. P. Cong, X. C. Ren, P. Wang, S. H. Yu, 2012, *ACS Nano*, 2012, **6**, 2693.
25. H. Bai, C. Li, X. L. Wang, G. Q. Shi, *Chem. Commun.*, 2010, **46**, 2376.
26. P. Schexnailder, G. Schmidt, *Colloid Polym. Sci.*, 2009, **287**, 1.
27. Y. Nishio, R. S. J. Manley, *Macromolecules*, 1988, **21**, 1270.
28. L. Zhang, Z. P. Wang, C. Xu, Y. Li, J. P. Gao, W. Wang, Y. Liu, *J. Mater. Chem.*, 2011, **21**, 10399.
29. H. Bai, K. X. Sheng, P. F. Zhang, C. Li, G. Q. Shi, *J. Mater. Chem.*, 2011, **21**, 18653.
30. S. T. Sun, P. Y. Wu, *J. Mater. Chem.*, 2011, **21**, 4095.
31. C. W. Lo, D. F. Zhu, H. R. Jiang, *Soft Matter*, 2011, **7**, 5604.
32. N. N. Zhang, R. Q. Li, L. Zhang, H. B. Chen, W. C. Wang, Y. Liu, T. Wu, X. D. Wang, W. Wang, Y. Li, Y. Zhao, J. P. Gao, *Soft Matter*, 2011, **7**, 7231.
33. Q. Y. Cheng, D. Zhou, Y. Gao, Q. Chen, Z. Zhang, B. H. Han, *Langmuir*, 2012, **28**, 3005.
34. B. Adhikari, J. Nanda, A. Banerjee, *Chem. Eur. J.* 2011, **17**, 11488.
35. C. C. Huang, H. Bai, C. Li, G. Q. Shi, *Chem. Commun.*, 2011, **47**, 4962.
36. Y. X. Xu, Q. Wu, Y. Q. Sun, H. Bai, G. Q. Shi, *ACS Nano*, 2010, **4**, 7358.
37. B. Adhikari, A. Banerjee, *Soft Matter*, 2011, **7**, 9259.
38. H. Bai, C. Li, G. Q. Shi, *Adv. Mater.*, 2011, **23**, 1089.
39. Y. Q. Sun, Q. Wu, G. Q. Shi, *Energy Environ. Sci.*, 2011, **4**, 1113.
40. Y. X. Xu, K. X. Sheng, C. Li, G. Q. Shi, *ACS Nano*, 2010, **4**, 4324.
41. L. Zhang, G. Q. Shi, *J. Phys. Chem. C* 2011, **115**, 17206.
42. H. N. Lim, N. M. Huang, S. S. Lim, I. Harrison, C. H. Chia, *Int. J. Nanomed.*, 2011, **6**, 1817.
43. Y. Q. Sun, Q. Wu, G. Q. Shi, *Phys. Chem. Chem. Phys.*, 2011, **13**, 17249.
44. S. Stankovich, R. D. Piner, X. Q. Chen, N. Q. Wu, S. T. Nguyen, R. S. Ruoff, *J. Mater. Chem.*, 2006, **16**, 155.

45. T. Cassagneau, J. H. Fendler, *J. Phys. Chem. B* 1999, **103**, 1789.
46. G. Wang, J. Yang, J. Park, X. Gou, B. Wang, H. Liu, J. Yao, *J. Phys. Chem. C*, 2008, **112**, 8192.
47. X. Fan, W. Peng, Y. Li, X. Li, S. Wang, G. Zhang, F. Zhang, *Adv. Mater.*, 2008, **20**, 449.
48. K. X. Sheng, Y. X. Xue, C. Li, G. Q. Shi, *New Carbon Mater.*, 2011, **26**, 9.
49. Z. Y. Sui, X. T. Zhang, Y. Lei, Y. J. Luo, *Carbon*, 2011, **49**, 4314.
50. K. X. Sheng, Y. Q. Sun, C. Li, W. J. Yuan, G. Q. Shi, *Sci. Rep.* 2012, **2**, 247.

CHAPTER 3

Electrically Conductive Polymer–Graphene Composites Prepared Using Latex Technology

NADIA GROSSIORD*[1], MARIE-CLAIRE HERMANT[2] AND EVGENIY TKALYA[3]

[1] Sabic Innovative Plastics, Plasticslaan 1, PO Box 117, 4600 AC Bergen op Zoom, The Netherlands; [2] BASF SE, Global Polymer Research, GKT/U–B 001, 67056 Ludwigshafen, Germany; [3] Laboratory of Polymer Chemistry, Eindhoven University of Technology, SPC–Helix, Den Dolech 2, 5600 MB Eindhoven, the Netherlands
*E-mail: nadia.grossiord@sabic-ip.com

3.1 Introduction

The low-cost production of high-performance conductive polymeric composites is eagerly sought after due to the rising cost of materials currently used in the preparation of conductive components, especially in electronics. Factors including the scalable production of polymer-based materials, their low density, easy processing and flexibility of use allow designers to apply electronics in new and innovative platforms.

With the discovery of novel inorganic (nano)fillers (such as fullerenes, nanotubes, quantum dots, *etc.*) with exceptional properties, the progress in the (nano)composite field has been rapid. In order for these materials to compete with current technologies, low cost and high performance are two key parameters that need to be achieved. Graphene has emerged from the dust

RSC Nanoscience & Nanotechnology No. 26
Polymer–Graphene Nanocomposites
Edited by Vikas Mittal
© The Royal Society of Chemistry 2012
Published by the Royal Society of Chemistry, www.rsc.org

settling after the carbon nanotube (CNT) hype as a filler that could achieve these goals. With a high electrical conductivity[1] coupled to an expected low raw material cost, all that is required now is a means to prepare polymeric composites that alter these two characteristics as little as possible.

A methodology aimed at processing cheap materials with a minimum number of steps is ideal. If most of the materials utilized are environmentally friendly and/or recyclable, it is an added bonus. Latex technology is one such technique, developed initially for use with CNTs, which has now been applied to graphene with promising results.[2]

In this chapter, the basics of the latex-based process to produce nanocomposites, with examples of its first application to produce CNT–polymer, are given. This is followed by a short overview of the strategies developed to interface graphene and polymers using aqueous-based methodologies. In particular, the adaptation of latex-based technology to ensure a homogeneously dispersion of graphene sheets in a polymer matrix to produce technologically performing, electrically conductive nanocomposites, is explained. Finally, the performance of graphene-based nanocomposites prepared by latex technology is compared with the state-of-the-art of graphene–polymer composites prepared by other methods.

3.2 Fundamentals of Latex Technology

Intimately mixing inorganic fillers and polymer matrices is often a major challenge, as this mixing step must overcome the initial aggregated state of the fillers. This tendency to aggregate is amplified for conductive nanofillers such as CNTs and graphene that are entangled and/or tightly bound into bundles[3] and stacks[4] respectively. Individualizing these specific fillers has, however, been achieved through the use of surfactants, an aqueous media and ultrasonication.[5,6] These aqueous nanofiller dispersions can be saturated with water-soluble polymers and subsequently dried to yield a composite.[7] A more versatile approach utilizes the so-called *latex technology*. The main steps of the process are the following (see Figure 3.1): (i) preparation of an aqueous colloidal dispersion of the nanofiller; (ii) mixing with a polymer latex to form a two-component colloidal mixture; (iii) drying (lyophilization) of the colloidal mixture and (iv) processing in order to yield a composite in which the fillers are homogeneously dispersed in the polymer matrix. This process has already been applied to a variety of polymeric latexes,[8–10] as well as fillers.[11,12]

The conductive nature of (nano)composites made using conductive fillers, such as CNTs, dispersed in an insulating matrix, is most of the time exclusively due to the formation of conductive paths of filler particles within the matrix. The formation of conductive paths is governed not only by geometrical factors, such as the aspect ratio and state of aggregation of the fillers, but also by filler–matrix interactions. Keeping this in mind, it is important to understand how the characteristics of the as-produced filler can dramatically be modified during the composite manufacturing process. For example, the

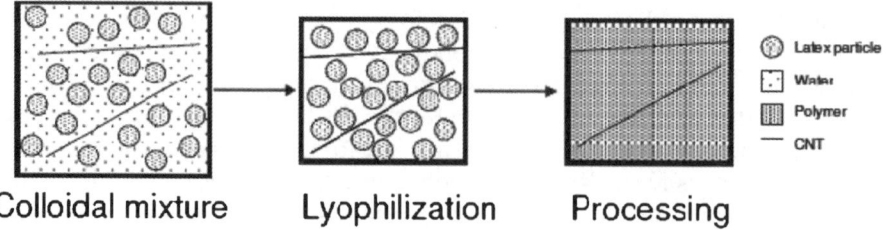

Colloidal mixture   Lyophilization   Processing

TEM and SEM micrographs of the three main steps:

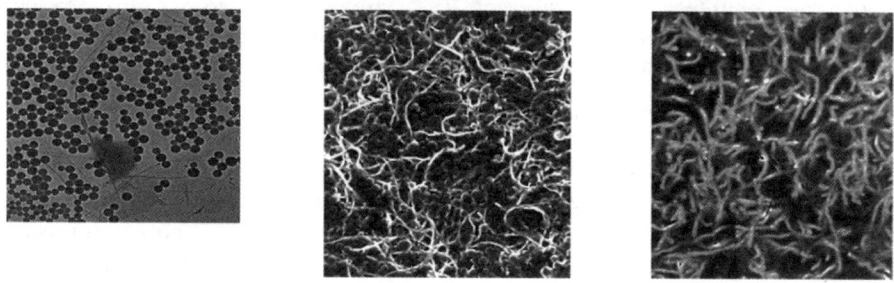

Figure 3.1   Schematic and pictorial representation of the latex technology process as applied to CNT composites.

CNT aspect ratio is ultimately determined by the synthesis method but may be dramatically reduced by sonication-assisted preparation of an aqueous dispersion of the filler (first step of latex technology).[13,14]

Considering the impact that the geometry/configuration or state of aggregation of the filler has on the final nanocomposite's performance, it is important to understand the impact of the various steps in the latex technology process. Broadly speaking, the following steps are most influential:

- Sonication-driven CNT individualization in an aqueous system
- Mixing of the two aqueous colloidal dispersions (step 1 of Figure 3.1) since this step can potentially lead to a new equilibrium of the system
- Melt-processing, because the filler diffusion and reorientation in the polymer melt are governed by the filler–polymer chain interaction.

On the contrary, the lyophilization step does not significantly modify the state of aggregation of the fillers compared to the previous solvated state (aqueous colloidal dispersions of fillers and polymer latex).

In the case of electrically conductive composites, certain properties are evaluated so as to enable the comparison and rating of various processing methods. The *ultimate electrical conductivity* (maximum value reached) and the measured *electrical percolation threshold* are typically used to enable such a comparison.[15] The second parameter corresponds to a critical filler concentration which coincides with the formation of a conductive network of filler particles in the continuous polymer matrix. In practice, it can be evidenced by

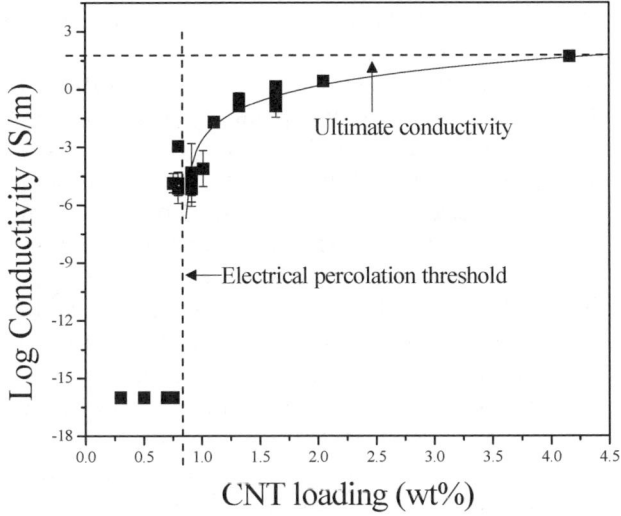

Figure 3.2 Determining the percolation threshold and ultimate conductivity for composites prepared with CNTs.

a drastic increase of conductivity of the nanocomposite upon a very small increase of filler loading. Additionally, although difficult to characterize accurately and quantitatively, the degree of dispersion of the filler in the polymer matrix is commonly considered too. The determination of the values of the ultimate conductivity and electrical percolation threshold is illustrated for a set of conductivity measurements made on CNT–polymer composites in Figure 3.2.

In recent years, the applicability of latex technology to various polymeric systems has been illustrated, as can be seen in Table 3.1. The industrial scalability has also been demonstrated by applying latex technology in the preparation of a composite master batch.[16] When examining this large variety of composite types, a range in percolation thresholds is observed for composites prepared with different polymeric matrices. This even holds true

Table 3.1 Percolation thresholds for polymer–CNT composites prepared with different latexes.

Polymer latex	Percolation threshold $(SW/MW)^a$ (wt%)	Ref.
PS	0.30–0.50	8,17
PMMA	0.80	8
PS–PMMA blends[b]	0.70	18
iPP-g-MA	0.04/0.11	9
PPO–PS blend[c]	0.30	16
PMA	0.30–0.50	19
PVAc	0.04	10

[a]Single-walled CNTs (SW) and multiwalled CNTs (MW). [b]50/50 (v/v) ratio. [c]90/10 (v/v) ratio.

when one nanofiller type, more specifically one CNT type, is chosen and held constant and the matrix is varied.[17]

Similar variations seen for a variety of other preparation methodologies have been linked to various physical phenomena including the dielectric constant of the polymer matrix,[16] polymer viscosity[17,20] and crystallisation behaviour,[21] among others. The differences in surface tension between the nanofillers and the polymeric matrix can also influence mixing and absorption of the matrix on the nanofiller surface.[16] Separating these variables and trying to understand them individually is very difficult experimentally, and even after many years of research, few hard-and-fast rules have been laid down.

Ultimately, low percolation thresholds combined with high levels of conductivity are desired due to cost and processing considerations. Moreover, low filler loadings imply higher composite transparency, which is extremely advantageous for applications requiring the use of transparent thin films, such as transparent coatings or flexible transparent electrodes of (organic) photovoltaic cells and light-emitting diodes.[22]

Many advances in latex technology targeting lower percolation thresholds have been made since the process was introduced in 2004. By tuning the polymer matrix, the filler characteristics, and the preparation and processing conditions, it is possible to reduce the percolation threshold and/or to increase the maximum conductivity of the nanocomposites prepared. For example, extremely low percolation threshold values of the order of 0.15–0.20 wt% were achieved for MWCNT–PS composites when a filler with very high aspect ratio was employed,[23] as the percolation threshold is inversely proportional to the filler aspect ratio (length/diameter). The use of CNTs with enhanced aspect ratios simultaneously led to an increase in ultimate conductivity (10^3 *versus* 10^2 S m^{-1} for shorter MWCNTs), likely due to the reduction in contacts per percolated CNT path.[24] Other manipulation strategies used include the use of conductive surfactants to disperse the CNTs,[25] an intrinsically conductive polymer matrix,[26] and a polymer matrix with low viscosity (inducing slight nanoparticle aggregation).[17]

As for CNT–polymer nanocomposites, the distribution of the graphene layers in the polymer matrix and the filler–matrix interactions are predominant factors determining the properties of the nanocomposite. For this reason, choosing a composite preparation strategy that gives the desired level of filler distribution is of great importance. Strategies that are aqueous-based have many benefits; a few strategies are summarised here.

3.2.1   *In Situ* Polymerization and Heterocoagulation Strategies

Various techniques to interface graphene with aqueous polymeric systems have been reported.[12,27–34] In all instances, an initial dispersion of pristine oxidized or reduced graphene is prepared. The successful preparation of dispersions of individual graphene sheets in water is crucial, as it is a prerequisite for the preparation of nanocomposites with non-aggregated nanofillers homogeneously dispersed in a polymer matrix.

Pristine graphene is incompatible with water and thus remains in an aggregated state within pure water. In order to circumvent this problem, this material needs to be modified by covalent or non-covalent surface modification. Covalent modification involves functionalization of the sheet surface whereas a non-covalent route is based on short-range interactions between the graphene sheets and molecules such as surfactants.

The preparation of aqueous graphene dispersions starting from untreated graphite has been reported.[6,35] Dispersions in water using common surfactants (such as sodium dodecyl sulfate (SDS), sodium dodecyl benzene sulfonate (SDBS) or sodium cholate) and aggressive ultrasound treatments yield a fairly low number of single graphene sheets.[36] These surfactant-covered graphene aqueous dispersions have been used in combination with *in situ* microemulsion polymerizations.[29] Low conductivities (10^{-5} S m^{-1}) were reached, most likely due to the thin insulating polymer layers surrounding the graphene sheets which can significantly increase the junction–junction resistance of the graphene network.[37] A percolation threshold of over 5 wt% was observed. This system was expanded to include MWCNTs as a second conductive filler, unfortunately without much increase in conductivity.[38]

An expansion of this concept was reported for graphene oxide (GO), which consists of heavily oxidized graphene sheets. Dispersed GO was modified with reactive surfactants called *surfmers*.[30] The functional groups physisorb to the GO surface, allowing the encapsulation of these flakes in the subsequent miniemulsion polymerization. It is likely that, once again, this optimal encapsulation will have deleterious consequences on the composite's conductivity, as has often been seen for CNTs.[39]

Unlike untreated graphene-based aqueous dispersions, GO can easily be dispersed in an aqueous medium.[40] GO consists of graphene sheets bearing a large density of functional groups such as hydroxyl, epoxide, diols, ketones and carboxyls. The presence of these groups promotes graphene sheet exfoliation. In addition, carbonyl and hydroxyl groups are usually located on the edges of the sheets, rendering GO strongly hydrophilic and easily dispersible in water. Furthermore, functionalization of graphene sheets by oxidation can subsequently enhance the interaction with the polymer matrix. However, the presence of functional groups disrupts the graphitic network of sp^2 carbon bonds, rendering the material electrically insulating. Consequently, in order to ultimately prepare technologically interesting graphene–polymer nanocomposites, the GO must undergo a reduction step in order to restore the graphene sheet conductivity. As soon as the functional groups, which both maintain the electrostatic repulsion between the graphene sheets and render them water soluble, are removed by chemical reduction, the sheets immediately agglomerate.[41] This particle aggregation can be circumvented by using surfactants,[42] conjugated organic acids[43] or water-soluble polymers,[44] yielding stable reduced graphene dispersions. Without the use of additional acids and surfactants, stable graphene dispersions have also been prepared by maintaining a solution at pH 10 during controlled reduction of GO colloids.[45]

Performing *in situ* reduction of GO in the presence of the final matrix material can also lead to a good dispersion of the graphene in the final composite. Homogeneous systems using the water-soluble matrix poly(vinyl alcohol),[34] and heterogeneous systems using latexes[27,28] have been reported. In these three instances, a coagulation step is required to amalgamate the solids. Percolation thresholds of 0.5 wt% have been achieved,[34] but in general this technique leads to composites with fairly poor conductivities ($\sim 10^{-6}$ S cm^{-1}).[27] A similar heterocoagulation was also used in systems where no GO reduction step was used.[46] Once again, the percolation threshold lies above 0.5 wt%. It appears that reducing the contact between adjacent graphene sheets with polymer layers deposited either during polymerization or heterocoagulation negatively influences electrical percolation.

3.3 Graphene–Polymer Composites via Latex Technology

Latex technology fundamentally starts with a colloidal system comprising an aqueous dispersion of a polymer and a filler; importantly, it is stated that the "state" or configuration of this colloidal state is conserved during the subsequent processing steps.[47] Additionally, due to the fact that a homogeneous compaction step or volume reduction (the *lyophilization* step shown in Figure 3.1) is performed before the intimate mixing of the graphene and polymer matrix, it is expected that the contact between adjacent graphene sheets could be improved.

The latex technology process has been applied to graphene with much success.[2] Like many other techniques, the exfoliation of as-received, bundled graphite to graphene sheets with as few layers as possible, is the first crucial step. For latex technology this exfoliation is performed in water.

Firstly, graphite oxide is prepared following the Hummers method, yielding a layered structure.[48] It is then exfoliated in water by a gentle sonication process. This method yields graphite oxide platelets with a thickness less than 1 nm, corresponding to 2–3 atomic layers, and an average surface area of 1–3 mm^2. A compromise has to be reached between the sonication time and the final graphene surface area and surface/thickness ratio obtained after reduction of the graphite oxide. A sonication time of around 12–15 min should prevent extensive breaking and destruction of the sheets, but still provide a good exfoliation. A reduction step is subsequently performed by using hydrazine at 120 °C and the water-soluble polymer poly(styrene sulfonate) (PSS). After filtering and washing the exfoliated and chemically reduced graphene sheets, a stable colloidal dispersion of graphene nanoplatelets stabilized by PSS is prepared in water using sonication. This process can be followed using UV-vis spectroscopy, much like the dispersion of CNTs in surfactant solutions,[5] as long as the concentrations of dispersed nanofiller are kept low to avoid contributions due to scattering.[49] This dispersion is finally

Figure 3.3   Latex technology process as applied to graphene-based composites. (Reproduced by permission of the Royal Society of Chemistry).

mixed with a polymer latex of choice. The whole process is shown graphically in Figure 3.3.

This process yields composites with a homogeneous dispersion of graphene in the polymeric matrix. This can be visualised using the electrical contrast mode in scanning electron microscopy (SEM; Figure 3.4).[50]

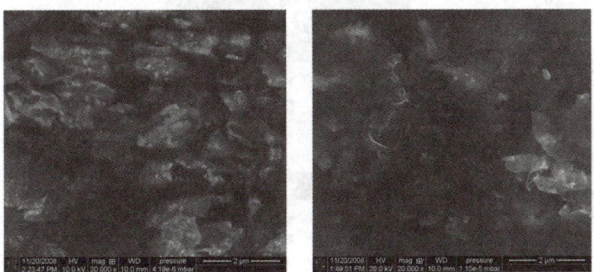

Figure 3.4   SEM micrographs revealing the embedded graphene sheets and their connected network. Scale bar: 2 μm for both pictures. (Reproduced by permission of the Royal Society of Chemistry).

The organization of the graphene sheets in the conductive nanocomposites and their conductivity distribution can also be analysed with nanometre resolution using conductive atomic force microscopy (C-AFM). Using a C-AFM probe, the local electrical conductivity, topography and phase contrast imaging can all be obtained simultaneously. The C-AFM tip measures the current throughout the volume of the nanocomposite specimen at a given voltage which runs *via* the graphene network to the ground contacts. Only platelets that are connected with the ground contacts can be measured, and the observed differences in current are determined by the intranetwork graphene junctions with highest resistivity. Graphene contributing to subnetworks without connection to the ground contacts shows no current. In this way, a current distribution image is obtained and the conductive platelets can be distinguished from the insulating polymer matrix. In Figure 3.5 an example of such an image is shown; the bright (white) areas correspond to graphene in the cross-section topographic image (Figure 3.5a) fit with the higher current level seen on the conductivity image (Figure 3.5b), indicating the presence of conductive pathways. Moreover, information regarding the 3D local organization of the graphene network inside the nanocomposite can be gained by combining electrostatic force microscopy (EFM) and C-AFM. This study enables the 3D reconstruction (see Figure 3.5d) and the drawing of a *xz* cross-section (up to 10 nm in the *z* direction, see Figure 3.5c) of a graphene cluster.[51]

Conductivities of graphene-based nanocomposites prepared using latex technology compete well with those reported for CNT-based composites made in a similar fashion. A typical conductivity profile is shown in Figure 3.6.

The composites exhibit minimal conductivity below 0.6 wt%. The onset of electrical percolation lies just above 0.6 wt%. At concentrations between 0.9

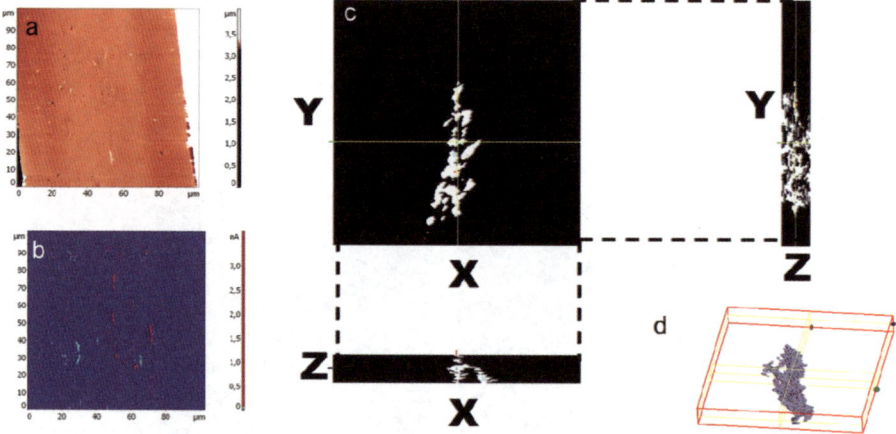

Figure 3.5   C-AFM images of composites containing 1.9 wt% graphene obtained in (a) topography mode and (b) as an electrical current distribution image. A (c) *xz* cross-section and (d) 3D reconstruction of a graphene cluster (see reference 51 for details regarding the 3D reconstruction).

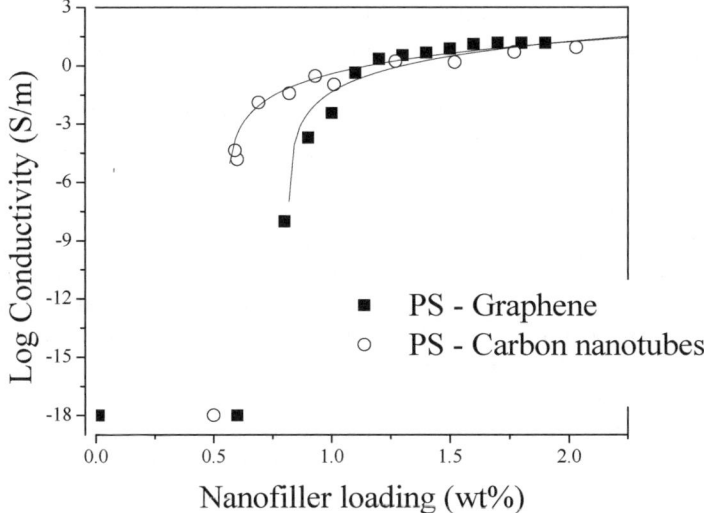

Figure 3.6 Conductivity profile for polystyrene–graphene composites prepared using latex technology.

and 1.2 wt% the conductivity increases drastically up to 2 S m^{-1}. At graphene loadings of about 2 wt%, the conductivity level is ~15 S m^{-1}. For composites prepared with single-walled CNTs and the exact same latex, a percolation threshold slightly above 0.5 wt% is observed. Interestingly, the ultimate conductivity achieved is similar for the two types of composites.

A similar latex-based route has also been reported for polycarbonate–graphene composites.[52] Percolation thresholds for such latex-based techniques were compared to solution blending strategies. It was shown that using latexes led to a threefold reduction of the percolation threshold from 0.38 to 0.14 vol%. This clearly demonstrates the advantages of using a latex-based processing route.

3.4 Graphene–Polymer Composite Production: An Overview

At this point, it makes sense to compare graphene–PS nanocomposites prepared by latex technology with similar nanocomposites manufactured by alternative routes. This type of comparison is arduous, as systems referred to in the literature as "graphene–polymer nanocomposites" as a general definition correspond to extremely diverse types of graphene-based materials manufactured by various methodologies. As stated earlier, as for any nanofiller dispersed in an insulating polymer matrix, the state of dispersion, the geometry, the abundance and the intrinsic properties of the nanofillers, as well as the type and strength of polymer interactions, substantially affect the electrical properties of the resulting nanocomposite. While reviewing the existing body of literature dealing with the preparation of electrically conductive graphene–polymer nanocomposites, one cannot help noticing that most so-called graphene–polymer nanocomposites (including those made *via* latex technology) do not exclusively contain individual pristine graphene sheets dispersed in a polymer matrix, but rather a mixture of individual (sometimes functionalized) graphene sheets and stacks of at least a few sheets. From this point onwards, we use the term *graphene nanofiller particle* to refer to the latter.[53,54]

Whatever the exact type of graphene-related nanofiller, it is generally agreed that, for graphene–polymer nanocomposites, high aspect ratio values are necessary to lower the percolation threshold. Another crucial parameter to take into account is the degree of exfoliation of the filler, or, in other words, the number of graphene nanofiller particles per volume unit. As for the ultimate electrical conductivity, it was reported that a high degree of exfoliation, a low density of defects of the sp^2 graphitic sheet network and doping by intercalation of dopants between graphene planes are beneficial.[53]

Table 3.2 gives an overview of the performances reported in literature of nanocomposites made of various graphene-based nanofillers dispersed in a PS matrix. Information about the type of filler used, as well as the fabrication method of the nanocomposites, is given.

The large spread in reported percolation threshold values of graphene–PS nanocomposites indicates that the state of dispersion and the properties of the graphene, as well as the conditions in which the composite has been processed, can strongly influence the electrical properties of the resulting nanocomposite. In this respect, it is challenging to evaluate the effect of, for example, the state of dispersion and properties of the graphene on the electrical behaviour of the composites based on the body of existing literature.

Nevertheless, due to the reproducibility of the preparation of nanocomposites produced using latex technology, dispersing graphene with specific characteristics into the same polymer matrix enabled us to draw conclusions regarding the influence of the state of aggregation of the graphene nanofiller in the PS matrix. In a work reported in more detail elsewhere,[66] aqueous graphene dispersions prepared in similar ways and differing only in terms of states of aggregation were used to prepare graphene–PS nanocomposites by latex technology. It was found that the composites prepared with dispersions exhibiting a lower degree of exfoliation, *i.e.* containing clusters, displayed lower percolation threshold values than those prepared with graphene dispersions with a higher degree of exfoliation (about 2 wt% and 3 wt% for lower and higher degrees of exfoliation, respectively), see Figure 3.7. Note that these results were perfectly in agreement with those of studies done on CNT–polymer nanocomposites.[67–70]

3.5 Industrial Relevance

For the preparation of conductive polymer nanocomposites via latex technology, several types of graphene produced from graphite can be used in

Table 3.2 Electrical properties of graphene/graphite-based PS nanocomposites reported in literature.

Fabrication method[a]	Filler	Percolation threshold	Max cond reached in S m^{-1} (+ loading)	Reference
Solution processing	Graphene	0.15 vol% (0.31 wt%)[b]	0.1 S m^{-1} at 1 vol% 1 S m^{-1} at 2.5 vol%	55
Solution processing	Graphene nanosheets	0.38 vol%	13.84 S m^{-1} at 4.19 vol%	56
Solution processing	Functionalized graphene sheets	Not reported	Conductivity 1–24 S m^{-1} for 10 wt% loading	57
In situ polymerization (PS emulsion polymerization)	Graphene nanosheets – PS particles (90–150 nm) chemically grafted to the graphene surface (mostly to the edges of the sheets)	Not reported – all measurements performed for filler loading above percolation threshold – application transistors	2.9 × 10^{-2} S m^{-1} with 2 wt% GNS	58
In situ polymerization	GNP – foliated graphite (5–10 μm diameter/80–150 nm thickness)	1.1 vol% (~2 wt%)[c]	Conductivity of the pure GNS: 4.85 × 10^{2} S m^{-1} ~1 S m^{-1} at 0.03 vol% (about 6 wt%) ~10^{-1} S m^{-1} at 0.02 vol% (about 4 wt%)	59
In situ polymerization	Expanded graphite – nanosheets 10–30 nm thick × 60–400 nm long	1.8 wt%	1 S m^{-1} at 2.8–3.0 wt%	60
In situ polymerization	Expanded graphene	1 vol%	10^{-2} S m^{-1} at 3.0 vol%	61
In situ polymerization	Potassium intercalated graphite – expanded graphene – different stage of potassium tested – determined that stage IV or higher is needed for high conductivity applications	Not reported	10 S m^{-1} at 10 wt%	62

Table 3.2 (*Continued*)

Fabrication method[a]	Filler	Percolation threshold	Max cond reached in S m^{-1} (+ loading)	Reference
In situ polymerization	Polymerization carried out in presence of sonicated expanded graphite – graphene sheets/ starting material: thickness 30–80 µm and 0.5–20 µm diameter/in polymer matrix: graphene nanosheets of 1–5 nm thickness – space of ~5 nm between them)	1 wt%	10^{-1} S m^{-1} at 4 wt%	63
In situ polymerization (anionic polymerization in THF)	System of potassium–THF– graphite intercalation compound (lamellae thickness <100 nm)	<8.2 wt%	~10 S m^{-1} at 10 wt%	64
Mixed techniques – kind of master batch approach 1.Cationic grafting polymerization of styrene initiated by CO^{+}ClO$_4^{-}$ on the surface of expanded graphite 2.Mixing with PS – solution blending	Grafted expanded graphite	2.6 wt%	1 S m^{-1} at 13.4 wt%	65
Latex technology	Stacks of a few graphene sheets (<1 nm thick) – large polydispersity in size	0.8 wt% (~0.4–0.5 vol%)[c]	15 S m^{-1} at 1.6–2wt%	2

GNS, grafted nanosheets. [a]Melt-processing was also reported as an alternative process for other polymers than PS. [b]Density for the phenyl isocyanate-treated graphite oxide sheets of 2.2 g cm^{-3} used for the conversion from vol% to wt%.[55] [c]Conversion from wt% to vol% assuming a density value of 1.75 g cm^{-3} for expanded graphite, as mentioned in reference.[59]

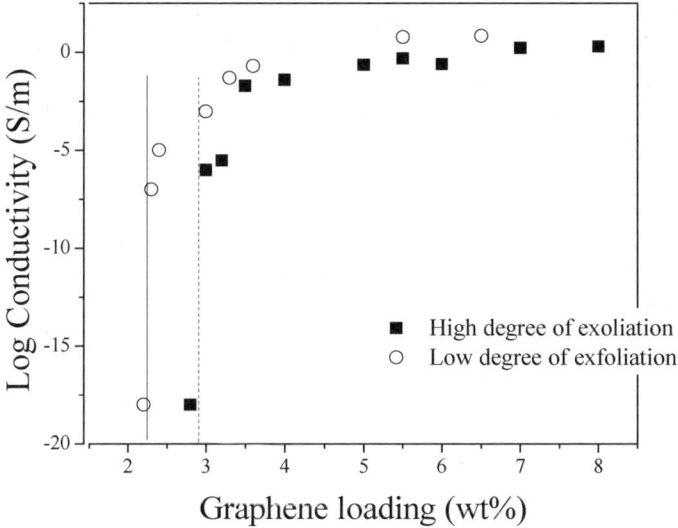

Figure 3.7　Conductivity profiles for graphene–PS composites prepared with a graphene dispersion exhibiting a high (squares) and low (circles) degree of exfoliation.

addition to the one previously described in the present chapter and others reported elsewhere.[2] in other words, the oxidation of graphite and subsequent chemical reduction of graphite oxide in the presence of PSS is not a prerequisite. Alternative routes consist of: (1) the oxidation of graphite and subsequent thermal reduction of graphite oxide and (2) liquid-phase exfoliation of as-produced graphite by bath sonication in the presence of surfactant (sodium cholate).[71] In general, these three methods are suitable for large-scale graphene production required for industrial polymer nanocomposite applications. Starting from graphite, or its derivatives, and using a solvent-based processing method offers significant economic advantages over methods such as mechanical exfoliation, chemical vapour deposition and epitaxial growth, as graphite is a commodity material with an annual global production of over 1.1 million tons at $825/ton in 2008.[72] With a view to industrial application one can say that the production of graphene via oxidation of graphite with subsequent thermal reduction of graphite oxide is more attractive than the other two methods due to the shorter time needed for the preparation of graphene. Oxidation of graphite to graphite oxide via the method described by Hummers requires about 3 hours; the subsequent thermal reduction and exfoliation of the generated GO requires only a few minutes.[73,74] To complete the exfoliation and to disperse the reduced graphene in water using horn-sonication in the presence of surfactant, another 3–4 hours are needed. In contrast, chemical reduction and exfoliation of GO can take up to 70 hours depending on the level of reduction needed.[2,75] Another disadvantage of the chemical reduction is that it requires a large excess of surfactant (namely a weight ratio of surfactant to graphite oxide of up to 10:1), which in turn introduces a filtration step in order to remove any excess surfactant. Depending on the concentration of graphene needed, the graphite liquid-phase exfoliation method requires at least 100 hours of bath sonication of graphite in water followed by centrifugation.[76,77]

Alternatively, a significant reduction in the time needed to process a complete nanocomposite can be achieved by the *in situ* reduction of GO during the last step of the latex technology process, *i.e.* during compression moulding. There have been few reports utilizing *in situ* thermal reduction of graphite oxide, *i.e.* during polymerization of monomers for composite preparation in the presence of graphite oxide at high temperatures.[78–80] Liu *et al.* prepared polyester–reduced GO composites via the *in situ* polymerization of terephthalic acid and ethylene glycol at elevated temperature containing well-dispersed graphite oxide. The composites exhibited a maximum conductivity of 0.5 S m^{-1}.[78] A similar strategy when applied to latex technology yielded highly conductive composites. Well-dispersed graphite oxide (without any surfactant) was stirred into a PS latex to promote the homogenous distribution of the hydrophilic GO in the water-based, colloidal system. After freeze-drying of the mixture, composite films were obtained by compression moulding at 180 °C. This high-temperature moulding step can partially reduce the graphite oxide. The conductivity curve of the resulting series of composite is shown in Figure 3.8.

The *in situ* reduced graphite oxide platelets in the composites percolate at 0.7 wt% of GO loading. A maximum conductivity of 0.3 S m^{-1} is reached for about 3.5 wt% loading. This percolation threshold is lower than the one

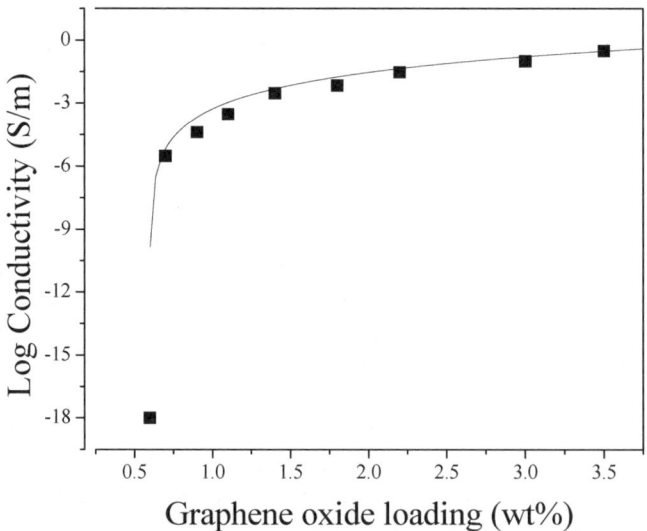

Figure 3.8   Electrical conductivity of *in situ* reduced graphite oxide–PS composites as a function of graphite oxide weight fraction.

obtained for the composites utilizing graphene obtained by chemical reduction of graphite oxide (see Figure 3.6). This difference can be attributed to the higher aspect ratio of the *in situ* reduced graphite oxide sheets in comparison to graphene platelets used for the preparation of graphene–PS composites described earlier in this chapter. The process of incorporating graphite oxide fillers into the nanocomposites prepared *in situ* requires required less sonication energy. Longer sonication times can break graphene sheets and reduce their aspect ratio, thereby leading to an increase in percolation threshold. On the other hand, the maximum conductivity achieved for the composites based on *in situ* reduced graphite oxide is much lower than that achieved for the composites based on the chemically reduced graphite oxide. This is probably caused by the lower conductivity of the filler, which results from the incomplete reduction of the *in situ* reduced graphite oxide. The conductivity of a bucky-paper made of graphite oxide is approximately 1×10^{-6} S m^{-1}, *versus* 5.5×10^{3} S m^{-1} for the chemically reduced graphite oxide covered by PSS. It is likely that higher conductivity values can be obtained by using the *in situ* reduction process at elevated temperatures, provided that the polymer used as a matrix can stand higher processing temperatures. Nevertheless, the conductivity levels achieved (0.15 S m^{-1}) might be sufficient for antistatic, and even for electromagnetic interference (EMI) shielding applications.

To sum up, the *in situ* nanocomposite preparation method is not only advantageous in terms of reduced processing time and number of steps, but it also does not require the use of surfactant which may be detrimental to mechanical properties of the final nanocomposites, among others. Additionally, this method does not require sonication of the nanofiller at high sonication power, which (1) reduces the total energy consumption and (2) prevents excessive damage to the graphene flakes. Higher maximum conductivity values are expected when higher compression moulding temperatures are used during the final processing step of the composite.

Finally, freeze-drying was chosen to remove water since it does not require high drying temperatures while preventing filler aggregation from occurring, as it does if regular, thermal-vapour routes are chosen.[81] On the other hand, this drying technique is a relatively lengthy and expensive process in terms of operating costs and energy consumption.[82,83] In an industrial environment, alternative dehydration strategies, such as a flash-drying technique for example, would certainly be preferred.

3.6 Conclusion

Several methods have been developed over the years to achieve the incorporation of graphene into a polymer matrix in order to obtain electrically conductive nanocomposites. The key factors for producing such technologically and industrially relevant nanocomposites with low graphene loadings and high electrical conductivity values are easy processing and process up-scaling.

Latex technology fulfils all these requirements. This technology is highly versatile, as any filler that can be dispersed to yield an aqueous colloidal dispersion can be used. Similarly, any polymer that can be synthesized by emulsion polymerization or can artificially be brought into the form of a polymer latex is suitable.

After having been successfully used to produce CNT–polymer nanocomposites, latex technology was utilized to prepare graphene–PS nanocomposites. The resulting composites possessed percolation thresholds as low as 0.8 wt%, as well maximum conductivity values comparable to those of CNT–polymer nanocomposites (15 S m^{-1} for 1.6–2 wt% graphene loading). These results are among the lowest values of percolation threshold and the highest values of maximum conductivity reported in the existing literature for graphene–PS nanocomposites prepared by various techniques. Furthermore, it was demonstrated that controlled clustering of the graphene filler favours the lowering of the percolation threshold. From an industrial point of view, *in situ* oxidation of the graphite oxide filler during the final compression moulding appears to be very attractive since it paves the way to reduced processing time and steps.

References

1. X. Du, I. Skachko, A. Barker, E. Y. Andrei, *Nature Nanotechnol.*, 2008, **3**, 491–495.
2. E. Tkalya, M. Ghislandi, A. Alekseev, C. E. Koning, J. Loos, *J. Mater. Chem.*, 2010, **20**, 3035–3039.
3. C. A. Dyke, J. M. Tour, *J. Phys. Chem. A*, 2004, **108**, 11151–11160.
4. A. K. Geim, K. S. Novolselov, *Nature Mater.*, 2007, **6**, 183–191.
5. N. Grossiord, O. Regev, J. Loos, J. Meuldijk, C. E. Koning, *Anal. Chem.*, 2005, **77**, 5135–5139.
6. M. Lotya, Y. Hernandez, P. J. King, R. J. Smith, V. Nicolosi, L. S. Karlsson, F. M. Blighe, S. De, Z. M. Wang, I. T. McGovern, G. S. Duesberg, J. N. Coleman, *J. Am. Chem. Soc.*, 2009, **131**, 3611–3620.
7. M. S. P. Shaffer, A. H. Windle, *Adv. Mater.*, 1999, **11**, 937–941.
8. N. Grossiord, J. Loos, H. E. Koning *J. Mater. Chem.*, 2005, **15**, 2349–2352.
9. K. Lu, N. Grossiord, C. E. Koning, H. E. Miltner, B. van Mele, J. Loos, *Macromolecules*, 2008, **41**, 8081–8085.
10. J. C. Grunlan, A. R. Mehrabi, M. V. Bannon, J. L. Bahr, *Adv. Mater.*, 2004, **16**, 150–153.
11. L. Raka, G. Bogoeva-Gaceva, K. Lu, J. Loos, *Polymers*, 2009, **50**, 3739–3746.
12. E. Tkalya, M. Ghislandi, A. Alekseev, C. E. Koning, J. Loos, *J. Mater. Chemistry*, 2010, **20**, 3035–3039.
13. N. Grossiord, J. Loos, L. van Laake, M. Maugey, C. Zakri, C. E. Koning, A. J. Hart, *Adv. Funct. Mater.*, 2008, **18**, 3226–3234.

14. S. Badaire, P. Poulin, M. Maugey, C. Zakri, *Langmuir*, 2004, **20**, 10367–10370.
15. W. Bauhofer, J. Z. Kovacs, *Compos. Sci. Tech.*, 2008, **69**, 1486–1498.
16. N. Grossiord, PhD. Thesis, Technische Universiteit Eindhoven, The Netherlands, 2007.
17. M. C. Hermant, N. M. B. Smeets, R. C. F. van Hal, J. Meuldijk, H. P. A. Heuts, B. Klumperman, A. M. van Herk, C. E. Koning, *e-Polymers*, 2009, **022**, 1–13.
18. N. Grossiord, M. C. Hermant, C. E. Koning, *Polymer Carbon Nanotube Composites: The Polymer Latex Concept*, Pan Standford Publishing, Singapore, 2012.
19. Hermant, M. C. Thesis, Technische Universiteit Eindhoven, The Netherlands, 2009.
20. M. Micušik, M. Omastova, I. Krupa,, J. Prokeš, P. Pissis, E. Logakis, C. Pandis, P. Pötschke, J. Piontec, *J. Appl. Polym. Sci.*, 2009, **113**, 2536 – 2551.
21. K. Jeon, S. Warnock, C. Ruiz-Orta, A. Kismarahardja, J. Brooks, R. G. Alamo, *J. Polym. Sci. B*, 2010, **48**, 2084–2096.
22. E. Kymakis, *NanoTechn. Law Business*, 2006, **3**, 405–410.
23. N. Grossiord, J. Loos, L. van Laake, M. Maugey, C. Zakri, C. E. Koning, A. J. Hart, *Adv. Func. Mater.*, 2008, **18**, 3226–3234.
24. D. Hecht, L. Hu, G. Grüner, *Appl. Phys. Lett.*, 2006, **89**, 1331121–1331123.
25. M. C. Hermant, B. Klumperman, A. V. Kyrylyuk, P. Van der Schoot, C. E. Koning, *Soft Matter*, 2009, **5**, 878–885.
26. N. Grossiord, M. E. L. Wouters, H. E. Miltner, K. Lu, J. Loos, B. Van Mele, C. E. Koning, *Eur. Polym. J.*, 2010, **46**, 1833–1843.
27. R. Wissert, P. Steurer, S. Schopp, R. Thomann, R. Mulhaupt, *Macromol. Mater. Eng.*, 2010, **295**, 1107–1115.
28. Y. Zhan, J. Wu, H. Xia, N. Yan, G. Fei, G. Yuan, *Macromol. Mater. Eng.*, 2011, **296**, 590–602.
29. A. S. Patole, S. P. Patole, H. Kang, J.-B. Yoo, T.-H. Kim, J.-H. Ahn, *J. Colloid Interface Sci.*, 2010, **350**, 530–537.
30. H. M. Etmimi, R. D. Sanderson, *Macromolecules*, 2011, **44**, 8504–8515.
31. M. Yoonessi, J. R. Gaier, *ACS Nano*, 2010, **4**, 7211–7220.
32. M. Gudarzi, F. Sharif, *Soft Matter*, 2011, **7**, 3432–3440.
33. H. B. Lee, A. V. Raghu, K. S. Yoon, H. M. Jeong, *J. Macromol. Sci. B*, 2010, **49**, 802–809.
34. H. J. Salavagione, G. Martinez, M. A. Gomez, *J. Mater. Chem.*, 2009, **19**, 5027–5032.
35. M. Lotya, P. J. King, U. Khan, S. De, J. N. Coleman, *ACSNano*, 2010, **4**, 3155.
36. A. A. Green, M. C. Hersam, *Nano Lett.*, 2009, **9**, 4031.
37. I. Balberg, D. Azulay, D. Toker, O. Millo. *Int. J. Mod. Phys. B*, 2004, **18**, 2091.

38. A. S. Patole, S. P. Patole, S.-Y. Jung, J.-B. Yoo, J.-H. An, T.-H. Kim, *Eur. Polym. J.*, 2012, **48**, 252–259.
39. E. K. Hobbie, J. Obrzut, S. B. Kharchenko, E. A. Grulke, *J. Chem. Phys.*, 2006, **125**, 044712–044713.
40. S. Park, R. S. Ruoff, *Nat. Nanotechnol.*, 2009, **4**, 217.
41. S. Stankovich, D. A. Dikin, K. Piner A. Kolhaas, A. Kleinhammes, Y. Jia, Y. Wu, S. T. Nguyen, R. S. Ruoff, *Carbon*, 2007, **45**, 1558.
42. J. R. Lomeda, C. D. Doyle, D. V. Kosynkin, W. F. Hwang, J. M. Tour, *J. Am. Chem. Soc.*, 2008, **130**, 16201.
43. Y. X. Xu, H. Bai, G. W. Lu, C. Li, G. Shi, *J. Am. Chem. Soc.*, 2008, **130**, 5856.
44. S. Stankovich, R. D. Piner, X. Q. Chen, N. Q. Wu, S. T. Nguyen, R. S. Ruoff, *J. Mater. Chem.*, 2006, **16**, 155.
45. D. Li, B. Muller, S. Gilje, R. B. Kaner, G. G. Wallace, *Nat. Nanotechnol.*, 2008, **3**, 101.
46. J. S. Kim, S. Hong, D. W. Park, S. E. Shim, *Macromol. Res.*, 2010, **18**, 558–565.
47. N. Grossiord, P. J. J. Kivit, J. Loos, J. Meuldijk, A. V. Kyrylyuk, P. van der Schoot, C. E. Koning, *Polym.*, 2008, **49**, 2866–2872.
48. W. S. Hummers, R. E. Offeman, *J. Am. Chem. Soc.*, 1985, **80**, 1339.
49. O. N. Torrens, D. E. Milkie, J. M. Zheng, J. M. Kikkawa, *Nano Lett.*, 2006, **6**, 2864–2867.
50. J. Loos, A. Alexeev, N. Grossiord, C. E. Koning, O. Regev, *Ultramicroscopy*, 2005, **104**, 160–167.
51. A. Alekseev, D. Chen, E. E. Tkalya, M. G. Ghislandi, Y. Syurik, O. Ageev, J. Loos, G.de With, *Adv. Funct. Mater.*, 2012, **22**, 1311–1318.
52. M. Yoonessi, J. R. Gaier, *ACS Nano*, 2010, **4**, 7211–7220.
53. B. Li, W.-H. Zhong, *J. Mater. Sci.*, 2011, **46**, 5595–5614.
54. T. Kuilla, S. Bhadra, D. Yao, N. H. Kim, S. Bose, J. H. Lee, *Prog. Polym. Sci.*, 2010, **35**, 1350–1375.
55. S. Stankovich, A. D. Dikin, G. H. B. Dommett, K. M. Kohlhaas, E. J. Zimney, E. A. Stach, R. D. Piner, S. T. Nguyen, R. S. Ruoff, *Nature*, 2006, **442**, 282–286.
56. N. Liu, G. Luo, H. Wu, Y. Liu, C. Zhang, J. Chen, *Adv. Funct. Mater.*, 2008, **18**, 1518–1525.
57. G. Eda, M. Chhowalla, *Nano Lett.*, 2009, **9**, 814–818.
58. H. Hu, X. Wang, J. Wang, L. Wan, F. Liu, H. Zheng, R. Chen, C. Xu, *Chem. Phys. Lett.*, 2010, **484**, 247–253.
59. N. K. Srivastava, R. M. Mehra, *J. Appl. Polym. Sci.*, 2008, **109**, 3991–3999.
60. G.-H. Chen, D.-J. Wu, W.-G. Weng, B. He, W.-L. Yan, *Polym. Int.*, 2001, **50**, 980–985.
61. J. F. Zou, Z. Z. Yu, Y. X. Pan, X. P. Fang, Y. C. Ou, *J. Polym. Sci. B.*, 2002, **40**, 954–963.

62. H. Kim, H. T. Hahn, L. M. Viculis, S. Gilje, R. B. Kaner, *Carbon*, 2007, **45**, 1578–1582.
63. G. Chen, C. Wu, W. Weng, D. Wu, W. Yan, *Polym.*, 2003, **44**, 1781–1784.
64. M. Xiao, L. Sun, J. Liu, Y. Li, K. Gong, *Polym.*, 2002, **43**, 2245–2248.
65. W.-P. Wang, C.-Y. Pan, *Polym.*, 2004, **45**, 3987–3995.
66. E. E. Tkalya, M. G. Ghislandi, R. Otten, M. Lotya, A. Alekseev, P. van der Schoot, J. Coleman, G. de With, C. E. Koning, *in preparation*, 2012.
67. J. Li, P. C. Ma, W. S. Chow, C. K. To, B. Z. Tang, J. K. Kim, *Adv. Funct. Mater.*, 2007, **17**, 3207–3215.
68. C. A. Martin, J. K. W. Sandler, M. S. P. Shaffer, M. K. Schwarz, W. Bauhofer, K. Schulte, A. H. Windle, *Comp. Sci. Technol.*, 2004, **64**, 2309–2316.
69. J. J. Hernandez, M. C. Garcia-Gutierrez, A. Nogales, D. R. Rueda, M. Kwiatkowska, A. Szymczyk, Z. Roslaniec, A. Concheso, I. Guinea, T. A. Ezquerra, *Comp. Sci. Techno.*, 2009, **69**, 1867–1872.
70. J. O. Aguilar, J. R. Bautista-Quijano, F. Aviles, *Express Polym. Lett.*, 2010, **4**, 292.
71. E. E. Tkalya, PhD Thesis, Eindhoven University of Technology, The Netherlands, 2012.
72. H. Kim, A. A. Abdala, C. W. Macosko,, *Macromolecules*, 2010, **43**, 6515–6530.
73. W. S. Hummers, R. E. Offeman, *J. Am. Chem. Soc.*, 1958, **80**, 1339.
74. M. J. McAllister, J. L. Li, D. H. C Adamson, H. C. Schniepp, A. A. Abdala, J. Liu, M. Herrera-Alonso, D. L. Milius, R. Car, R. K. Prud'homme, I. A. Aksay, *Chem. Mater.*, 2007, **19**, 4396–4404.
75. S. Stankovich, R. D. Piner, X. Q. Chen, N. Q. Wu, S. T. Nguyen, R. S. Ruoff, *J. Mater. Chem.*, 2006, **16**, 155–158.
76. M. Lotya, Y. Hernandez, P. J. King, R. J. Smith, V. Nicolosi, L. S. Karlsson, F. M. Blighe, S. De, Z. M. Wang, I. T. McGovern, G. S. Duesberg, J. N. Coleman, *J. Am. Chem. Soc.*, 2009, **131**, 3611–3620.
77. M. Lotya, P. J. King, U. Khan, S. De, J. N. Coleman, *ACS Nano*, 2010, **4**, 3155–3162.
78. K. Liu, L. Chen, Y. Chen, J. Wu, W. Zhang, F. Chen, Q. Fu, *J. Mater. Chem.*, 2011, **21**, 8612–8617.
79. Z. Xu, C. Gao, *Macromolecules*, 2010, **43**, 6716–6723.
80. D. Chen, H. Zhu, A. Liu, *Appl. Mater. Interfaces*, 2010, **2**, 3702–3706.
81. M. Maugey, W. Neri, C. Zakri, A. Derre, A. Penicaud, L. Noe, M. Chorro, P. Launois, M. Monthioux, P. Poulin, *J. Nanosci. Nanotech.*, 2007, **7**, 2633–2639.
82. I. Alibas, *Biosyst. Eng.*, 2007, **96**, 495–502.
83. Y. Xu, J. Zhang, D. Tu, J. Sun, L. Zhou, A. S. Mujumdar, *Int. J. Food Sci. Technol.*, 2005, **40**, 589–595.

CHAPTER 4

Polymer–Graphene Nanocomposites by Living Polymerization (RAFT) in Miniemulsion

HUSSEIN M. ETMIMI* AND RON D. SANDERSON

UNESCO Associated Centre for Macromolecules & Materials/Department of Chemistry and Polymer Science, University of Stellenbosch, South Africa
*E-mail: hussein@sun.ac.za or hmetmimi@gmail.com

4.1 Introduction

Recent rapid growth in nanoscience has led to the preparation of polymer nanocomposites with enhanced properties compared to pure polymers. Currently, one of the most advanced research areas of nanotechnology focuses on the inclusion of nanoparticle fillers into polymers in order to enhance the functional and physical properties of the polymers. Some of the nanoparticles used to date include nanofibres, silica nanoparticles, carbon nanotubes (CNTs), clays and graphene.[1] Perhaps the most studied is the use of clay, due to its ease of modification and availability. However, because of its unique properties, graphene has become the material of choice in the field of nanoscience and nanotechnology in recent years.[2,3] Not only do graphene nanosheets provide most of the advantages offered by the nanometre-size fillers, but they can be also incorporated in both hydrophilic and hydrophobic polymers.[4–6] Moreover, graphene is naturally available and thus its use is

RSC Nanoscience & Nanotechnology No. 26
Polymer–Graphene Nanocomposites
Edited by Vikas Mittal
© The Royal Society of Chemistry 2012
Published by the Royal Society of Chemistry, www.rsc.org

generally cost-effective. The low cost of this material, together with its good mechanical, thermal and barrier properties, offers new possibilities for material development of polymer nanocomposites using graphene nanoparticles.

In the past, most researchers have focused mainly on the synthesis and characterization of polymer–graphene nanocomposites (PGNs) using conventional free radical polymerization.[7] Very few studies on the use of controlled/ living radical polymerization (CLRP) for the synthesis of polymer nanocomposites, such as the reversible addition-fragmentation chain transfer (RAFT) method, focus on the use of clay[8,9] and CNTs.[10,11] However, the use of CLRP for the synthesis of polymer nanocomposites based on graphene has not been fully investigated. In particular, the use of RAFT method for the synthesis of PGNs has never been investigated, except by our own group.[12]

Since its discovery in the late 1990s,[13] the RAFT method has become one of the most effective and versatile methods of CLRP.[14] The method operates *via* a degenerative transfer mechanism in which a thiocarbonylthio compound acts as a chain transfer agent. Its suitability over a wide range of reaction conditions required for the RAFT process and its versatility for use with different monomers make this method among the most useful of all the controlled polymerization techniques for designing molecular architectures.[15,16] By attaching the RAFT agent to the surface of graphene, polymer nanocomposites based on graphene can be obtained in a controlled manner. The use of an anchored RAFT agent will result in controlled living radical polymer growth from the graphene surface, leading to PGNs with controlled molar masses and dispersity (Đ).

In general, various synthesis techniques are now available and have been widely used for the preparation of these PGNs. These include solution blending,[4] exfoliation–adsorption,[17] *in situ* intercalative polymerization[18] and melt intercalation.[19] Although great success has been achieved in the preparation of such nanocomposites using *in situ* polymerization of the monomer in the presence of graphene nanosheets,[18,20,21] reports on the preparation of these composites in emulsion systems are rare. Particularly, the use of miniemulsion polymerization for the synthesis of these nanocomposites has not been investigated. In this chapter, the use of miniemulsion polymerization for the synthesis of PGNs using the RAFT method is discussed.

4.2 Synthesis of PGNs Based on Functionalized Graphene

The properties of polymer nanocomposites based on graphene depend strongly on how well the graphene nanosheets are dispersed in the final nanocomposites.[22] Graphene is often obtained as reinforcement nanoplatelets with thicknesses as little as 2–10 nm through a process of intercalation and exfoliation of natural graphite.[23] Unfortunately, the preparation of PGNs based on parent natural graphite is difficult to achieve, as most monomers and polymers cannot be easily intercalated between graphene nanosheets in the pristine graphite. This is mainly because there are no reactive groups on the surface of pristine natural graphite. Therefore, natural graphite lacks both the space and affinity for polymer molecules (or monomers) to be intercalated into its galleries. Furthermore, the graphene layers are bound together by Van der Waals forces, which make the interlayer distance in natural graphite very narrow. However, the synthesis of graphene oxide (GO) from natural graphite by oxidation[24] creates many oxygen functionalities on its surface, which could greatly facilitate the interaction of monomer and polymer molecules into graphene galleries.[7] Therefore, GO has been widely used instead of pristine graphite in the synthesis of PGNs.[25]

In contrast to natural graphite, GO is very hydrophilic and soluble in aqueous and organic media.[26,27] In addition, the presence of hydrophilic polar groups (*e.g.* –OH and –COOH) will facilitate physical and chemical interaction between the graphite and other organic molecules, such as monomers and polymers, which can be then loaded onto its surface. The nanometre-scale sheets and galleries in the GO as well as the polar groups generated by chemical oxidation create favourable conditions allowing for suitable polymers to intercalate and form PGNs. In recent years, the structure of GO has been widely studied in a number of theoretical[28–30] and experimental[31–35] investigations. Similar to graphite, which contains stacks of graphene, GO is composed of graphene oxide nanoplatelets with expanded interlayer spacing. Several authors have proposed that the epoxy and hydroxyl functional groups lie on the surface of each graphene layer, while the carboxyl groups are located near the layers' edges, as shown in Figure 4.1.[31,36,37] The oxygen functionalities alter the chemistry of the graphene nanoplatelets in GO and render them hydrophilic in nature, thus facilitating their hydration and exfoliation in aqueous media. As a result, GO readily disperses in water and forms stable colloidal dispersions of thin GO nanosheets in water.[38]

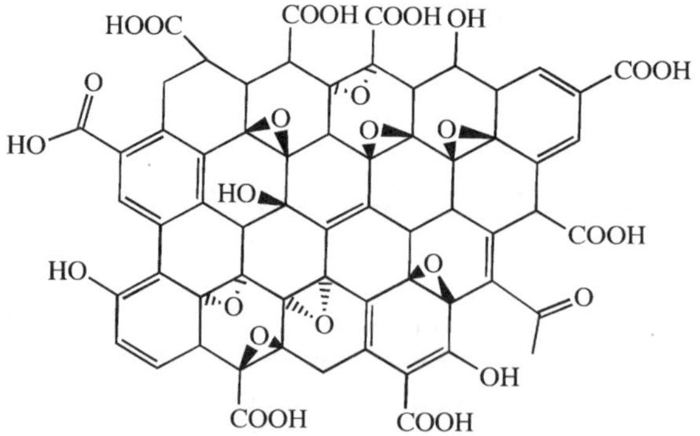

Figure 4.1   Chemical structure of GO with different oxygen-based groups on the basal plane and around the edges of a graphene layer.

In recent years, various studies have focused on the synthesis of PGNs using chemically modified GOs.[39,40] Modification of GO sheets is expected to play a vital role in tailoring the structure and properties of GO, and improving the solubility and compatibility of GO sheets in polymer systems. The functionalization of GO will also enable us to prepare novel PGNs with enhanced functional properties. In 2010, Pramoda *et al.*[41] reported the synthesis of poly(methyl methacrylate) (PMMA)–graphite nanocomposites using the in situ polymerization of methylmethacrylate (MMA) monomer in the presence of modified GO. First, the GO was functionalized with octadecylamine, and then reacted with methacryloyl chloride to incorporate polymerizable groups at the graphene nanoplatelets. The modified GO was then employed in the polymerization of MMA to obtain covalently bonded PMMA–graphite nanocomposites. The authors indicated that the nanocomposites thus obtained showed a significant enhancement in thermal and mechanical properties compared with neat PMMA. The functionalization of graphene with these oxygen-containing groups could also lead to direct attachment of a RAFT agent to the surface of graphene. This process presents a new approach for the preparation of polymer nanocomposites based on graphene in a controlled manner.[12]

4.3   Miniemulsion Polymerization

Miniemulsions contain submicron-size monomer droplets, ranging from 50 to 500 nm.[42] The droplets are formed by shearing a premixed system containing water, monomer, surfactant and a hydrophobe (also referred to as a costabilizer). The surfactant prevents the droplets from coalescence, and the hydrophobe prevents Ostwald ripening. Coalescence occurs upon the collision of droplets while Ostwald ripening is caused by degradation of the droplets *via* diffusion. In a system susceptible to Ostwald ripening, larger monomer droplets will grow in size at the expense of the smaller ones due to the difference in the chemical potential between droplets of different radii.[43] The low molecular weight molecules of the hydrophobe can diffuse only very slowly from one droplet to the other due to their highly hydrophobic nature, therefore they are trapped in the droplets. This will lead to the creation of an osmotic pressure in every droplet, which will suppress monomer diffusion from smaller to bigger droplets.

A well-designed miniemulsion formulation therefore relies greatly on a suitable choice of surfactant(s) and hydrophobe. The amount of surfactant used allows control over particle size of the final latex particles.[44] An increase in the surfactant concentration will lead to a decrease in the particle size. In addition to a variation in surfactant concentration, the size of the droplets can be controlled through changes in the shear rate and time.[45] Different surfactant/hydrophobe systems can be used for miniemulsion formulations. The most common model systems employ sodium dodecyl sulfate (SDS) in combination with cetyl alcohol (CA) or hexadecane (HD).

A characteristic feature of miniemulsion polymerization is that droplet nucleation is the predominant mechanism of particle formation.[46] The nanometre-size monomer droplets formed by the application of high shear to the system have a sufficiently large surface area to compete effectively with the micelles or particles for radical capture.[47] The large droplet surface area is stabilized by the adsorption of an additional amount of surfactant from the water phase, which leads to a decrease in surfactant concentration in the water phase. Thus, there are usually no micelles present in a well-prepared miniemulsion.

The first report on miniemulsion polymerization dates back to 1973, when Ugelstad *et al.*[46] reported the polymerization of styrene (St) in the presence of a mixed emulsifier system of SDS and CA. For comparison, polystyrene (PS) emulsion made with SDS alone was also prepared. Results showed that the prepared emulsion was unstable and phase separated within a few minutes when the stirring was stopped. On the other hand, when CA was used in addition to SDS, the stability of the PS emulsions was very good and the average droplet size was small. At that time the term miniemulsion had not been used; however, the polymerization features fit the general definition of miniemulsion polymerization: monomer droplets smaller than 1 µm were obtained by simple mixing of the monomer into an aqueous solution of a surfactant and a costabilizer. The reduction in their average size makes the monomer droplets more competitive in capturing radicals generated in the aqueous phase, which provides the basis of miniemulsion polymerization, *i.e.* monomer droplet nucleation.

4.3.1 Miniemulsion *Versus* Emulsion Polymerization

Several authors have studied the differences between conventional emulsion and miniemulsion polymerization.[48,49] The difference in size of the monomer droplets in emulsion and miniemulsion polymerization is the key in distinguishing between the two systems. The size of the dispersed droplets in a miniemulsion is quite small (50–500 nm) relative to the size of monomer droplets in an emulsion system (1–100 µm).[42] This significant difference in the droplet size is liable for the different mechanisms of particle nucleation operating in the two systems.

Emulsion polymerization normally consists of water-insoluble monomer(s), a dispersing medium (usually water), a suitable surfactant and a water-soluble initiator. The surfactant plays an important role in the stability, rheology and control of particle size of the resulting lattices. When the concentration of the surfactant is above its critical micelle concentration (cmc), the unabsorbed surfactant molecules remain in the aqueous phase and form micelles. The polymerization process commences with radicals, generated by the thermal decomposition (or otherwise) of the initiator, reacting with the monomer in the aqueous phase to form oligomeric radical chains.

In an emulsion system, there are three possible nucleation mechanisms for the growing oligomeric radical species: micellar, homogeneous (water phase), and (less often) droplet nucleation.[50] In homogenous nucleation, oligomers growing in the aqueous phase begin to precipitate from solution as they reach a degree of polymerization that exceeds their solubility limit (critical length). The oligomeric radicals will then form precursor particles, which are stabilized by adsorbing surfactant molecules. These primary particles can then absorb monomer for further propagation, to form polymer particles. Droplet nucleation occurs when radicals formed in the aqueous phase enter monomer droplets and propagate to form polymer particles.

Micellar nucleation, on the other hand, occurs when sufficient surfactant is present in the system to exceed the cmc. As a result of Ostwald ripening in the emulsion system, monomer molecules tend to diffuse from smaller monomer droplets to larger ones to minimize the total interfacial energy of the system. The droplets are consequently large and the total interfacial area is unable to accommodate all of the surfactant molecules. The desorbed surfactant molecules remain in the aqueous phase and form micelles if the concentration of the surfactant is above the cmc. The hydrophobic tail of the aggregates will then be swollen by monomer, forming monomer-swollen micelles. Initiator radicals (or oligomeric radicals) generated in the aqueous phase can then enter the monomer-swollen micelles to form monomer-swollen polymeric particles. These swollen polymeric particles will grow further by propagation reactions until monomer and surfactant are depleted from unentered micelles.

All three of the above-mentioned mechanisms can occur in classical emulsion polymerization. However, due to the large size (small surface area) of the monomer droplets, they cannot effectively compete with micellar and homogeneous nucleation. Droplets merely act as reservoirs for monomer that diffuses through the water phase to the growing latex particles. Therefore, droplet nucleation is insignificant for most emulsion polymerizations. On the other hand, in miniemulsion polymerization, droplet nucleation is the predominant mechanism of particle formation due to the small size of monomer droplets and the presence of little or no micelles in the system.[51] These submicron droplets have a large interfacial area and are capable of capturing most of the oligomeric free radicals; thus the droplets become the locus of nucleation.

4.3.2 Typical Miniemulsion Formulations

A typical miniemulsion formulation includes water, a monomer (or monomer mixture), a surfactant, a hydrophobe and a suitable initiator system. Different monomers, with a wide range of water solubilities, including vinyl acetate (VAc),[52] MMA,[53,54] *n*-butyl acrylate (BA)[55] and St,[56,57] have been polymerized by means of miniemulsion polymerization. In other cases, formulations that contain more than one monomer have also been prepared, including miniemulsions in which small quantities of very water-soluble monomers, such as acrylic acid[58] and methacrylic acid,[59] have been used.

An important factor for the formulation of a stable miniemulsion is the choice of an appropriate water-insoluble compound, or so-called hydrophobe. In most of the early work, authors investigated the miniemulsion polymerization of St stabilized with CA as a hydrophobe.[60] It was found that although the nucleation period was rather long, most of the particles were nucleated at low conversion. As proposed by Landfester *et al.*,[61] the most efficient hydrophobes are very water-insoluble, surface-inactive reagents. The authors found that the predominant requirement for the hydrophobe is an extremely low water solubility ($<10^{-7}$ mL mL^{-1}), independent of its chemical nature. It was also found that regardless of the amount and type of the hydrophobe, stable miniemulsions with similar structural characteristics were obtained.

The water-insoluble compound is usually a fatty alcohol or a long-chain alkane. The addition of the hydrophobe, such as a long-chain alkane (*e.g.* HD)[62,63] or a long-chain alcohol (*e.g.* CA),[46,57] can efficiently retard the destabilization of the nanodroplets by Ostwald ripening. It should be noted that both linear and branched molecules can be used provided that they have very low water solubility. Other costabilizers that have been used include dodecyl mercaptan[54] or reactive alkyl methacrylates (*e.g.* dodecyl methacrylate).[64]

Another important formulation variable in miniemulsion polymerization is the use of an emulsifier or surfactant system to prevent the degradation of particles by collision. Many different surfactants can be used for miniemulsion formulations, including anionic,[57] cationic,[65] non-ionic,[66] non-reactive surfactants and reactive surfactants.[67] The surfactant provides stability against physical degradation (*i.e.* coalescence). This is due to the trend towards a minimal interfacial area between the dispersed phase and the dispersion medium. The surfactants used in miniemulsion polymerization should meet the same requirements as in conventional emulsion polymerization.[68] These requirements are the following: (1) their structure must have polar and non-polar groups, (2) they must be more soluble in the aqueous phase than the oil phase so as to be readily available for adsorption on the oil droplet surface, (3) they must adsorb strongly and not be easily displaced when two droplets collide, (4) they must be effective at low concentrations, and (5) they should be relatively inexpensive, non-toxic and safe to handle.

4.3.3 Preparation of Miniemulsions

In principle, miniemulsion preparation can be carried out by dissolving a suitable surfactant system in water and dissolving the hydrophobe in monomer (or monomer mixture), followed by premixing under stirring (see Scheme 4.1). The mixture is then subjected to a highly efficient homogenization process called miniemulsification. This can be achieved by using a high-shear dispersion device to disperse the premixed solution into small droplets. Various homogenization techniques can be used for the preparation of stable

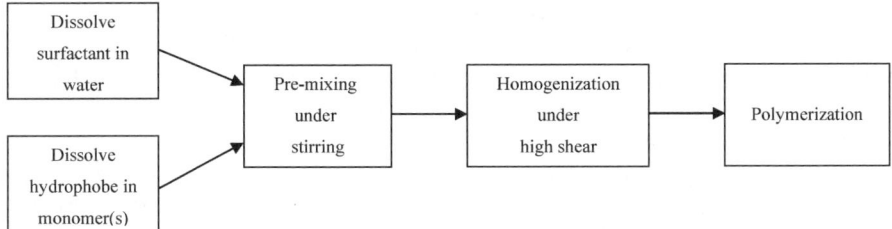

Scheme 4.1 Schematic representation of miniemulsion preparation.

miniemulsions. Stirring, used in the earlier work on miniemulsions,[46] has now been replaced with high-shear mechanical agitation and ultrasonication. The energy transferred by simple stirring is not sufficient to prepare small, well-distributed particles,[69] so a much higher-energy device such as a sonifier is required to create smaller droplets. According to Asua,[68] the following devices are the most commonly used to achieve homogenization: rotor–stator devices, sonifiers and high-pressure homogenizers. Today miniemulsification by ultrasound, first reported in 1927,[70] is most frequently used, especially for small latex quantities. Rotor–stator devices, which rely on turbulence to produce the miniemulsion, are used to prepare large quantities of latex. High-pressure homogenizers such as microfluidizers are also used to prepare such stable miniemulsions, and can be scaled up.

4.3.4 Initiators Used in Miniemulsions

In miniemulsions, polymerization can be initiated by using either a water-soluble or oil-soluble initiator. In the case of a water-soluble initiator polymerization commences in the aqueous phase; the initiator generates free radicals, by thermal decomposition, in the aqueous phase. This is similar to the case in conventional emulsion polymerization, where mainly water-soluble initiators are used. Polymerization involves the formation of oligomeric radicals, which will enter the monomer droplets when they reach a certain critical chain length. In this case, the initiator is added after the miniemulsification process takes place. Bechthold and Landfester[71] studied the miniemulsion polymerization of St using the water-soluble initiator, potassium persulfate (KPS). They found that the reaction rate was slightly increased by increasing the initiator concentration. However, increasing the initiator concentration caused a significant reduction of the average degree of polymerization.

On the other hand, an oil-soluble initiator can be mixed with the oil phase (monomer and hydrophobe) before premixing with the surfactant–water solution. Because of the small size of monomer droplets, radical recombination is then often a problem. Oil-soluble initiators are preferred when water-soluble monomers such as MMA and vinyl chloride are used. This is because nucleation can take place in the water phase (also referred to as secondary or homogeneous nucleation).[72] Oil-soluble initiators are also preferred when monomers with extremely low water solubility, such as lauryl methacrylate, need to be polymerized. Here the monomer concentration in the water phase is not high enough to frequently create oligomeric radicals which can enter the droplets. The possibility of nucleation in the water phase can also be minimized by using a redox initiation system, which contains two components (*e.g.* $(NH_4)_2S_2O_8/NaHSO_3$). In this case one component is in the aqueous phase and the other is in the oil phase.[73] Hence, the initiation is restricted to the interfacial layer of monomer droplets with the water phase.

4.3.5   Miniemulsion Polymerization for the Synthesis of PGNs

Miniemulsion is a convenient one-step technique that can be used for the incorporation of nanolayered filler materials such as clay[8] and CNTs[74] in polymer matrices. In the miniemulsion process, the oil phase, which consists of the monomer and the filler, can be dispersed in the water phase by a high-shear device such as a sonicator.[42] Similarly, the initial dispersion of the graphene nanosheets within the monomer droplets can be achieved by using the high-shear device during the miniemulsification process. The use of the high-shear device will also lead to the exfoliation of graphene nanoplatelets, which is followed by the *in situ* polymerization of monomers in the presence of these exfoliated graphene nanosheets. This will lead to the formation of monomer droplets containing the filler particles, and stabilized by the surfactant and the hydrophobe, from which polymer particles will develop during the polymerization step.[8,12,75]

Miniemulsion also offers many advantages over other polymerization methods: it is environmentally friendly, lattices with high solids content and high conversion can be obtained, polymers with high molar masses can be prepared, high rates of polymerization are achieved, during polymerization the viscosity remains low, and it is compatible with highly hydrophobic monomers. These advantages of miniemulsion polymerization make it even more attractive to use for the synthesis of PGNs. In a recent study, miniemulsion polymerization was successfully used for the incorporation of modified graphene nanoplatelets within a polymer matrix.[75] The study showed that miniemulsion could be used as an effective method for the synthesis of polymer nanocomposites based on graphene with improved properties.

A schematic representation of the formation of PGNs by miniemulsion polymerization is shown in Scheme 4.2. The graphene nanosheets can be added to a mixture of monomer and a hydrophobe for swelling. Surfactant solution is then added, followed by the miniemulsification process by sonication. The sonication step (*i.e.* miniemulsification) will lead to the exfoliation of graphene nanosheets to thinner graphene nanoplatelets. Upon polymerization, these graphene nanoplatelets will be finely distributed within the polymer matrix, resulting in an intercalated or exfoliated polymer nanocomposite system.

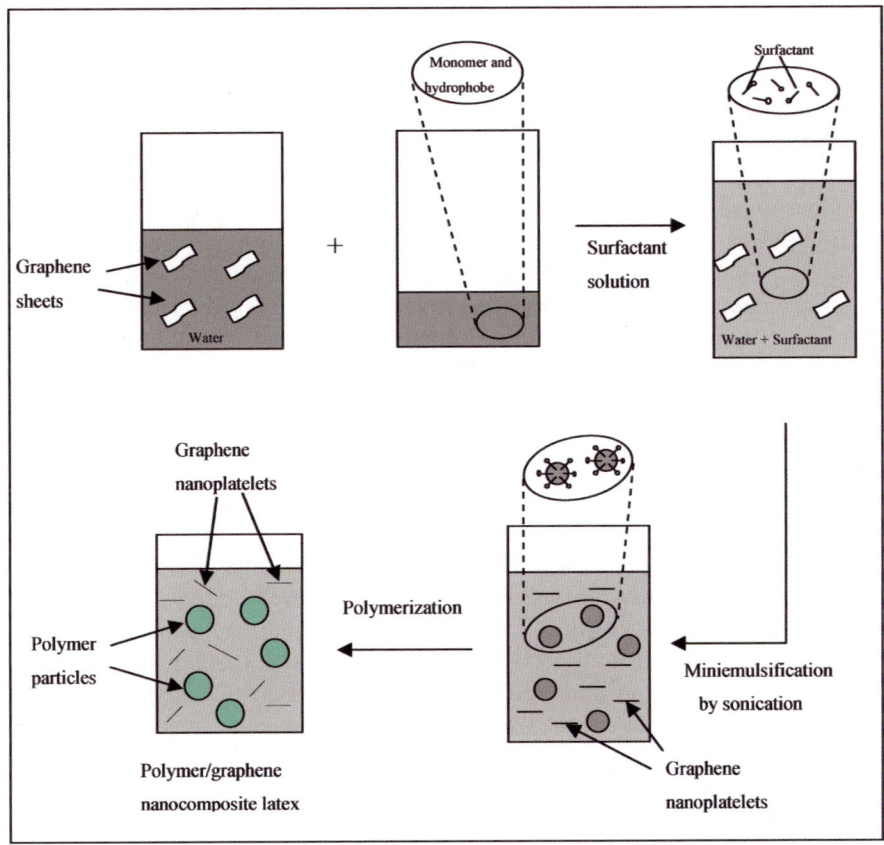

Scheme 4.2 The formation of polymer nanocomposite lattices based on graphene using miniemulsion polymerization.

4.4 Conventional Free Radical Polymerization *Versus* Controlled/Living Radical Polymerization

Free radical polymerization is a chain addition reaction in which polymer chains are formed by monomer molecules adding to free radicals. The radicals are created by the thermal decomposition (or otherwise) of an initiator, usually organic peroxides or azo compounds. These free radicals have single unpaired electrons, which are highly reactive. Therefore, they tend to take part in addition reactions such as polymerization processes.[76] This polymerization is one of the most important and versatile methods used for the synthesis of high molecular weight polymers on a commercial scale.[77] It is a powerful and inexpensive technique that can be easily applied, in comparison to other polymerization methods. In addition, a wide range of monomers and functional groups, including methacrylates, styrenics, acrylamides, and butadiene, can be polymerized under different reaction conditions.

Furthermore, using this process, polymerizations such as bulk, solution, emulsion, miniemulsion, and suspension have been successfully implemented.[78]

However, conventional free radical polymerization offers very little control over the macromolecular structure, such as the molecular weight distribution, composition and architecture of the polymers. This can be attributed to a constant radical generation followed by the occurrence of irreversible termination reactions throughout the polymerization process.[79,80] In 1955, the first report of living polymerization by an anionic process was published by Szwarc.[81] The author referred to the polymers formed (*i.e.* PS) as 'living polymers' because they were able to grow whenever additional monomer was supplied.[82] This had a tremendous impact on polymer science, and several controlled radical polymerization techniques have since been reported. A detailed description of the mechanistic developments in the field of CLRP is given in a review by Braunecker and Matyjaszewski.[80]

CLRP could also provide new synthesis methods for PGNs that allow very precise control over the polymerization process while retaining much of the versatility of conventional free radical polymerization. Generally speaking, CLRP enables the synthesis and design of new polymer architectures with predictable molecular weights, controlled molecular weight distribution (*i.e.* Đ) and well-defined end groups. The prediction and control of the molecular weight is achieved through a complex series of reactions and intermediates. Under appropriate conditions, the termination reaction will be reduced and the polymerization will behave as a living system. In a typical living polymerization, polymer chains must increase in molecular weight upon addition of new monomer units and the degree of polymerization should increase linearly with conversion.

4.5   Fundamentals of CLRP

In general, CLRP methods are based on the creation of a rapid equilibrium between the growing polymer radicals and dormant species. The irreversible chain termination reaction is suppressed by the presence of reagents that react with the propagating radicals in a activation/deactivation reaction (*i.e.* chain transfer process).[83] This allows the slow and simultaneous growth of all chains while keeping the concentration of radicals low enough to minimize termination. In the ideal case, the active growing polymer chains will continue growing as long as there is monomer present in the system. Only two processes (initiation and propagation), should occur, hence all growing polymer chains in the system should be active. Thus, all polymer chains, which were initiated at the beginning of the polymerization process, grow at the same rate, resulting in a better control of the polymerization.

The technique provides a simple method for the synthesis of advanced complex polymer architectures such as star, block and branched copolymers, which are more difficult to obtain by other synthetic methods.[84] CLRP is generally more compatible with the functional groups of the monomers than classical living polymerization methods, and can be carried out in many solvents over a wide temperature range. Several fundamental characteristics for an ideal CLRP can be summarized as follows:[83]

- Polymerization of monomers should proceed until all monomer molecules are consumed.
- The addition of new monomer results in the growth of the polymer chains without any new ones being initiated.
- During the polymerization process, the molar mass of polymers should be predictable.
- The number of living species must remain constant during the entire polymerization process and the molar mass should increase linearly with conversion.
- Polymers with narrow molecular weight distribution can be obtained.
- The end groups of the chain transfer compound are preserved at the ends of the resulting polymer chains.

4.6 Common CLRP Techniques

Various free radical processes offering controlled growth of polymer chains have been developed. Many new techniques, including nitroxide-mediated polymerization (NMP), atom transfer radical polymerization (ATRP) and RAFT-mediated polymerization have recently emerged and are now available for the production of new polymers with well-defined structures. These methods are discussed in detail in the following sections, with emphasis on the RAFT method.

4.6.1 NMP

NMP, first introduced by Solomon and Rizzardo in the 1980s, is based on the reversible trapping of carbon-centred radicals by nitroxides.[85,86] Scheme 4.3

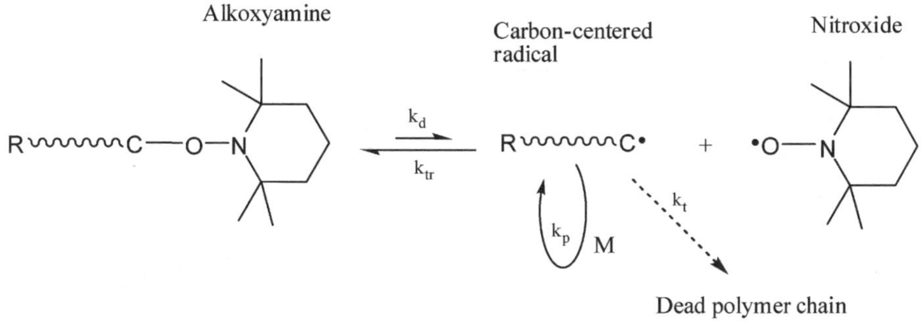

Scheme 4.3 Nitroxide-mediated polymerization: k_d is the dissociation rate coefficient, k_{tr} is the trapping rate coefficient, k_p is the propagation rate coefficient and k_t is the termination rate coefficient.

shows a general mechanism of the NMP method. At high temperature, the carbon–oxygen bond of the alkoxyamine species can be cleaved to form a nitroxide and a carbon-centred radical (*i.e.* equilibrium exists). The carbon-centred radical reacts with the monomer (M) units present to form propagating radicals. The radical can propagate or undergo a termination reaction until it is trapped by a nitroxide again. In an ideal case, the equilibrium lies close to the alkoxyamine, resulting in a low concentration of radicals, and therefore minimizing the termination rate for the polymerization. There are some disadvantages of NMP, for example, the elevated reaction temperatures that are often required for polymerization ($\sim$ 120 °C) and the limited monomer range that can be polymerized in a controlled way.

4.6.2 ATRP

This technique was first independently reported by Wang and Matyjaszewski[87] and Kato *et al.*[88] in 1995. The process is based on a well-known reaction in organic chemistry that is referred to as atom transfer radical addition. In this reaction, an organic radical is produced from an alkyl halide initiator (R–X), which can be transferred to a transition metal complex in a higher oxidation state. The radical can react with monomer to form polymer chains, which can be reversibly deactivated by transfer of the halogen back from the metal complex, or undergo termination or chain transfer reactions. If the correct halide is used, the exchange of the halogen atom between the alkyl group and the propagating radical polymer chain is fast and selective, resulting in a controlled polymerization. Scheme 4.4 illustrates the ATRP method.

Although the range of monomer types in ATRP is broader than that in NMP, the contamination of the polymer with the metal catalyst is a major drawback of ATRP. Removal of the toxic transition metal from the final polymer is thus required. Halogens such as chlorine, bromine and iodine have been found to be suitable for such migration reaction with various transition

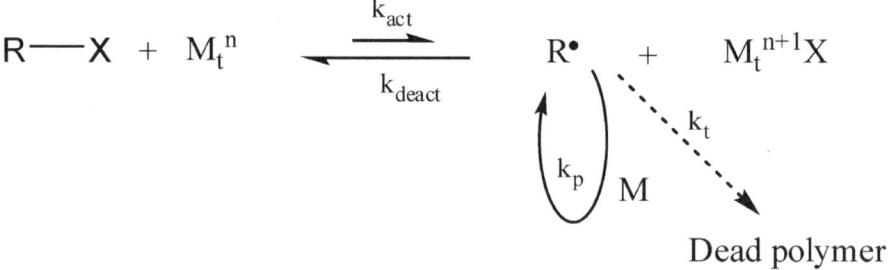

Scheme 4.4   Atom transfer radical polymerization: M_t^n is a transition metal complex, k_{act} is the activation rate coefficient, k_{deact} is the deactivation rate coefficient, k_p is the propagation rate coefficient and k_t is the termination rate coefficient.

metal systems for specific monomers. Transition metal systems used in ATRP include Cu(I)/Cu(II), Ru(II)/Ru(III), Fe(II)/Fe(III) and Ni(II)/Ni(III).

4.6.3 RAFT-Mediated Polymerization

Since its discovery in the late 1990s, the RAFT method has become one of the most effective and versatile methods of CLRP.[79,89] As a CLRP technique, this process allows the construction of polymers having targeted molecular weights with very low dispersity.[90–92] In principle, the RAFT method operates via a degenerative transfer mechanism in which a thiocarbonylthio compound (*e.g.* dithioesters, xanthates, dithiocarbamates and trithiocarbonates) acts as a chain transfer agent.[93] Other thiocarbonylthio compounds such as phosphoryl dithioesters and dithiocarbazates have also been employed in RAFT polymerization.[14] The preserved end groups of the RAFT agent can be reactivated, allowing the incorporation of additional monomer molecules to produce a variety of polymer architectures, including stars,[94] grafts,[95] brushes and branches.[96]

The key factor in a successful RAFT polymerization is the appropriate choice of the RAFT agent. This agent is a simple organic compound that possesses the thiocarbonylthio moiety (S=C–S) that imparts the living behaviour to free radical polymerization. The general molecular structure of a RAFT agent is illustrated in Figure 4.2, where Z denotes the stabilizing group and R is the free radical homolytic leaving group. The Z-group controls the reactivity of the C=S bond toward radical addition and fragmentation, and the R-group is responsible for re-initiating the polymerization.

Scheme 4.5 shows all the reactions involved in the RAFT process. The RAFT agent is added to the reaction medium in the presence of monomer (M) and radical initiator (I). Once the polymerization has commenced the initiator decomposes, generating free radicals, which can react with monomer to produce propagating radicals (P_n°) (Equation 4.1). The propagating radicals can react with the RAFT agent to give dormant chains, as shown in Equation 4.2. The leaving group radical then reacts with another monomer species, starting another active polymer chain (Equation 4.3). Equation 4.4 shows the main chain equilibrium reaction. This is the fundamental step in the RAFT process, which traps the majority of the active propagating species into the

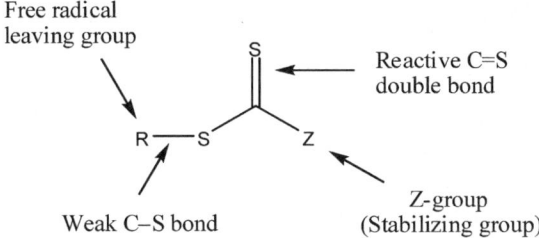

Figure 4.2   Basic structure of a typical RAFT agent.

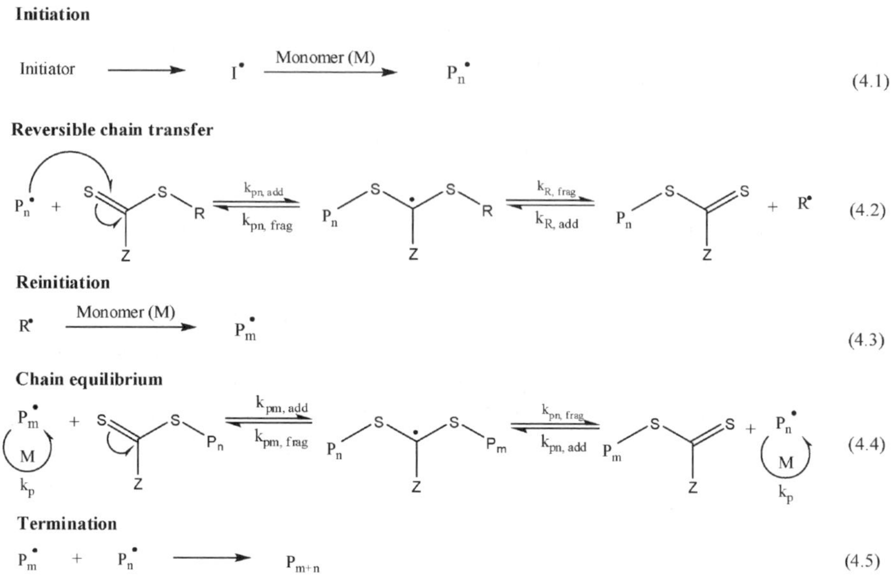

Initiation

$$\text{Initiator} \longrightarrow I^{\bullet} \xrightarrow{\text{Monomer (M)}} P_n^{\bullet} \tag{4.1}$$

Reversible chain transfer

$$\tag{4.2}$$

Reinitiation

$$R^{\bullet} \xrightarrow{\text{Monomer (M)}} P_m^{\bullet} \tag{4.3}$$

Chain equilibrium

$$\tag{4.4}$$

Termination

$$P_m^{\bullet} + P_n^{\bullet} \longrightarrow P_{m+n} \tag{4.5}$$

Scheme 4.5 The RAFT mechanism.

dormant thiocarbonyl compound. This limits the possibility of chain termination. By controlling the initiator/RAFT ratio, it is possible to produce polymer chains in a controlled manner with very narrow molecular weight distribution. The number of dead chains that are terminated by radical coupling (Equation 4.5) corresponds to the amount of decomposed initiator and the living polymerization features will be observed.

The RAFT process has many advantages over other CLRP techniques (*i.e.* NMP and ATRP), such as its versatility towards different monomers and functional groups (*e.g.* methacrylates, acrylates and styrenics) as well as the suitability to a wide range of reaction conditions. NMP is usually carried out at a high temperature. The ATRP catalyst, on the other hand, tends to bind strongly to the functional groups in the monomers used. A major drawback of this contamination is that the removal of the toxic transition metal from the final polymer is necessary.

4.7 RAFT-Mediated Emulsion Polymerization *Versus* Miniemulsion Polymerization

In the past, most research groups have focused on the RAFT method in homogeneous systems such as bulk[97] and solution[98] polymerization, where good understanding of the polymerization mechanism has been achieved. As CLRP in aqueous medium is industrially preferred, applying the RAFT process in emulsion or miniemulsion systems using water as the medium will be most useful. If RAFT-mediated polymerization can be successfully carried out in aqueous systems, the application and versatility of this process will be greatly enhanced.

The use of RAFT-mediated polymerization for the synthesis of polymer particles is, however, difficult to achieve by conventional emulsion polymerization, where colloidal instability is a major problem. This can be attributed to the poor transport of the RAFT agent from the monomer droplets to the polymer particles, which is necessary in such systems. This is due to the high hydrophobicity and low water solubility of most RAFT agents.[15,89] This means that in conventional emulsion systems, no transport of the fairly water-insoluble RAFT components into micelles can take place, leading to phase separation (*i.e.* a RAFT-rich and a polymer-rich phases are observed). Thus, polymer lattices with poor colloidal stability, loss of molecular weight control, slow polymerization rates and broad molecular weight distribution are obtained.[99,100] However, due to the initial dispersion of the hydrophobic components (RAFT agent and monomer), miniemulsion polymerization can be a powerful technique for the preparation of latex particles using RAFT-mediated polymerization.[101]

In comparison with emulsion polymerization, in miniemulsion polymerization most monomer droplets are, in principle, directly converted into particles, since the droplets are regarded as the locus of the initiation and propagation reaction.[102] Therefore, the transport of the monomer or other hydrophobic compounds from a reservoir to the polymerization locus, as in the case for emulsion polymerization, is unnecessary. This feature makes miniemulsion polymerization quite efficient as a convenient one-step nano-incorporation technique for hydrophobic compounds. In addition, in the miniemulsion process, the monomer droplets are directly polymerized, thus the resulting polymer particles are often one-to-one copies of the monomer droplets.[103] The RAFT agent can be equally distributed in the droplets at the beginning of the polymerization (*i.e.* during miniemulsification process) and the transport of the RAFT agent is not required during the polymerization.

4.8 Synthesis of PGNs Using the RAFT Process in Miniemulsion

The use of chain transfer agents in CLRP makes it possible to achieve control of the polymerization process. Among the CLRP methods, the discovery of the RAFT-mediated polymerization has been an outstanding achievement.[14,104] RAFT-mediated polymerization allows the preparation of polymers with low dispersity and predetermined molecular weights. In addition, the compatibility over a wide range of reaction conditions required for the RAFT process and its versatility toward different monomers make this method the most useful of all the CLRP techniques in designing macromolecular architectures. Thus, use of a combination of RAFT technology and graphite nanosheets for the synthesis of PGNs by RAFT-mediated polymerization is expected to allow for the preparation of a wide range of tailor-made polymer composites with enhanced properties.

The attachment of the RAFT agent to a solid support such as graphene nanosheets can be performed using either the R-group or the Z-group approach. In the R-group approach, the RAFT agent is attached to the solid support *via* the leaving and re-initiating R-group. In the Z-group approach, the RAFT agent is attached to the solid support through the stabilizing Z-group. Recently, Stenzel *et al.*[105] showed that in the R-group approach attachment *via* the R-group will lead to detachment of the RAFT agent during the polymerization, which may result in the loss of immobilized functionalities. In the Z-group approach these side reactions can be prevented, and controlled growth of polymer chains can be achieved.[106] Recently, in our group, RAFT-grafted graphene was dispersed in water in the presence of St monomer, a surfactant and a hydrophobe, to from miniemulsions.[12] The obtained miniemulsions were polymerized to produce polystyrene–graphite oxide (PS–GO) nanocomposites in a controlled manner. Scheme 4.6 presents the overall synthesis route for the preparation of these RAFT-immobilized graphene nanosheets.

As can be seen in Scheme 4.6, the RAFT agent (dodecyl isobutyric acid trithiocarbonate (DIBTC)), which has a carboxylic end group in the re-initiating group (R), can be successfully anchored to the graphene oxide nanosheets in the GO surface *via* an esterification reaction. GO is prepared by the oxidation of pristine natural graphite using a strong oxidizing agent such as potassium permanganate in the presence of concentrated mineral acids (*e.g.*

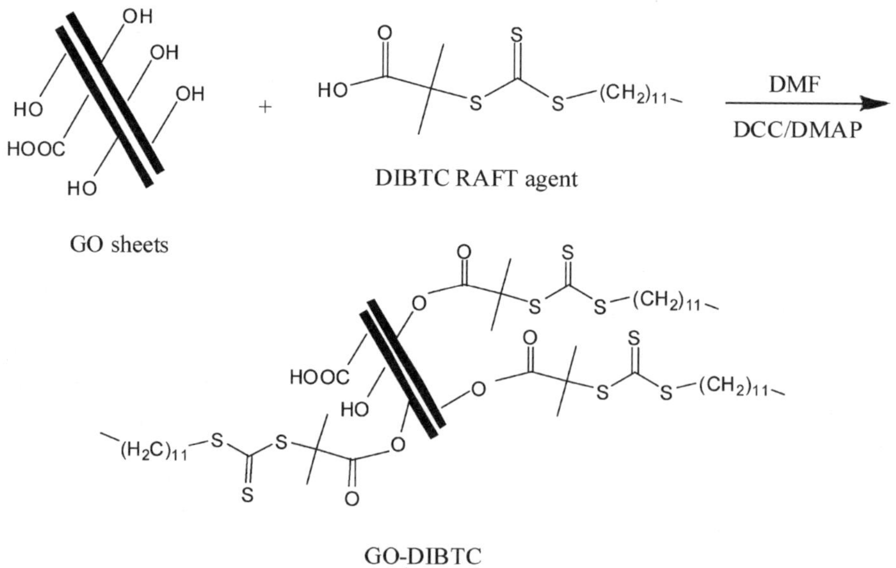

Scheme 4.6   Overall synthesis route for the preparation of RAFT-immobilized GO nanosheets. DMF, *N,N*-dimethylformamide; DCC, 1,3-dicyclohexyl carbodiimide; DMAP, 4-dimethylaminopyridine.

H_2SO_4).[24] The resultant RAFT-modified graphene can be used for the preparation of PGNs using miniemulsion polymerization similar to the procedure shown in Scheme 4.2. The RAFT-grafted GO is dispersed in the monomer phase and the resultant mixtures can be sonicated in the presence of a surfactant and a hydrophobe to form miniemulsions. The stable miniemulsions thus obtained are polymerized using suitable initiator to yield PGNs in a controlled manor. The intercalation of graphene nanoplatelets within the polymer matrix will be achieved in situ during the miniemulsion process without a prior exfoliation step.

4.9 Characterization of PGNs Synthesized by the RAFT Method

Generally speaking, the structure of PGNs can be determined using X-ray diffraction (XRD) and transmission electron microscopy (TEM) analysis. The structure of a PGN (either intercalated or exfoliated) may be identified by using XRD to monitor the position of the basal reflections from the graphene layers. TEM allows a qualitative understanding of the internal structure of PGNs through direct visualization at the nanometre level. However, special care is required when TEM is used to guarantee a representative part of the sample. Other analytical techniques such as dynamic mechanical analysis (DMA), thermogravimetric analysis (TGA), size exclusion chromatography (SEC) and Fourier-transform infrared (FT-IR) spectroscopy can be used to fully characterize the obtained PGNs. DMA gives the mechanical response for the composite material while TGA shows the chemical degradation behaviour as a function of temperature. FT-IR could be used to determine the chemical structure of the RAFT-attached graphene and the final PGNs. SEC could also be used to determine the molar masses and Đ of the polymer in the PGNs. These analytical techniques are discussed in detail in the following sections.

4.9.1 FT-IR and Solubility Analysis

In order to determine the attachment of the RAFT agent on the GO surface, the RAFT-attached-GO product is characterized by FT-IR (KBr discs). The amount of GO sample added to the KBr has to be strictly controlled because the black GO can absorb most of the infrared rays if too high concentration of GO is used. Figure 4.3 shows the FT-IR analysis of GO, RAFT and RAFT-attached GO. Compared with the FT-IR spectrum of crude GO (Figure 4.3b), the FT-IR spectrum of GO-RAFT (Figure 4.3c) shows the characteristic peaks of the RAFT agent, such as C−H, C=O, C−S and C=S stretching vibrations, centred at 2921 and 2853, 1701, 815 and 1069 cm^{-1} respectively (see Figure 4.3a). Using this information, and knowing that all free RAFT was removed, it can be concluded that the RAFT agent is covalently bound to the surface of the GO.

Figure 4.4 shows digital images of the GO and the RAFT-grafted GO dispersed in St monomer. Is it apparent that GO is insoluble in St, and there is

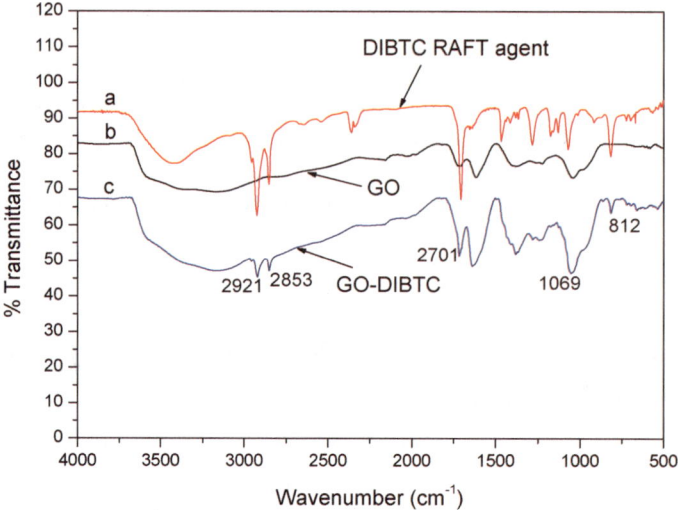

Figure 4.3   FT-IR spectra of (a) DIBTC RAFT agent, (b) GO and (c) GO-DIBTC.

much sedimentation of GO at the bottom of the vial (see Figure 4.4a). This can be attributed to the hydrophilic nature of GO, which cannot be easily dispersed in hydrophobic monomers such as St. However, as shown in Figure 4.4b, the RAFT-functionalized GO is soluble in St and forms a homogeneous solution (no sedimentation observed). The anchored RAFT agent led to an increase in the hydrophobic character of the GO nanosheets, subsequently leading to better compatibility between GO and the water-insoluble monomer St.

Figure 4.4   Digital photographs of GO and GO-RAFT dispersed in St monomer: (a) 1 wt% GO relative to monomer; (b) 5 wt% GO-RAFT relative to monomer.

4.9.2   TEM Analysis

TEM can be used to visualize the latex particles and to determine the morphology of the films prepared from the obtained lattices. Figure 4.5 shows TEM images of the latex particles prepared using 4 and 5 wt% GO-RAFT relative to the monomer using miniemulsion polymerization. Individual polymer/graphene core-shell particles with diameters ranging from 150 to 180 nm were obtained. In addition, the particle size distribution is fairly narrow, which is an indication that little to no secondary particle nucleation occurred during the polymerization process. The lighter areas are representative of the polymer shell, while the darker areas represent the core RAFT-modified GO. This is due to the difference in contrast between the core and the shell domains as a result of the different path lengths and material densities of the constituent materials. This resulted in increased scattering of the incident electron beam from the core material (GO-RAFT), resulting in a darker region in the TEM images.

The GO nanoplatelets that have smaller particle dimensions could be incorporated into the polymer particles. This is due to the effect of the RAFT agent, which made GO more hydrophobic, allowing for better compatibility between the monomer and the GO nanosheets. As indicated in Figure 4.4, GO is hydrophilic and does not mix with monomer (St), while GO-RAFT is hydrophobic and disperses in monomer. The GO nanosheets can also be seen in the TEM images of the dried film that was embedded into epoxy resin. Figures 4.6a and b show TEM images of films made with 3 and 5 wt% of GO-RAFT relative to monomer, respectively. Most of the graphene nanoplatelets were of exfoliated morphology, with the exception of some areas that contained a few intercalated GO nanosheets (in agreement with the XRD measurements).

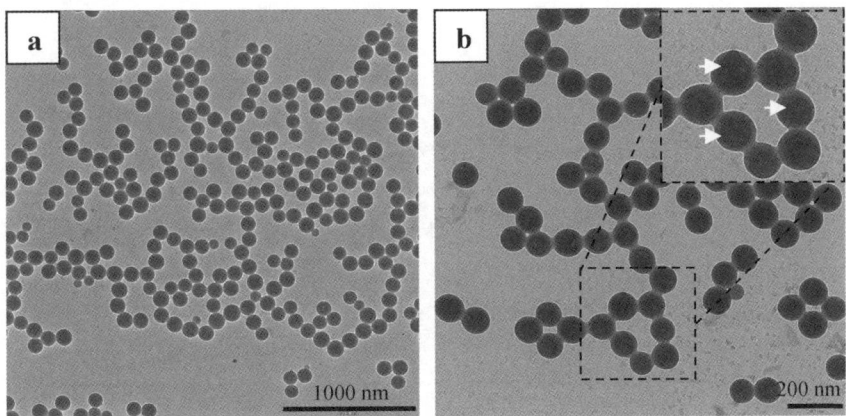

Figure 4.5   TEM images of PS–GO nanocomposite latex particles made with different amounts of GO-RAFT: (a) 5 wt% GO-RAFT (at low magnification); (b) 4 wt% GO-RAFT (at higher magnification).

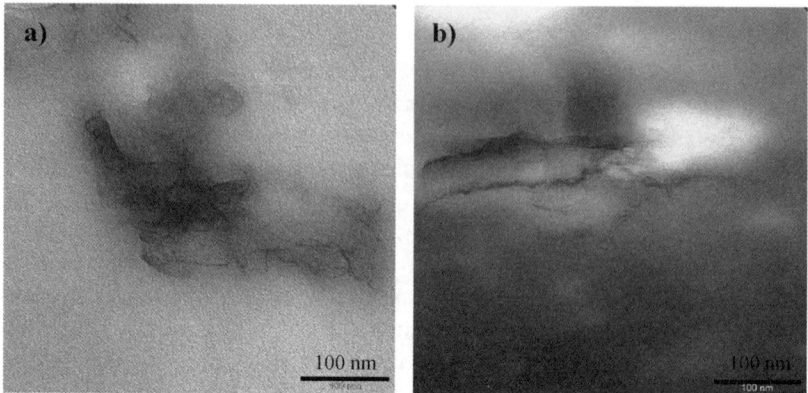

Figure 4.6 TEM images of microtomed films cast from PS-GO nanocomposite latices prepared with different amount of GO-RAFT: (a) 3 wt% GO-RAFT; (b) 5 wt% GO-RAFT.

4.9.3 XRD Analysis

The XRD patterns of PS–GO nanocomposites synthesized with different RAFT-attached GO loadings are shown in Figure 4.7. The measurements were performed on nanocomposites containing 1, 5 and 7 wt% GO-RAFT relative to monomer. Complete exfoliation of the GO sheets was obtained (no diffraction peak of GO nanoplates was observed). The absence of the characteristic peak of GO suggested that the layered GO had been exfoliated in the nanocomposites.[107] The broad peak observed at a 2θ value of 20° is due to PS, as reported in literature.[108]

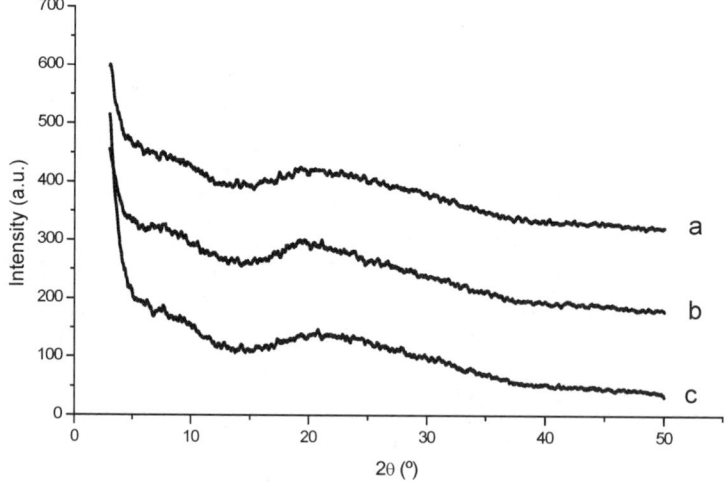

Figure 4.7 XRD results of PS-GO nanocomposites with different GO-RAFT loadings: (a) 1 wt% GO-RAFT; (b) 5 wt% GO-RAFT; (c) 7 wt% GO-RAFT.

4.9.4 SEC Analysis

Table 4.1 summarizes the molecular weights (weight average molecular weight, $\bar{M}_W$ and number average molecular weight, $\bar{M}_W$) and Đ values of PS in the nanocomposites prepared with different quantities of GO-RAFT in miniemulsion. An increase in GO-RAFT loading resulted in a decrease in the molar masses of the PS chains. This was expected, because the concentration of the RAFT agent increases with an increase in the amount of RAFT-modified GO incorporated into the polymer nanocomposites. It is well known that an increase in the RAFT agent concentration results in a decrease in the molar mass of polymers.[98,109]

From Table 4.1, it can be seen that the best control was achieved with 7 wt% GO-RAFT loading, which seems to be the threshold concentration at which good control begins to be observed. The higher Đ values recorded at low GO-RAFT content may be attributed to the fact that the system is highly heterogeneous. In this system the monomer is emulsified in the water phase and the modified GO, on which the RAFT agent is attached, is suspended in the monomer phase. Thus the RAFT agent is not homogeneously distributed in the monomer phase, resulting in regions of low and high RAFT agent concentration. Therefore, the probability of a growing polymeric radical to encounter a RAFT molecule, and thus control of polymerization, increases with an increase in GO-RAFT concentration (*i.e.* an increase in RAFT agent concentration).

Furthermore, at lower GO-RAFT loadings the number of GO-RAFT particles might vary from one polymer particle to another, leading to different target chain lengths, and thus higher Đ is observed. In addition, it was observed that at low GO-RAFT loading (*i.e.* low RAFT concentration) polymer chains with high $\bar{M}_n$ were obtained (see Table 4.1). This could result in a larger Đ value due to the effect of the more newly formed PS chains from the initiator. As the concentration of RAFT-grafted GO increases, such variation is expected to be smaller, resulting in lower Đ values.

Table 4.1 Molar masses and Đ of the PS–GO nanocomposites and PS reference

GO-RAFT relative to monomer (wt%)	$\bar{M}_n$ (g/mol)	$\bar{M}_W$ (g/mol)	Đ
–	96 700	162 600	1.68
1	177 400	287 700	1.62
2	116 600	185 000	1.58
3	74 600	106 800	1.43
4	84 600	126 400	1.49
5	61 700	87 900	1.42
6	71 100	93 900	1.32
7	53 600	67 500	1.26

4.9.5 Mechanical Properties

The enhancement in mechanical properties of PGNs is caused by the strong interaction between polymer chains and graphene nanoplatelets, which have a high aspect ratio. This will lead to a significant increase in the mechanical properties of polymers in the presence of graphene, compared to pure polymers. The interaction between polymer and graphene layers could suppress the mobility of the polymer segments near the polymer–graphene interface, leading to improved mechanical properties of polymers.[110,111] Figures 4.8 and 4.9 show the mechanical properties of the PS–GO nanocomposites made with the RAFT process as evaluated by DMA. It can be seen that the nanocomposites with high GO-RAFT content had enhanced storage and loss modulus in the glassy state relative to the neat PS reference.

At RAFT-modified GO loadings of 1 wt% relative to monomer, the storage modulus of the nanocomposite was lower than that of the PS reference (5.7 × 10^7 Pa). However, samples with higher GO-RAFT content (2–6 wt%) had storage modulus values higher than that of the pure PS. Furthermore, at low GO-RAFT content (1–3 wt%) the loss modulus of the nanocomposites was lower than that of the PS reference (1.6 × 10^7 Pa). However, when the RAFT-modified GO content reached 6 wt% relative to monomer the loss modulus was higher than that of the pure PS reference (see Figure 4.9). These results also show that the modulus was simply a function of filler content in the nanocomposite. Both the storage and the loss modulus of PS–GO nanocomposites increase with increasing RAFT-modified GO content in the sample. The enhancement in storage and loss modulus is caused by the strong

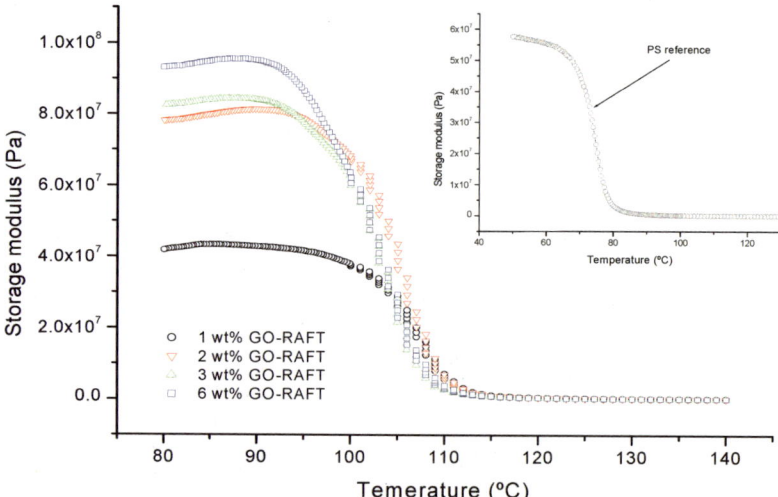

Figure 4.8   Storage modulus as function of temperature of PS–GO nanocomposites at GO loadings of 1, 2, 3 and 6 wt%. The insertion shows the storage modulus of PS reference.

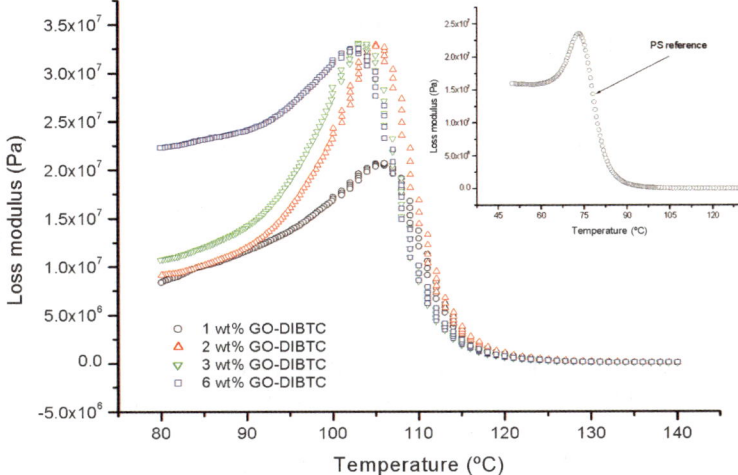

Figure 4.9 Loss modulus as function of temperature of PS–GO nanocomposites, at GO-RAFT loadings of 1, 2, 3 and 6 wt%. The insertion shows the loss modulus of PS reference.

interaction between polymer chains and GO nanoplatelets, which have a high aspect ratio. This results in a decrease in the polymer segments' mobility near the polymer–graphene interface, leading to a higher modulus.[110,111]

The T_g of the PS polymer in the nanocomposite was determined from the onset temperature of the tan δ curve in the DMA scan. Figure 4.10 shows the

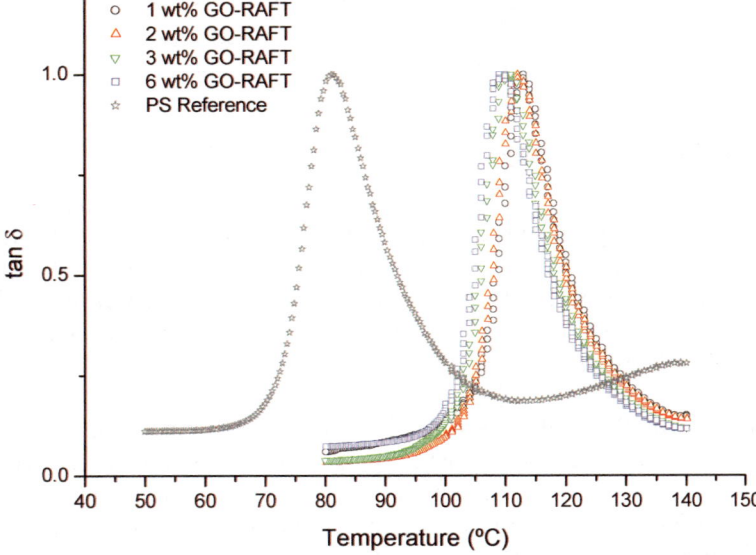

Figure 4.10 Tan δ as a function of temperature of PS–GO nanocomposites at GO-RAFT loadings of 1, 2, 3 and 6 wt%.

Table 4.2 T_g values of PS–GO nanocomposites and PS reference obtained from the onset temperature of the tan δ curve in the DMA scan

GO-RAFT content relative to monomer (wt%)	T_g (°C)
0	74
1	105
2	103
3	102
6	101

variation of tan δ of the PS–GO nanocomposites with temperature. Table 4.2 shows the T_g of PS–GO nanocomposites containing different loading of GO-RAFT. A shift of the tan δ peaks of the nanocomposites to higher temperatures relative to the PS reference was recorded. This indicates that the PS–GO nanocomposites have higher T_g values, ranging from 101 to 105 °C, compared to the value of the pure PS ($T_g = 74$ °C) (see Table 4.2). This was due to restricted chain mobility of the polymer caused by the presence of GO nanosheets.

However, as the RAFT-modified GO loading increased, a slight shift of the tan δ peaks to lower temperatures (lower T_g values) was recorded. This was attributed to the change in molar masses of the PS in the nanocomposites prepared with different quantities of RAFT-modified GO. It was shown in Table 4.1 that an increase in the RAFT-functionalized GO loading resulted in a significant decrease in the molar mass of the PS chains. This led to a significant decrease in the T_g of PS in the nanocomposites. It is well known that T_g increases with increasing $\bar{M}_n$ which can be attributed to a reduction in the relative number of polymer chain ends.[112]

4.9.6   Thermal Stability

The onset temperature of degradation for the PS in the nanocomposites increased noticeably in the presence of GO-RAFT, and all synthesized nanocomposites are more thermally stable relative to the neat PS. This indicates that the incorporation of GO-RAFT into the PS leads to better thermal stability of the polymer. However, recent results indicate that improvement in thermal stability is not simply a function of GO-RAFT loading (see Figure 4.11). This was attributed to the effect of GO-RAFT concentration in the molar masses of the PS in the nanocomposite. It was found that the molar masses of the polymer in the nanocomposites decreased markedly as the RAFT-functionalized graphene loading increased (see Table 4.1). The effect of this change in molar mass could counteract the effect of the increased graphene content on the thermal stability.

The improvement in thermal stability of PS in the presence of GO-RAFT can be attributed to the intercalation of PS into the lamellae of graphene. The

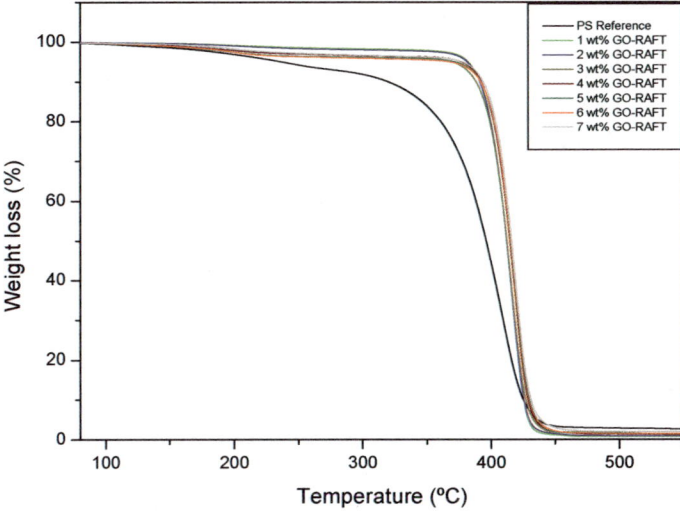

Figure 4.11 TGA thermograms of PS–GO nanocomposites and a PS reference.

PS chains are trapped between the graphene nanoplatelets in GO-RAFT, which may act as an insulator between the heat source and the surface area of polymer, where the combustion occurs.[113] The presence of graphene nanoplatelets may also hinder the diffusion of volatile decomposition products within the nanocomposites by promoting char formation. The char-formed layer acts as a mass transport barrier that retards the escape of the volatile products generated as the PS decomposes.[107] The enhancement of the nanocomposites' thermal stability has also been attributed to the movement restriction of the polymer chains inside the graphene nanogalleries.[107]

4.10 Conclusions

PGNs are a new class of polymer materials that exhibit unique functional and physical properties. The ability to synthesize such nanocomposite materials can be of great scientific and industrial importance due to their potential properties and applications. The usual method for their preparation involves the use of exfoliation of graphene nanoplatelets in monomers followed by polymerization, mainly under bulk conditions. This process has led to various degrees of success in forming the true dispersed nanocomposites. The use of miniemulsion polymerization as the polymerization method allows the formation of polymer latices, containing the graphene nanosheets, which can be exfoliated during the miniemulsion process. Miniemulsion will also promote the intercalation of monomers into the graphene nanosheets, leading to the formation of exfoliated structure. This presents a new approach for the preparation of polymer nanocomposites based on graphene nanosheets.

In the past, most researchers in the field of nanotechnology have focused mainly on the synthesis and characterization of PGNs using conventional free radical polymerization. Only a few studies on the use of CLRP, such as ATRP and the RAFT method, focus on the use of clay and CNTs. The modification of graphene with direct functionalization using the RAFT process, could lead to a controlled living radical polymer growth from the graphene surface. This process present a new approach for the preparation of polymer nanocomposites based on graphene in a controlled manor. This opens the possibility for the synthesis of a wide range of PGNs based on modified graphene nanosheets due to the versatility of the RAFT polymerization process.

References

1. Y.-W. Mai, Z.-Z. Yu, *Polymer Nanocomposites*, Woodhead Publishing, Cambridge, 2nd edition, 2006.
2. S. Stankovich, D. A. Dikin, G. H. B. Dommett, K. M. Kohlhaas, E. J. Zimney, E. A. Stach, R. D. Piner, S. T. Nguyen, R. S. Ruoff, *Nature*, 2006, **442**(7100), 282–286.
3. A. K. Geim, K. S. Novoselov, *Nat. Mater.*, 2007, **6**(3), 183–191.
4. W. Zheng, S.-C. Wong, *Compos. Sci. Technol.*, 2003, **63**(2), 225–235.
5. W. Zheng, S.-C. Wong, H.-J. Sue, *Polymer*, 2002, **73**(25), 6767–6773.
6. J. Xu, Y. Hu, L. Song, Q. Wang, W. Fan, G. Liao, Z. Chen, *Polym. Degrad. Stab.*, 2001, **73**(1), 29–31.
7. H. Kim, A. A. Abdala, C. W. Macosko, *Macromolecules*, 2010, **43**(16), 6515–6530.
8. A. Samakande, R. D. Sanderson, P. C. Hartmann, *J. Polym. Sci, Part A: Polym. Chem.*, 2008, **46**(21), 7114–7126.
9. A. Samakande, R. D. Sanderson, P. C. Hartmann, *Eur. Polym. J.*, 2009, **45**(3), 649–657.
10. G.-J. Wang, S.-Z. Huang, Y. Wang, , L. Liu, J. Qiu, Y. Li, *Polymer*, 2007, **48**(3), 728–733.
11. X. Pei, J. Hao, W. Liu, *J. Phys. Chem. C*, 2007, **111**(7), 2947–2952.
12. H. M. Etmimi, M. P. Tonge, R. D. Sanderson, *J. Polym.Sci, Part A: Polym. Chem.*, 2011, **49**(7), 1621–1632.
13. T. P. Le, G. Moad, E. Rizzardo, S. H. Thang, PCT Int. Appl. WO, 9801478 A1, 980115.
14. J. Chiefari, Y. K. Chong, F. Ercole, J. Krstina, J. Jeffery, T. P. T. Le, R. T. A. Mayadunne, G. F. Meijs, C. L. Moad, G. Moad, E. Rizzardo, S. H. Thang, *Macromolecules*, 1998, **31**(16), 5559–5562.
15. E. Rizzardo, M. Chen, B.. Chong, G. Moad, M. Skidmore, S. H. Thang, *Macromol. Symp.*, 2007, **248**, 104–116.
16. S. Perrier, P. Takolpuckdee, *J. Polym. Sci, Part A: Polym. Chem.*, 2005, **43**(22), 5347–5393.
17. S. Jing-Wei, C. Xiao-Mei, H. Wen-Yi, *J. Appl. Polym. Sci.*, 2003, **88**(7), 1864–1869.

18. P.-G. Liu, P. Xiao, M. Xiao, K.-c. Gong, *Chin. J. Polym. Sci.*, 2000, **18**(5), 413–418.

19. Z. Wenge, L. Xuehong, W. Shing-Chung, *J. Appl. Polym. Sci.*, 2004, **91**(5), 2781–2788.

20. X. S. Du, M. Xiao, Y. Z. Meng, *J. Polym. Sci., Part B: Polym. Phys.*, 2004, **42**, 1972–1978.

21. G. Zheng, J. Wu, W. Wang, C. Pan, *Carbon*, 2004, **42**, 2839–2847.

22. Meng Y. Polymer/graphite nanocomposites. In: *Polymer Nanocomposites*, ed. Y.-W. Mai and Z.-Z. Yu, Woodhead Publishing, Cambridge, 2nd edition, 2006, Vol. 19, pp. 510–539.

23. L. M. Viculis, J. J. Mack, O. M. Mayer, H. T. Hahn, R. B. Kaner, *J. Mater. Chem.*, 2005, **15**, 974–978.

24. W. S. Hummers, R. E. Offeman, *J. Am. Chem. Soc.*, 1958, **80**, 1339.

25. J. R. Potts, D. R. Bielawski, C. W. Dreyer, , R. S. Ruoff, *Polymer*, 2011, **52**(1), 5–25.

26. M. Hirata, T. Gotou, S. Horiuchi, M. Fujiwara, M. Ohba, *Carbon*, 2004, **42**(14), 2929–2937.

27. M. Hirata, T. Gotou, M. Ohba, *Carbon*, 2005, **43**(3), 503–510.

28. J. T. Paci, T. Belytschko, G. C. Schatz, *J. Phys. Chem. C*, 2007, **111**(49), 18099–18111.

29. D. W. Boukhvalov, M. I. Katsnelson, *J. Am. Chem. Soc.*, 2008, **130**(32), 10697–10701.

30. R. J. W. E. Lahaye, H. K. Park, C. Y. Jeong, Y. H. Lee, *Phys. Rev. B*, 2009, **79**(12), 125435.

31. A. Lerf, H. He, M. Forster, J. Klinowski, *J. Phys. Chem. B*, 1998, **102**(23), 4477–4482.

32. T. Szabo, O. Berkesi, P. Forgo, K. Josepovits, Y. Sanakis, D. Petridis, I. Dekany, *Chem. Mater.*, 2006, **18**(11), 2740–2749.

33. W. Cai, R. D. Piner, F. J. Stadermann, S. Park, M. A. Shaibat, Y. Ishii, D. Yang, A. Velamakanni, S. J. An, M. Stoller, J. An, D. Chen, R. S. Ruoff, *Science*, 2008, 321(5897), 1815–1817.

34. W. Gao, L. B. Alemany, L. Ci, P. M. Ajayan, *Nat. Chemi.*, 2009, **1**(5), 403–408.

35. K. A. Mikhoyan, A. W. Contryman, J. Silcox, D. A. Stewart, G. Eda, C. Mattevi, S. Miller, M. Chhowalla, *Nano Letters*, 2009, **9**(3), 1058–1063.

36. H. He, J. Klinowski, M. Forster, A. Lerf, *Chem. Phys. Lett.*, 1998, **287**(1–2), 53–56.

37. H. He, T. Riedl, A. Lerf, J. Klinowski, *J. Phys. Chem.*, 1996, **100**(51), 19954–19958.

38. G. I. Titelman, V. Gelman, S. Bron, R. L. Khalfin, Y. Cohen, H. Bianco-Peled, *Carbon*, 2005, **43**(3), 641–649.

39. B. Zhang, Y. Chen, X. Zhuang, G. Liu, B. Yu, E.-T. Kang, J. Zhu, Y. Li, *J. Polym. Sci, Part A: Polym. Chem.*, 2010, **48**(12), 2642–2649.

40. J. Liu, L. Tao, W. Yang, D. Li, C. Boyer, R. Wuhrer, F. Braet, T. P. Davis, *Langmuir*, 2010, **26**(12), 10068–10075.

41. K. P. Pramoda, H. Hussain, H. M. Koh, H. R. Tan, C. B. He, *J. Polym. Sci, Part A: Polym. Chem.*, 2010, **48**(19), 4262–4267.
42. E. Sudol, M. El-Aasser, Miniemulsion polymerization. In *Emulsion Polymerization and Emulsion Polymers*, ed. P. Lovell and M. El-Aasser, John Wiley & Sons, New York, 2nd edition, 1997, pp. 699–722.
43. J. Soma, K. Papadopoulos, *J. Colloid Interface Sci.*, 1996, **181**, 225–231.
44. M. Antonietti, K. Landfester, *Progr. Polym. Sci.*, 2002, **27**, 689–757.
45. V. Rodriguez, M. El-Aasser, J. Asua, C. Silebi, *J. Polym. Sci, Part A: Polym. Chem.*, 1989, **27**, 3659–3671.
46. J. Ugelstad, M. El-Aasser, J. Vanderhoff, *J. Polym. Sci., Polym. Lett. Ed.*, 1973, **11**, 503–513.
47. F. Hansen, J. Ugelstad, *J. Polym. Sci.: Polym. Chem. Ed.*, 1979, **17**, 3069–3082.
48. N. Kim, E. Sudol, , V. Dimonie, M. El-Aasser, *Macromolecules*, 2004, **37**, 2427–2433.
49. I. Aizpurua, M. Barandiaran, *Polymer*, 1999, **40**, 4105–4115.
50. M. El-Aasser, E. Sudol, Features of emulsion polymerization. In *Emulsion Polymerization and Emulsion Polymers*, ed. P. Lovell and M. El-Aasser, John Wiley & Sons, New York, 2nd edition, 1997, pp 37–58.
51. F. Schork, G. Poehlein, S. Wang, J. Reimers, J. Rodrigues, C. Samer, *Colloids Surf. A: Physicochemical and Engineering Aspects*, 1999, **153**, 39–45.
52. S. Wang, F. Schork, *J. Appl. Polym. Sci.*, 1994, **54**, 2157–2164.
53. J. Reimers, F. Schork, *J. Appl. Polym. Sci.*, 1996, **59**, 1833–1841.
54. D. Mouran, J. Reimers, F. Schork, *J. Appl. Polym. Sci.*, 1996, **34**, 1073–1083.
55. P. Tang, E. Sudol, M. Adams, M. El-Aasser, J. Asua, *J. Appl. Polym. Sci.*, 1991, **42**, 2019–28.
56. J. Alduncin, J. Forcada, J. Asua, *Macromolecules*, 1994, **27**, 2256–2261.
57. Y. Choi, M. El-Asser, E. Sudol, J. Vanderhoff, *J. Polym. Sci.: Polym. Chem. Ed.*, 1985, **23**, 2973–2987.
58. M. Unzue, J. Asua, *J. Appl. Polym. Sci.*, 1993, **49**, 81–90.
59. J. Masa, L. Arbin, J. Asua, *J. Appl. Polym. Sci.*, 1993, **48**, 205–213.
60. C. Miller, E. Sudol, C. Silebi, M. El-Aasser, *J. Polym. Sci, Part A: Polym. Chem.*, 1995, **33**, 1391–1408.
61. K. Landfester, N. Bechthold, F. Tiarks, M. Antonietti, *Macromolecules*, 1999, **32**, 5222–5228.
62. M.-S. Lim, H. Chen, *J. Polym. Sci, Part A: Polym. Chem.* 2000, **38**(10), 1818–1827.
63. J. Delgado, M. El-Aasser, J. Vanderhoff, *J. Polym.Sci, Part A: Polym. Chem.*, 1986, **24**, 861–874.
64. C. Chern, T. Chen, *Colloid Polym. Sci.*, 1997, **275**, 546–554
65. K. Landfester, N. Betchthold, F. Tiarks, M. Antonietti, *Macromolecules*, 1999, **32**, 2679–2683.
66. C. Chern, Y. Liou, *Polymer*, 1999, **40**, 3763–3772.
67. E. Kitzmiller, C. Miller, E. Sudol, M. El-Aasser, *Macromol. Symp.*, 1995, **92**, 157–168.
68. J. Asua, *Progr. Polym. Sci.*, 2002, **27**, 1283–1346.

69. B. Abismail, J. Canselier, A. Wilhelm, H. Delmas, C. Gourdon, *Ultrason. Sonochem.*, 1999, **6**, 75–83.
70. R. Wood, A. Loomis, *Phil. Mag.*, 1927, **4**, 417.
71. N. Bechthold, K. Landfester, *Macromolecules*, 2000, **33**, 4682–4689.
72. B. Saethre, P. Mork, J. Ugelstad, *J. Polym. Sci, Part A: Polym. Chem.*, 1995, **33**, 2951–2959.
73. C. Wang, N. Yu, C. Chen, J. Kuo, *J. Appl. Polym. Sci.*, 1996, **60**, 493–501.
74. H. F. Lu, B. Fei, J. H. Xin, R. H. Wang, L. Li, W. C. Guan, *Carbon*, 2007, **45**(5), 936–942.
75. H. M. Etmimi, R. D. Sanderson, *Macromolecules*, 2011, **44**(21), 8504–8515.
76. R. Morrison, R. Boyd, *Organic Chemistry*, Allyn & Bacon, New York, 3rd edition, 1973, pp. 46–50.
77. G. Moad, D. H. Solomon, *The Chemistry of Free Radical Polymerization*, Pergamon, Oxford, 2nd edition, 1995.
78. A. M. Van Herk, M. Monteiro, Heterogeneous systems. In: *Handbook of Radical Polymerization*, ed. K. Matyjaszewski and P. D. Thomas, John Wiley & Sons, New York, 2nd edition, 2003, Vol. 6, pp. 301–331.
79. G. Moad, E. Rizzardo, S. H. Thang, *Austral. J. Chem.*, 2005, **58**(6), 379–410.
80. W. A. Braunecker, K. Matyjaszewski, *Progr. Polym. Sci.*, 2007, **32**(1), 93–146.
81. M. Szwarc, *Nature*, 1956, 178(4543), 1168–1169.
82. M. Szwarc, *J. Polym. Sci, Part A: Polym. Chem.*, 1998, 36(1), IX–XV.
83. K. Matyjaszewski, General concepts and history of living radical polymerization. In: *Handbook of Radical Polymerization*, ed. K. Matyjaszewski and P. D. Thomas, John Wiley & Sons, New York, 2nd edition, 2003, Vol. 8, pp. 361–406.
84. Y. Ganou, D. Taton, Macromolecular engineering by controlled/living radical polymerization. In: *Handbook of Radical Polymerization*, ed. K. Matyjaszewski and P. D. Thomas, John Wiley & Sons, New York, 2nd edition, 2003, Vol. 14, pp 775–844.
85. D. H. Solomon, E. Rizzardo, P. Caciolo, U.S. Patent 4,581,429, 1986.
86. D. H. Solomon, *J. Polym. Sci, Part A: Polym. Chem.*, 2005, **43**(23), 5748–5764.
87. J.-S. Wang, K. Matyjaszewski, *J. Am. Chem. Soc.*, 1995, **117**(20), 5614–5615.
88. M. Kato, M. Sawamoto, M. Kamigaito, T. Higashimura, *Macromolecules*, 1995, **28**(5), 1721–1723.
89. G. Moad, J. Chiefari, J. Chong, J. Krstina, R. T. Mayadunne, A. Postma, E. Rizzardo, , S. H. Thang, *Polym. Int.*, 2000, **49**(9), 993–1001.
90. C. L. McCormick, A. B. Lowe, *Acc. Chem. Res.*, 2004, **37**(5), 312–325.
91. X. Yin, A. S. Hoffman, P. S. Stayton, *Biomacromolecules*, 2006, **7**(5), 1381–1385.
92. P. E. Millard, L. Barner, M. H. Stenzel, T. P. Davis, C. Barner-Kowollik, A. H. E., Müller, *Macromol. Rapid Commun.*, 2006, **27**(11), 821–828.
93. J. Chiefari, E. Rizzardo, Control of free-radical polymerization by chain transfer methods. In: *Handbook of Radical Polymerization*, ed. K.

Matyjaszewski and P. D. Thomas, John Wiley & Sons, New York, 2nd edition, 2003, pp. 629–690.

94. D. Boschmann, P. Vana, *Macromolecules*, 2007, **40**(8), 2683–2693.
95. Y. Zhao, S. Perrier, *Macromol. Symp.*, 2007, **248**(1), 94–103.
96. A. P. Vogt,S. R. Gondi, B. S. Sumerlin, *Austral. J. Chem.*, 2007, **60**(6), 396–399.
97. J. M. Lee, O. H. Kim, S. E. Shim, B. H. Lee, S. Choe, *Macromol. Res.*, 2005, **13**(3), 236–242.
98. A. B. Lowe, C. L. McCormick, *Progr. Polym. Sci.*, 2007, **32**(3), 283–351.
99. M. J. Monteiro, M. Hodgson, H. De Brouwer, *J. Polym.Sci, Part A: Polym. Chem*, 2000, **38**(21), 3864–3874.
100. M. J. Monteiro, J. de Barbeyrac, *Macromolecules*, 2001, **34**(13), 4416–4423.
101. X. Zhou, P. Ni, Z. Yu, *Polymer*, 2007, **48**(21), 6262–6271.
102. M. Antonietti, K. Landfester, *Progr. Polym. Sci.*, 2002, **27**(4), 689–757.
103. K. Landfester, N. Bechthold, S. Förster, M. Antonietti, *Macromol. Rapid Commun.*, 1999, **20**(2), 81–84.
104. T. P. Le, G. Moad, E. Rizzardo S. H. Thang, PCT Int. Appl. WO, 9801478, 1998.
105. H. S. Martina, P. D. Thomas, *J. Polym.Sci, Part A: Polym. Chem.*, 2002, **40**(24), 4498–4512.
106. H. S. Martina, Z. Ling, T. S. H. Wilhelm, *Macromol. Rapid Commun.*, 2006, **27**(14), 1121–1126.
107. R. Zhang, Y. Hu, J. Xu, W. Fan, Z. Chen, *Poly. Degrad. Stab.*, 2004, **85**(1), 583–588.
108. F. M. Uhl, C. A. Wilkie, *Poly. Degrad. Stab.*, 2002, **76**(1), 111–122.
109. A. Postma, T. P. Li, G. Davis, G. Moad, M. S. O'Shea, *Macromolecules*, 2006, **39**(16), 5307–5318.
110. J. Yang, M. Tian, Q.-X. Jia, J.-H. Shi, L.-Q. Zhang, , S.-H. Lim, Z.-Z. Yu, Y.-W. Mai, *Acta Mater.*, 2007, **55**(18), 6372–6382.
111. C. Donghwan, L. Sangyeob, Y. Gyeongmo, F. Hiroyuki T. D. Lawrence, *Macromol. Mater. Eng.*, 2005, **290**(3), 179–187.
112. J. M. G. Cown, *Eur. Polym. J.*, 1975, **11**(4), 297–300.
113. W. Jianqi, H. Zhidong, *Polym. Adv. Technol.*, 2006, **17**(4), 335–340.

CHAPTER 5

In Situ *Polymerization in the Presence of Graphene*

YUAN HU AND CHENLU BAO

State Key Laboratory of Fire Science, University of Science and Technology of China, Jinzhai Road 96, Hefei, Anhui 230026, China
E-mail: yuanhu@ustc.edu.cn or baocl@mail.ustc.edu.cn

5.1 Introduction

Polymer–graphene nanocomposites (PGNs) have triggered enormous interest in recent years. The combination of graphene and polymer results not only in dramatically enhanced properties, but also in novel functions. PGNs have been regard as a promising type of materials. In PGNs, the dispersion of graphene and the interface interaction between the graphene and the polymer matrix are two key factors.[1] The dispersion determines the effective specific surface area of graphene and the interface interaction directly influences the dispersion of graphene and the property enhancements of PGNs.

There are three main methods of preparing PGNs: (1) melt blending, (2) solvent blending, and (3) *in situ* polymerization. Due to the strong interactions among the graphene nanolayers, graphene usually re-aggregates easily, especially when it is dried. The strong tendency to re-aggregation makes dispersion of graphene in the polymer matrix difficult. Earlier researchers have pointed out that melt blending caused poorer dispersion than solvent blending and *in situ* polymerization.[2] As compared with solvent blending, *in situ* polymerization is obviously advantageous in the interface interaction when the polymer is copolymerized with the graphene (or derivative). *In situ*

RSC Nanoscience & Nanotechnology No. 26
Polymer–Graphene Nanocomposites
Edited by Vikas Mittal
© The Royal Society of Chemistry 2012
Published by the Royal Society of Chemistry, www.rsc.org

polymerization combines good dispersion and strong interface interaction, so it is widely used to prepare PGNs.

Pristine graphene, which consists of non-polar carbon, is seldom used in polymer nanocomposites due to its high price and small preparation quantities. Currently, the most promising preparation strategy to prepare graphene with low cost and in large quantities is the reduction of graphite oxide (GO). The reduced form of GO, which is usually called "reduced graphite oxide" (rGO), has several kinds of functional groups such as hydroxyl, carboxyl, ester, carbonyl and alkyl. These oxygen-containing groups offer a great chance for the chemical functionalization of graphene. The functionalization introduces reactive groups to the surface or the edge of graphene sheets, which can then be polymerized *in situ* with monomers or oligomers. In addition, graphene can also be incorporated into the polymer matrix by *in situ* polymerization without further functionalization.

Although Rouff's group reported PGNs prepared by solvent blending as early as 2006,[3] *in situ* polymerized PGNs were reported 3 years later[4] From then on, many kinds of PGNs have been reported with various enhancements and functions. In this chapter, the recent progress of *in situ* polymerized PGNs is introduced according to the kinds of polymeric matrix used.

5.2   Polyaniline (PANI)

PANI is well known as an electrically conducting polymer. Table 5.1 shows the published papers on *in situ* polymerized PANI–graphene nanocomposites, most of which are about electrochemical, electrical and electronic properties.

The most widely used preparation method is based on oxidation polymerization. Before the polymerization, graphene (or GO) and aniline are usually mixed to form a stable blend. Wang *et al.* prepared PANI–GO nanocomposites using H_2O_2, HCl and $FeCl_3.6H_2O$ in an aqueous medium. The electrochemical properties were investigated in 1 M H_2SO_4 using a three-electrode system. The highest initial specific capacitance was 746 F g^{-1} and the capacitance retention was 73% after 500 cycles.[7] Later, they prepared PANI–GO nanocomposites

Table 5.1   Details on the *in situ* polymerized PANI–graphene nanocomposites

First author	Main improvements	Reference
Murugan *et al.*	Electrochemical property	5–18
Zhang *et al.*	Electrical conductivity	19–22
Bao	Electroactivity	23
Majumdar	Electronic properties	24
Al-Mashat	Sensitivity of H_2 gas	25
Chang	Sensitivity of heavy metal ions	26
Luong	Mechanical properties	27
Basavaraja	Microwave absorption	28
Goswami	Electron field emission property	29
Zhang	Electrorheological response	30

using HCl and ammonium persulfate (APS) in ethylene glycol. The field emission scanning electron microscopy (FE-SEM) photograph of a nanocomposite in Figure 5.1b shows that PANI was homogenously attached to the GO surface. The nanocomposites exhibited excellent electrochemical properties. The initial specific capacitance reached 1126 F g^{-1} with good cycle stability (Figure 5.1a), the energy density reached 30–38 W h kg^{-1} and the power density reached about 140 h kg^{-1} (Figure 5.1c).[12]

Electropolymerization is another technique used to prepare PANI-based composites. Wang *et al.* fabricated graphene paper (G-paper) by filtration of the graphene suspension through a membrane. The G-paper was directly used as electrode and aniline was polymerized on the paper (Figure 5.2). The gravimetric capacitance of the obtained PANI–graphene nanocomposites was 233 F g^{-1} and the volumetric capacitance was 135 F cm^{-3}.[6]

PANI and graphene are both electroconductive, but GO usually has poor conductivity due to lattice defects. Chen *et al.* prepared PANI–GO

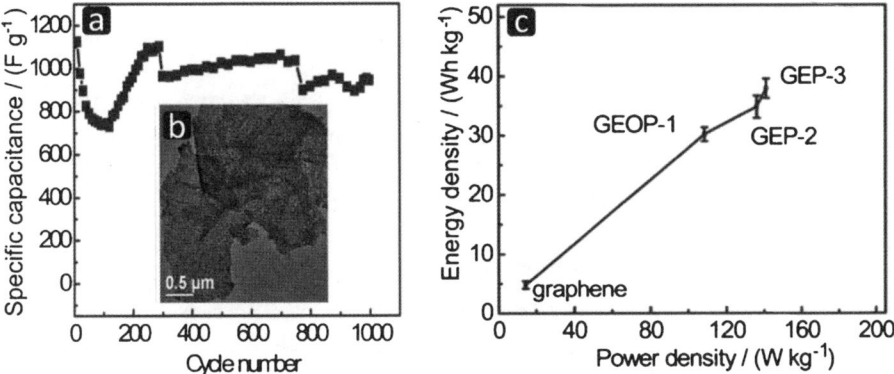

Figure 5.1 (a) Specific capacitance as a function of cycle number; (b) FE-SEM photograph of PANI–GO nanocomposites; (c) specific power and specific energy for graphene electrode. Reproduced with permission from ref. 12. Copyright 2012, Royal Society of Chemistry.

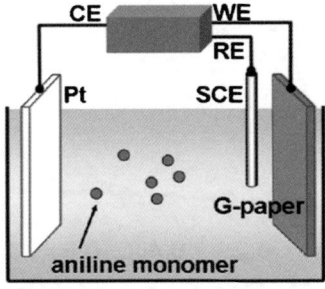

Figure 5.2 Electropolymerization of PANI in the presence of graphene paper. Reproduced with permission from ref. 6. Copyright 2012, American Chemistry Society.

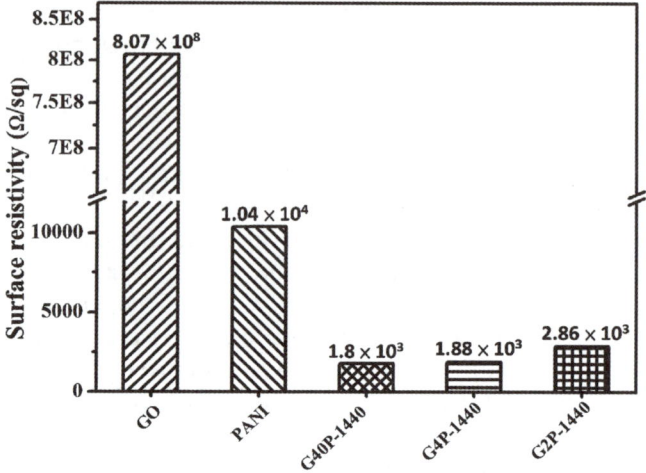

Figure 5.3 Surface resistivity of GO, PANI and PANI–GO nanocomposites.
Reproduced with permission from ref. 21. Copyright 2012, American
Chemistry Society.

nanocomposites by oxidation polymerization using HCl and APS. The
combination of PANI and GO significantly improved the electrical conductivity (Figure 5.3), because the GO served as an oxidizing agent for the
polymerization and an excellent electron-transfer path.[21]

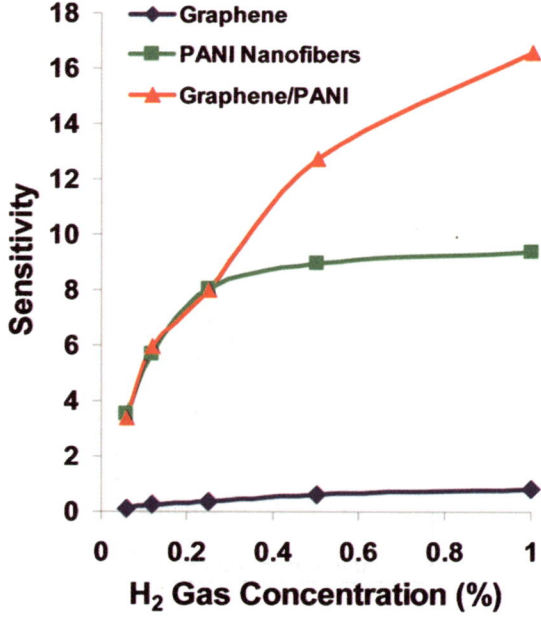

Figure 5.4 Sensitivities of H_2 gas sensors. Reproduced with permission from ref. 25.
Copyright 2012, American Chemistry Society.

PANI–graphene nanocomposites are also investigated in other fields besides the electron and electrochemical applications. Details are listed in Table 5.1. Al-Mashat *et al.* investigated the hydrogen sensing property of PANI–graphene nanocomposites prepared using HCl and APS. The sensitivity is shown in Figure 5.4. Graphene had fairly poor sensitivity and the sensitivity of PANI was about 9.38%. The sensor sensitivity of the PANI–graphene nanocomposites reached 16.57%. Moreover, the nanocomposites also showed improved response as compared with PANI.[25]

5.3 Polypyrrole (PPy)

PPy is another well-known type of electrical conductive polymer. Researches on *in situ* polymerized PPy–graphene nanocomposites is summarised in Table 5.2. As with PANI, most of the research relates to electrochemical applications.

The dispersion of graphene in the polymer matrix is important for property enhancements. In order to form a homogenous dispersion of graphene in PPy, Bose *et al.* reduced GO using hydrazine at the presence of poly(sodium 4-styrenesulfonate) (Na-PSS). Na-PSS was attached to the graphene surface, rendering further agglomeration of the graphene impossible. The functionalized graphene was then incorporated into PPy by *in situ* polymerization (Figure 5.5a). Due to the attachment of Na-PSS, the thickness of the functionalized graphene was about 2 nm, which is greater than the thickness of graphene prepared from the traditional hydrazine reduction method (usually 0.4–0.8 nm). This nanocomposite was used in a supercapacitor and the specific capacitance was about 267 F g^{-1} at a high scan rate of at 100 mV s^{-1}. As shown in Figure 5.5b, the PPy–graphene nanocomposites exhibited fairly good cycle stability, almost 90%.[33] Xu *et al.* prepared PPy–graphene nanocomposites using HCl and APS, and the nanocomposites obtained exhibited better cycle stability than in Bose's work, ~95% after 1000 cycles.[39] Zhang *et al.* prepared layered GO structures with sandwiched PPy (Figure 5.6). The morphology of PPy was controlled by controlling the shape of surfactant micelles. The specific capacitance was as high as 528 F g^{-1}.[32]

PPy–graphene nanocomposites were also investigated in other fields. Chandra *et al.* prepared PPy–graphene nanocomposites and investigated the selective removal of bivalent metallic ions. The nanocomposites presented high

Table 5.2 Research on *in situ* polymerized PPy–graphene nanocomposites

First author	Main improvements	Reference
Han *et al.*	Electrochemical properties	31–40
Bose	Electrically conductive	41
Zou	Phenol detection	42
Chandra	Selective removal of Hg^{2+}	43
Chandra	Selective adsorption of CO_2	44

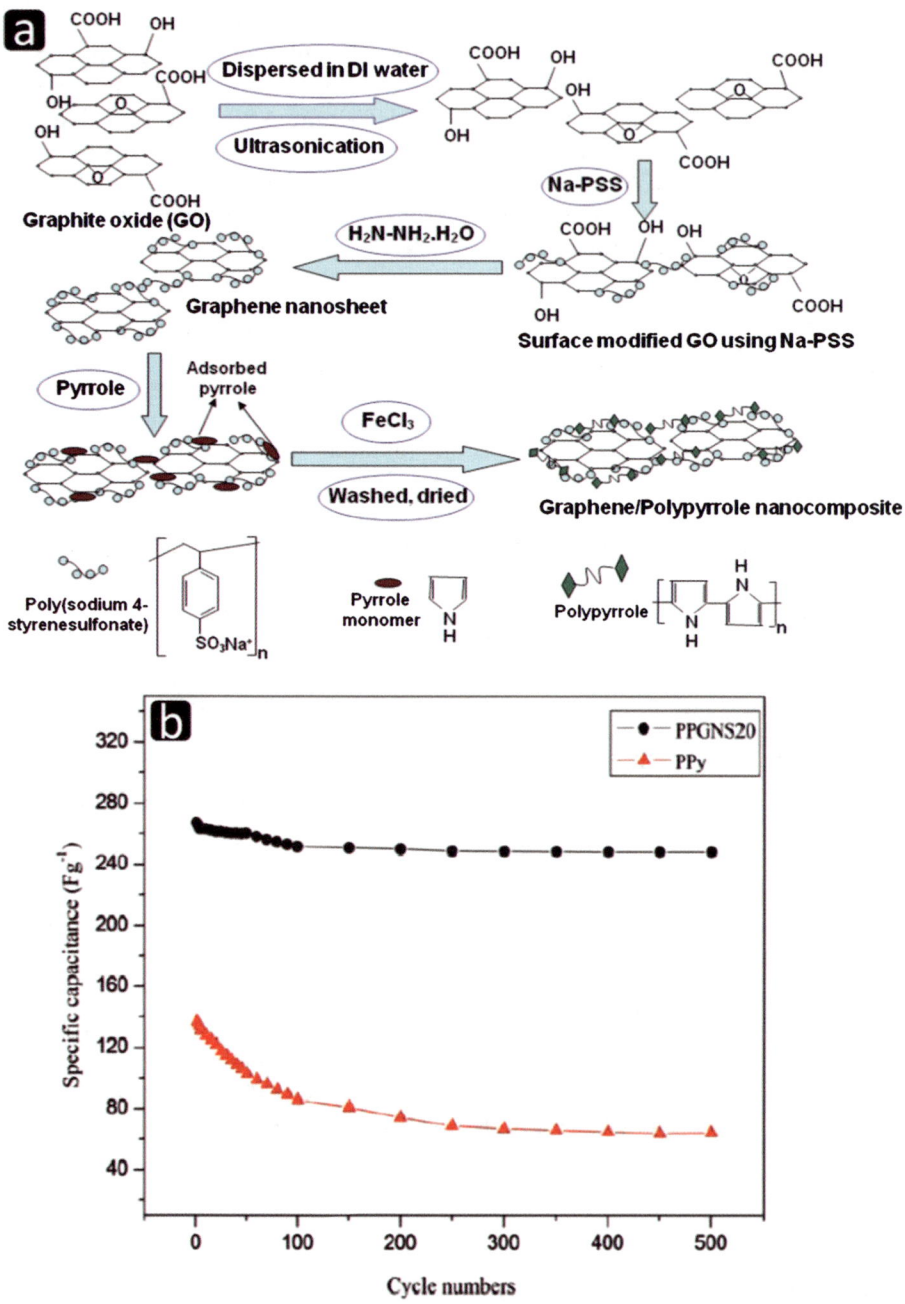

Figure 5.5 (a) Synthesis of PPy–graphene nanocomposites; (b) Specific capacitance of PPy and PPy–graphene nanocomposite measured at 100 mV s^{-1}. Reproduced with permission from ref. 33. Copyright 2012, IOP Publishing Ltd.

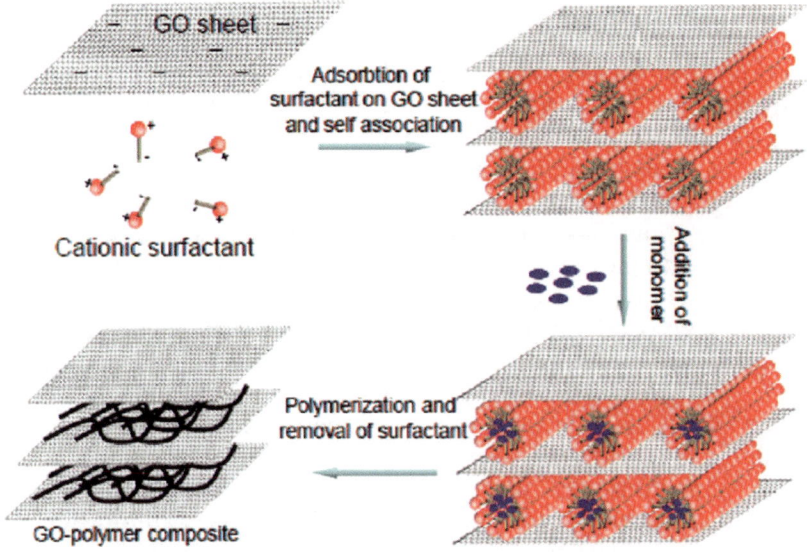

Figure 5.6 Preparation of PPy–GO nanocomposite in Zhang's work. Reproduced with permission from ref. 32. Copyright 2012, American Chemistry Society.

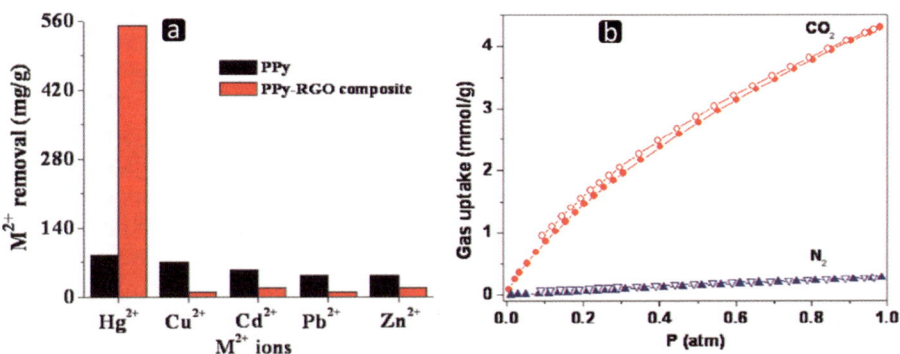

Figure 5.7 (a) M^{2+} removal properties of PPy and the PPy–graphene nanocomposites in reference.[43]; (b) Adsorption of CO_2 and N_2 in reference 44. Reproduced with permission from ref. 44. Copyright 2012, Royal Society of Chemistry.

selectivity for Hg^{2+} (Figure 5.7a).[43] Later, they reported N-doped porous nanocomposites produced via chemical activation (by KOH solution) of PPy and graphene. The adsorption of CO_2 was 4.3 mmol g^{-1} and the adsorption of N_2 was only 0.27 mmol g^{-1} (Figure 5.7a).[44] Zou *et al.* found that PPy–graphene nanocomposites exhibited fairly good detection of phenol.[42] This research indicates promising applications of PPy–graphene nanocomposites in selectivity and detection.

5.4 Epoxy

Epoxy is one of the most important polymers due to its useful properties and wide applications. As a kind of thermosetting plastic, epoxy is widely used in coatings, complexes, structural adhesives and electronic circuit board materials. Research on epoxy–graphene nanocomposites focuses on improving the mechanical, thermal, electrical, and fire safety properties, as shown in Table 5.3.

Graphene or functionalized graphene can be incorporated into epoxy by *in situ* curing. Due to the oxygen-containing functional groups, GO and graphene are usually functionalized to incorporate curable functional groups, such as amine and epoxide. Fang *et al.* functionalized the graphene surface with amines and then cured with epoxy matrix (Figure 5.8a). The thickness of graphene was increased from 0.97 nm (GO) to 4.83 nm (amine-functionalized graphene). The amine-functionalized graphene promoted the exfoliation and dispersion of graphene nanolayers in the matrix, served as a linker between graphene nanolayers and the matrix for improved load transfer and constructed a hierarchical structure to dissipate strain energy during fracture.[46] As a result, with only 0.6 wt% functionalized graphene, the fracture toughness and the flexural strength were increased by 93.8% and 91.5%, respectively.

Bao *et al.* reported an epoxide functionalization method to introduce epoxy groups on to the graphene surface using hexachlorocyclotriphosphazene (HCCP), a hexatomic ring containing six reactive chlorine groups in each molecule, and glycidol (Figure 5.9).[50] HCCP was attached to the GO surface by reaction with the hydroxyl groups, and the chlorine in HCCP was reacted with glycidol to introduce epoxy groups. Every hydroxyl group on the GO surface was bonded with one HCCP and then with six epoxy groups, which significantly extended the curable groups on GO surface. The functionalized graphene was *in situ* heat-cured with epoxy resin and good dispersion was obtained. The storage modulus, hardness, electrical conductivity and thermal stability of the nanocomposites were improved.

Epoxy is an easily combustible polymer, and its poor fire resistance restricts its applications. Wang *et al.* investigated the effect of graphene on the flame retardancy of epoxy by cone calorimetry.[53] As compared with neat epoxy, the epoxy–graphene nanocomposites exhibited longer ignition time, broadened

Table 5.3 Research on of epoxy–graphene nanocomposites

First author	Main improvements	Reference
Yang *et al.*	Mechanical properties	45–49
Bao	Mechanical, electrical and thermal properties	50
Kim	Electrical conductivity and storage modulus	51
Guo *et al.*	Fire safety	52–53
Zaman *et al.*	Mechanical and thermal properties	54–56
Qiu	Ease of the curing of epoxy	57
Teng	Thermal conductivity	58
Liu	Varistor effect	59

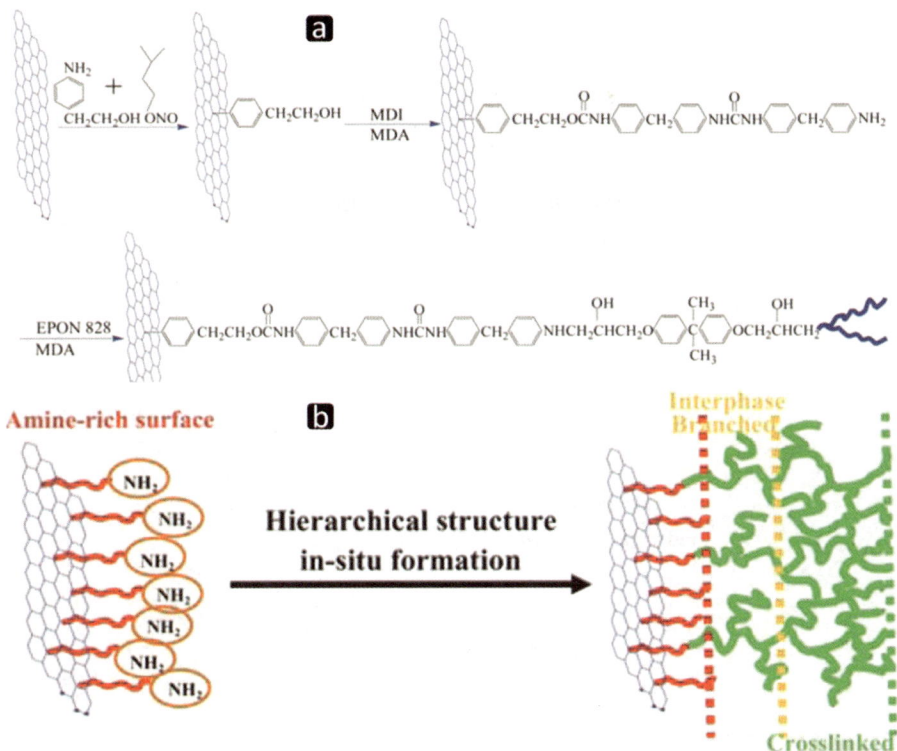

Figure 5.8 (a) Synthesis of amine-rich graphene and epoxy–graphene nanocomposites; (b) The surface of amine-rich graphene and hierarchical interphase. Reproduced with permission from ref. 46. Copyright 2012, Royal Society of Chemistry.

heat release rate curve and reduced peak heat release rate (PHRR). Guo *et al.* modified GO with flame-retardant elements including nitrogen and phosphorus, and then incorporated GO, graphene and the functionalized graphene into epoxy. The obtained nanocomposites showed obvious reductions in PHRR, total heat release rate and thermal degradation rate.[52]

Graphene also affects other properties of epoxy–graphene nanocomposites, such as thermal conductivity and glass transition temperature (T_g). Details are shown in Table 5.3. It is reasonable that epoxy–graphene is a promising family of PGNs due to the ease of the functionalization of GO, the convenient curing methods and obvious enhancements.

5.5 Poly(methyl methacrylate) (PMMA)

Research on *in situ* polymerized PMMA–graphene nanocomposites relate mainly to thermal and mechanical properties, similar to the research on epoxy–graphene. Table 5.4 gives details of the relevant work.

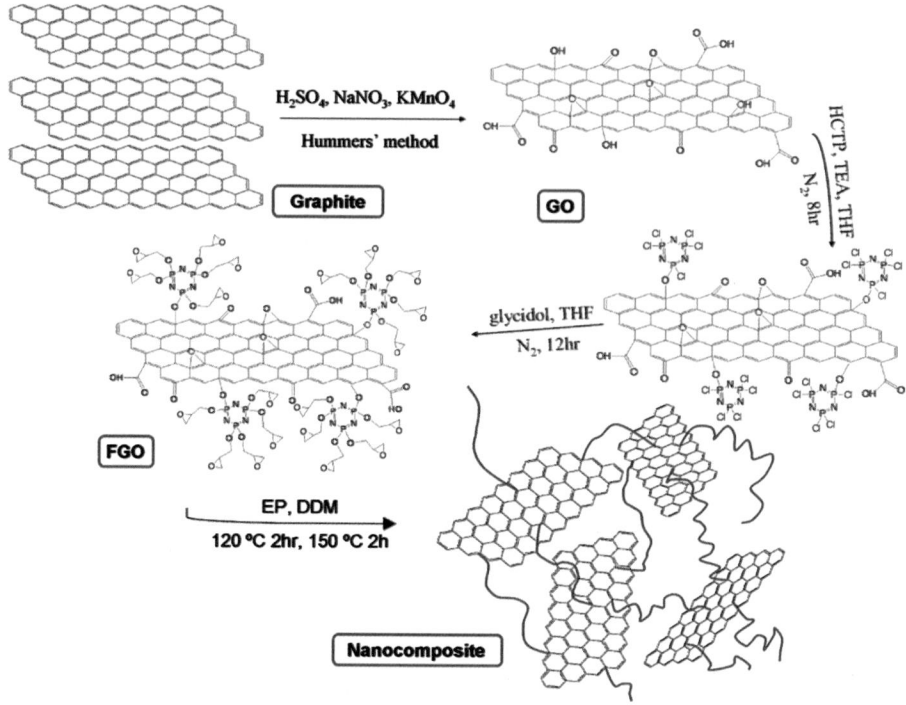

Figure 5.9   Preparation route of epoxide-functionalized graphene and the epoxy–
graphene nanocomposites. Reproduced with permission from ref. 50.
Copyright 2012, Royal Society of Chemistry.

Potts *et al.* prepared PMMA–GO nanocomposites using benzoyl peroxide
initiator and investigated their mechanical and thermal properties
(Figure 5.10).[64] The storage modulus, loss modulus, tensile strength and
thermal stability of PMMA–graphene nanocomposites are obviously
enhanced. Layek *et al.* prepared PMMA–graphene nanocomposites using
atom transfer radical polymerization, and the mechanical properties of the
obtained nanocomposites, including storage modulus (124%), stress at break
(157%) and Young's modulus (321%), are significantly improved.

5.6   Polystyrene (PS)

Research on *in situ* polymerized PS–graphene nanocomposites focuses mainly
on the preparation of novel PS–graphene nanostructures and the thermal,

Table 5.4   Statistics of *in situ* polymerized PMMA–graphene nanocomposites

First author	Main improvements	Reference
Layek *et al.*	Thermal and mechanical properties	60–64
Wang	Mechanical properties and electrical conductivity	65
Wang	Detection of capillary electrophoresis	66

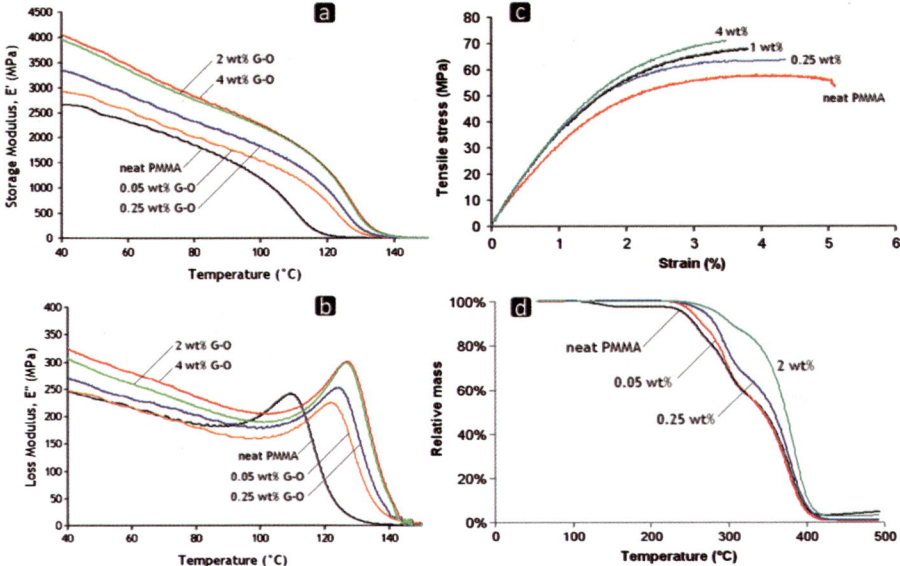

Figure 5.10   (a) Storage modulus, (b) loss modulus, (c) tensile properties and (d) thermogravimetric curves of PMMA and PMMA–GO nanocomposites. Reproduced with permission from ref. 64. Copyright 2012, Elsevier Ltd.

electrical and mechanical properties of the nanocomposites. Table 5.5 gives details.

There are several *in situ* polymerization methods for the preparation of PS–graphene nanocomposites. Wu *et al.* prepared PS–graphene nanocomposites by *in situ* free radical solution polymerization. Due to the covalent interface between the graphene and PS matrix, the graphene was well dispersed. The initial thermal degradation temperature was obviously increased by about 50 °C.[69] Hu *et al.* prepared PS–GO nanocomposites by *in situ* emulsion polymerization, then reduced the GO to graphene to obtain PS–graphene nanocomposites (Figure 5.11). The thermal degradation temperature of the nanocomposites was significantly increased (by ~100 °C), and the T_g was increased by 8 °C. The improved thermal stability was attributed to the restriction imposed on the mobility of PS macromolecules, which resulted in homogeneous heating and avoidance of heat concentration.[71] Patole *et al.* prepared PS–graphene nanocomposites by *in situ* microemulsion polymeriza-

Table 5.5   Details of *in situ* polymerized PS–graphene nanocomposites

First author	Main improvements	Reference
Fang *et al.*	thermal and electrical properties	67–71
Kassaee	Thermal stability and modulus	72
Das	Aggregation-resistance of graphene	73
Lu	Determination of bioactive constituents	74

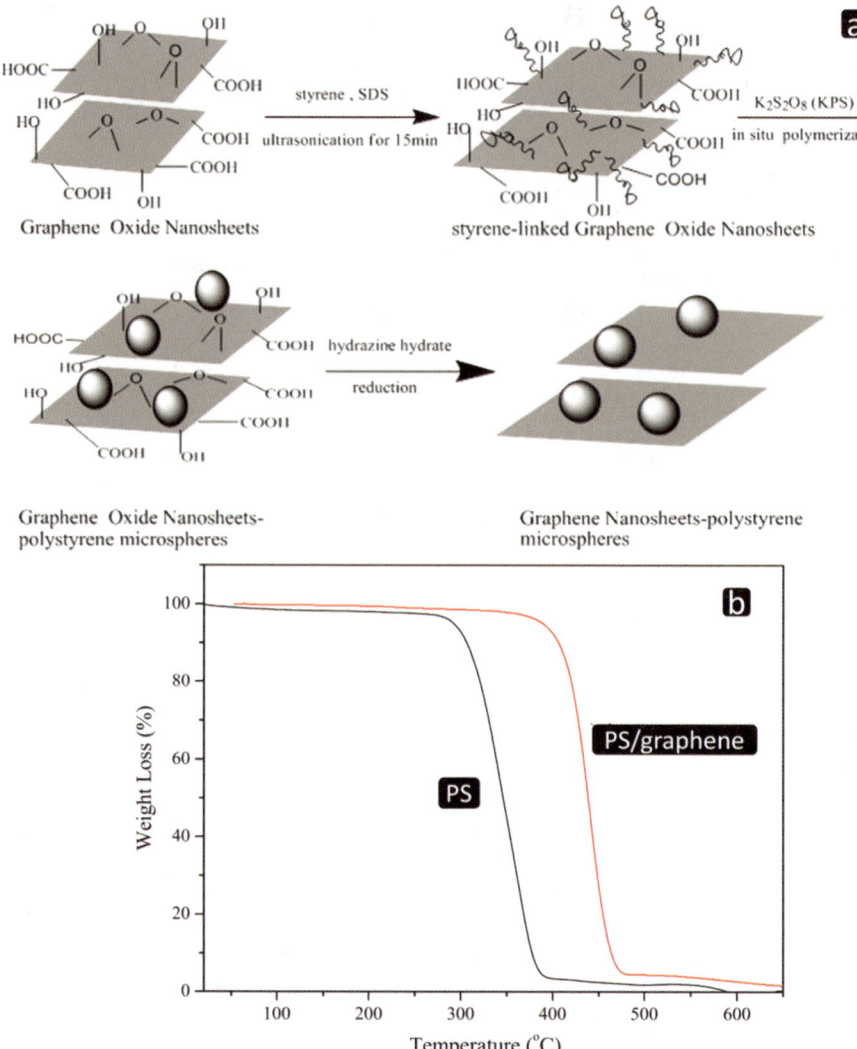

Figure 5.11 (a) Illustration of the synthesis of PS–graphene nanocomposites; (b) Thermogravimetric curves of PS and PS–graphene nanocomposites in reference. Reproduced with permission from ref. 71. Copyright 2012, Elsevier Ltd.

tion (Figure 5.12a). The obtained PS–graphene film exhibited good flexibility (Figure 5.12c–f), increased T_g and improved electrical conductivity.[68] As compared with other kinds of *in situ* polymerized PGNs, the *in situ* polymerized PS–graphene nanocomposites usually show more obvious improvements in thermal stability. This is probably because of the strong covalent and non-covalent interactions between the graphene and the PS matrix.

Kassaee *et al.* modified GO with nano-Fe_3O_4 and then incorporated it into PS. The obtained nanocomposites exhibited improved thermal stability and modulus, good flexibility and transparency (Figure 5.13b and c). Moreover,

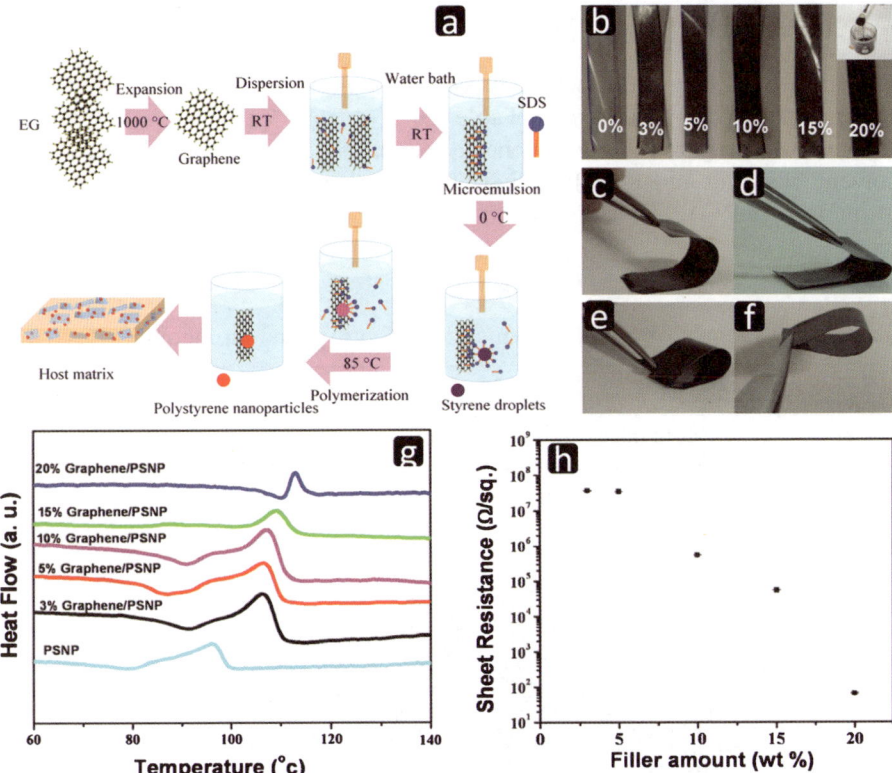

Figure 5.12 (a) Preparation of PS–graphene nanocomposites based films; (b) Photograph of PS–graphene films; (c–f) Photographs showing the flexibility of the PS–graphene films; (g) Dynamic differential scanning curves of the PS–graphene nanocomposites; (h) Sheet resistance of the films. Reproduced with permission from ref. 68. Copyright 2012, Elsevier Ltd.

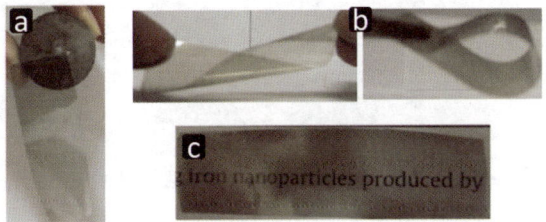

Figure 5.13 (a) PS–Fe_3O_4-GO film lifted by a magnet; (b) PS–Fe_3O_4-GO film exhibits good flexibility; (c) Transparency of the PS–Fe_3O_4-GO film. Reproduced with permission from ref. 72. Copyright 2012, Elsevier Ltd.

due to the attached nano-Fe_3O_4, the nanocomposites can be shifted by magnet (Figure 5.13a).[72] This work extended the application of graphene in polymer-based nanocomposites by incorporation of other functional materials.

5.7 Polyurethane (PU)

Research on PU–graphene mainly concerns the mechanical, electrical and thermal properties, as shown in Table 5.6.

Kim *et al.* prepared PU–graphene nanocomposites by *in situ* polymerization, solvent blending and melt blending. The graphene was prepared by thermal reduction (TRG). The dispersion of graphene was compared and the

Table 5.6 Details of *in situ* polymerized PS–graphene nanocomposites

First author	Main improvements	Reference
Lee *et al.*	Mechanical and electrical properties	4,75
Kim	Electrical, mechanical and gas barrier properties	2
Wang	Mechanical properties	76

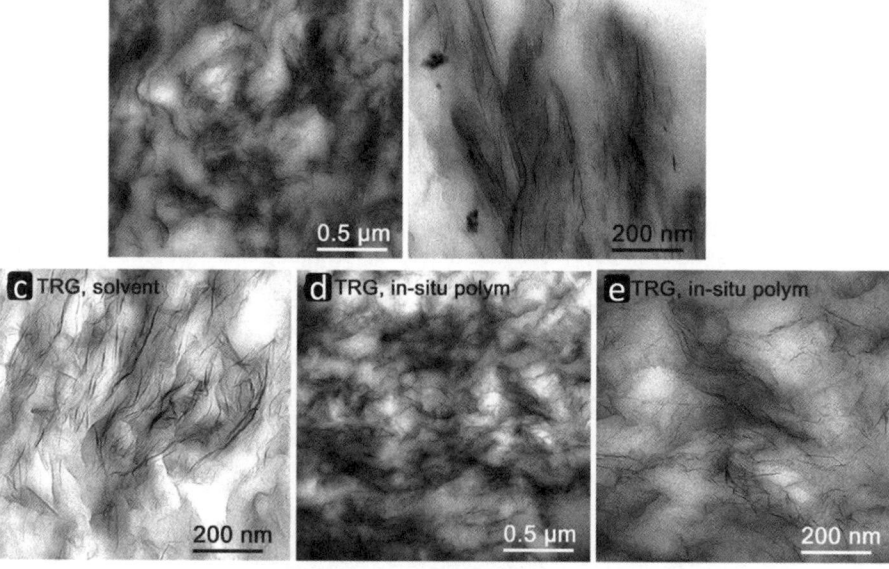

Figure 5.14 TEM images show the dispersion of TRG in PU matrix by: (a, b) melt blending; (c) solvent blending; (d, e) the *in situ* polymerization. Reproduced with permission from ref. 2. Copyright 2012, American Chemistry Society.

properties of the nanocomposites were investigated. It was found that *in situ* polymerization caused the best dispersion (Figure 5.14).[2] Wang *et al.* functionalized GO with 4,40-diphenylmethane diisocyanate and then polymerized it with poly(tetramethylene glycol) to prepare PU–functionalized-GO nanocomposites. The storage modulus of the nanocomposites was improved.[76]

5.8 Other Polymers

Besides the examples mentioned above, many other kinds of polymer are studied in the polymer–graphene field, as shown in Table 5.7.

PA6 (nylon) is one of the most widely used polymers. Xu prepared PA6–graphene nanocomposites by *in situ* ring-opening polymerization (Figure 5.15a). The nanocomposites exhibited reduced crystallinity and T_g, but the tensile property was obviously improved (Figure 5.15d).[78] Liu prepared polyester–graphene nanocomposites by a similar method, and obvious improvements in the tensile strength (72.2%) and elongation at break (54.6 %) were obtained.[85] Such enhancements are also reported in other work, as shown in Table 5.7.

Table 5.7 Other types of *in situ* polymerized PGNs

First author	Polymer	Main improvements	Reference
Huang	PP	Electrical conductivity	77
Xu	PA6	Mechanical properties	78
Alzari	PTGD	Mechanical and thermal properties	79
Alzari	PNPA	Swelling ratio	80
Wang	PNPA, PAA	Aqueous solution of GO	81
Yang	PNPA	Lowered critical solution temperature	82
Etmimi	PSBA	Thermal stability and modulus	83
Feng	Polyester	Electrical conductivity	84
Liu	Polyester	Tensile properties	85
Wang	PEAM	Photoresponsive properties	86
Wang	PI	Mechanical properteis	87
Liu	PI	Tribological properties	88
Wang	POPBI	Mechanical properties	89.
Zhang	PVK	Solubility in solvents and energy bandgap	90
Zhang	PAM	Mechanical properties	91
Oh	PEMA	Mechanical and electrical properties	92
Trang	PEDOT	Electrical conductivity, thermal stability	93
Sangerm-ano	PEG	Thermal properties	94

PA6, polyamide 6; PAA, poly(acrylic acid); PAM, polyacrylamide; PEAM, poly[2-(ethyl(phenyl)amino)ethyl methacrylate]; PEDOT, poly(3,4-ethylenedioxythiophene); PEG, poly(ethylene glycol); PEMA, poly(ethyl methacrylate); PI, polyimide; PNPA, poly(N-isopropylacrylamide; POPBI, poly[2,2'-(p-oxydiphenylene)-5,5'-bibenzimidazole]; PP, polypropylene; PSBA, poly(styrene-butyl acrylate); PTGD, poly(tetraethylene glycol diacrylate; PVK, poly(N-vinylcarbazole).

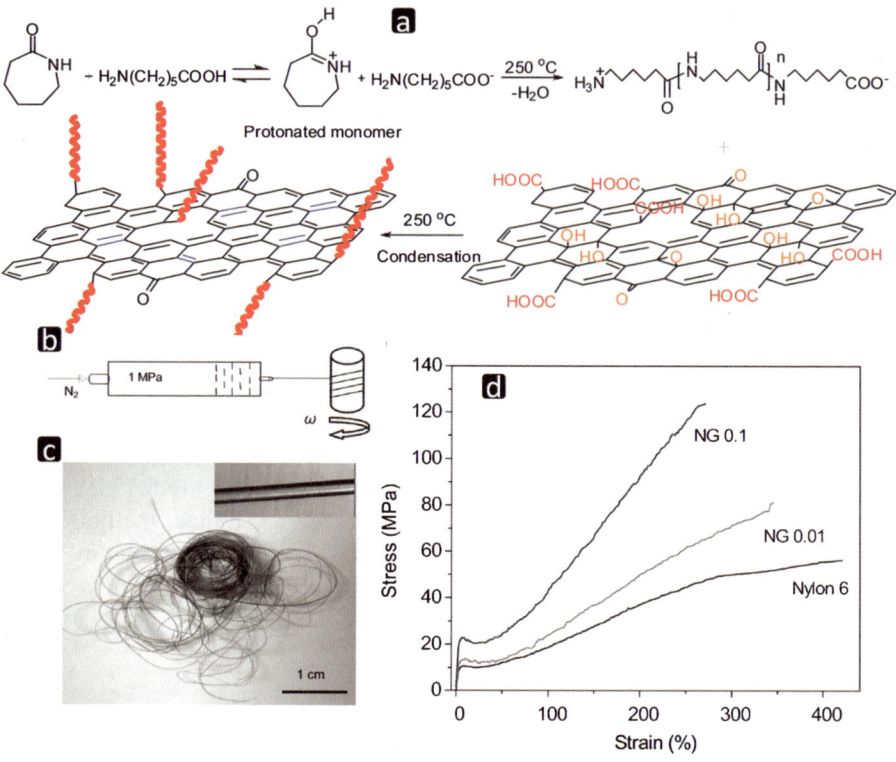

Figure 5.15   (a) Illustration for the preparation of PA6–graphene nanocomposites by *in situ* ring-opening polymerization; (b) Apparatus for melt spinning of PA6–graphene fibres; (c) Photograph of the fibres (0.5% graphene); (d) Stress–strain curves of PA6 and PA6–graphene nanocomposites with 0.01% and 0.1% graphene, respectively. Reproduced with permission from ref. 78. Copyright 2012, American Chemistry Society.

5.9   Summary

Many kinds of PGNs that are advantageous in the functionalization and dispersion of graphene have been prepared by *in situ* polymerization. The nanocomposites obtained usually exhibit obviously improved properties. The most widely studied polymers include PANI, PPy, epoxy, PMMA, PS and PU. Research on the properties of the PGNs has focused mainly on electro-chemical, mechanical, thermal and electrical properties. However, there is still a long way to go to reach a full understanding and practical application of *in situ* polymerized PGNs.

References

1. D. Y. Cai, M. Song, Recent advance in functionalized graphene/polymer nanocomposites. *J. Mater. Chem.*, 2010, **20**(37), 7906–7915.

2. H. Kim, Y. Miura, C. W. Macosko, Graphene/polyurethane nanocomposites for improved gas barrier and electrical conductivity. *Chem. Mater.*, 2010, **22**(11), 3441–50.

3. S. Stankovich, D. A. Dikin, G. H. B. Dommett, K. M. Kohlhaas, E. J. Zimney, E. A. Stach, *et al.*, Graphene-based composite materials. *Nature*, 2006, **442**(7100), 282–6.

4. Y. R. Lee, A. V. Raghu, H. M. Jeong, B. K. Kim, Properties of waterborne polyurethane/functionalized graphene sheet nanocomposites prepared by an *in situ* method. *Macromol. Chem. Phys.*, 2009, **210**(15), 1247–54.

5. A. V. Murugan, T. Muraliganth, A. Manthiram, Rapid, facile microwave-solvothermal synthesis of graphene nanosheets and their polyaniline nanocomposites for energy strorage. *Chem. Mater.*, 2009, **21**(21), 5004–6.

6. D. W. Wang, F. Li, J. P. Zhao, W. C. Ren, Z. G. Chen, J. Tan, *et al.*, Fabrication of graphene/polyaniline composite paper via *in situ* anodic electropolymerization for high-performance flexible electrode. *ACS Nano*, 2009, **3**(7), 1745–52.

7. H. L. Wang, Q. L. Hao, X. J. Yang, L. D. Lu, X. Wang, Effect of graphene oxide on the properties of its composite with polyaniline. *ACS Appl. Mater.Interfaces*, 2010, **2**(3), 821–8.

8. J. Yan, T. Wei, B. Shao, Z. J. Fan, W. Z. Qian, M. L. Zhang, *et al.*, Preparation of a graphene nanosheet/polyaniline composite with high specific capacitance. *Carbon*, 2010, **48**(2), 487–93.

9. X. B. Yan, J. T. Chen, J. Yang, Q. J. Xue, P. Miele, Fabrication of free-standing, electrochemically active, and biocompatible graphene oxide-polyaniline and graphene-polyaniline hybrid papers. *ACS Appl. Mater.Interfaces*, 2010, **2**(9), 2521–9.

10. J. Li, H. Q. Xie, Y. Li, J. Liu, Z. X. Li, Electrochemical properties of graphene nanosheets/polyaniline nanofibers composites as electrode for supercapacitors. *J. Power Sources*, 2011, **196**(24), 10775–81.

11. M. Sathish, S. Mitani, T. Tomai, I. Honma, MnO$_2$ assisted oxidative polymerization of aniline on graphene sheets: Superior nanocomposite electrodes for electrochemical supercapacitors. *J. Mater. Chem.*, 2011, **21**(40), 16216–22.

12. H. L. Wang, Q. L. Hao, X. J. Yang, L. D. Lu, X. Wang, A nanostructured graphene/polyaniline hybrid material for supercapacitors. *Nanoscale*, 2010, **2**(10), 2164–70.

13. J. Yan, T. Wei, Z. J. Fan, W. Z. Qian, M. L. Zhang, X. D. Shen, *et al.*, Preparation of graphene nanosheet/carbon nanotube/polyaniline composite as electrode material for supercapacitors. *J. Power Sources*, 2010, **195**(9), 3041–5.

14. J. J. Xu, K. Wang, S. Z. Zu, B. H. Han, Z. X. Wei, Hierarchical nanocomposites of polyaniline nanowire arrays on graphene oxide

sheets with synergistic effect for energy storage. *ACS Nano*, 2010, **4**(9), 5019–26.

15. K. Zhang, L. L. Zhang, X. S. Zhao, J. S. Wu, Graphene/polyaniline nanoriber composites as supercapacitor electrodes. *Chem. Mater.*, 2010, **22**(4), 1392–401.

16. X. M. Feng, R. M. Li, Y. W. Ma, R. F. Chen, N. E. Shi, Q. L. Fan, *et al.*, One-step electrochemical synthesis of graphene/polyaniline composite film and its applications. Adv Funct Mater., 2011, **21**(15), 2989–96.

17. L. Mao, K. Zhang, H. S. O Chan, J. S. Wu, Surfactant-stabilized graphene/polyaniline nanofiber composites for high performance supercapacitor electrode. *J. Mater. Chem.*, 2012, **22**(1), 80–5.

18. Q. L. Hao, H. L. Wang, X. J. Yang, L. D. Lu, X. Wang, Morphology-controlled fabrication of sulfonated graphene/polyaniline nanocomposites by liquid/liquid interfacial polymerization and investigation of their electrochemical properties. *Nano Res.*, 2011, **4**(4), 323–33.

19. W. L. Zhang, B. J. Park, H. J. Choi, Colloidal graphene oxide/polyaniline nanocomposite and its electrorheology. *Chem. Commun.*, 2010, **46**(30), 5596–8.

20. X. S. Zhou, T. B. Wu, B. J. Hu, G. Y. Yang, B. X. Han, Synthesis of graphene/polyaniline composite nanosheets mediated by polymerized ionic liquid. *Chem. Commun.*, 2010, **46**(21), 3663–5.

21. G. L. Chen, S. M. Shau, T. Y. Juang, R. H. Lee, C. P. Chen, S. Y. Suen, *et al.*, Single-layered graphene oxide nanosheet/polyaniline hybrids fabricated through direct molecular exfoliation. *Langmuir*, 2011, **27**(23), 14563–9.

22. S. H. Domingues, R. V. Salvatierra, M. M. Oliveirab, A. J. G. Zarbin, Transparent and conductive thin films of graphene/polyaniline nanocomposites prepared through interfacial polymerization. *Chem. Commun.*, 2011, **47**(9), 2592–4.

23. Y. Bao, J. X. Song, Y. Mao, D. X. Han, F. Yang, L. Niu, *et al.*, Graphene oxide-templated polyaniline microsheets toward simultaneous electrochemical determination of AA/DA/UA. *Electroanalysis*, 2011, **23**(4), 878–84.

24. D. Majumdar, M. Baskey, S. K. Saha, Epitaxial growth of crystalline polyaniline on reduced graphene oxide. *Macromol. Rapid Commun.*, 2011, **32**(16), 1277–83.

25. L. Al-Mashat, K. Shin, K. Kalantar-Zadeh, J. D. Plessis, S. H. Han, R. W. Kojima, *et al.*, Graphene/polyaniline nanocomposite for hydrogen sensing. *J. Phys. Chem. C*, 2010, **114**(39), 16168–73.

26. Y. H. Chang, B. Wang, H. Luo, L. J. Zhi, Polyaniline/graphene nanocomposite in detecting trace heavy metal ions. In: *Conference on Environmental Pollution and Public Health*, Scientific Research Publishing, Irvine, CA, 2010, pp. 1169–1172.

27. N. D. Luong, U. Hippi, J. T. Korhonen, A. J. Soininen, J. Ruokolainen, L. S. Johansson, *et al.*, Enhanced mechanical and electrical properties of

polyimide film by graphene sheets via *in situ* polymerization. *Polymers*, 2011, **52**(23), 5237–42.

28. C. Basavaraja, W. J. Kim, Y. D. Kim, D. S. Huh, Synthesis of polyaniline-gold/graphene oxide composite and microwave absorption characteristics of the composite films. *Mater. Lett.*, 2011, **65**(19–20), 3120–3.

29. S. Goswami, U. N. Maiti, S. Maiti, S. Nandy, M. K. Mitra, K. K. Chattopadhyay, Preparation of graphene-polyaniline composites by simple chemical procedure and its improved field emission properties. *Carbon*, 2011, **49**(7), 2245–52.

30. W. L. Zhang, Y. D. Liu, H. J. Choi, Fabrication of semiconducting graphene oxide/polyaniline composite particles and their electrorheological response under an applied electric field. *Carbon*, 2012, **50**(1), 290–6.

31. Y. Q. Han, B. Ding, X. G. Zhang, Preparation of graphene/polypyrrole composites for electrochemical capacitors. *J. New Mat. Electrochem. Syst.*, 2010, **13**(4), 315–20.

32. L. L. Zhang, S. Y. Zhao, X. N. Tian, X. S. Zhao, Layered graphene oxide nanostructures with sandwiched conducting polymers as supercapacitor electrodes. *Langmuir*, 2010, **26**(22), 17624–8.

33. S. Bose, N. H. Kim, T. Kuila, K. T. Lau, J. H. Lee, Electrochemical performance of a graphene-polypyrrole nanocomposite as a supercapacitor electrode. *Nanotechnology*, 2011, **22**(29), 295202.

34. A. Davies, P. Audette, B. Farrow, F. Hassan, Z. W. Chen, J. Y. Choi, *et al.*, Graphene-based flexible supercapacitors: pulse-electropolymerization of polypyrrole on free-standing graphene films. *J. Phys. Chem. C*, 2011, **115**(35), 17612–20.

35. M. Deng, X. Yang, M. Silke, W. M. Qiu, M. S. Xu, G. Borghs, *et al.*, Electrochemical deposition of polypyrrole/graphene oxide composite on microelectrodes towards tuning the electrochemical properties of neural probes. *Sens. Actuators, B*, 2011, **158**(1), 176–84.

36. S. Konwer, R. Boruah, S. K. Dolui, Studies on conducting polypyrrole/graphene oxide composites as supercapacitor electrode. *J. Electron. Mater.*, 2011, **40**(11), 2248–55.

37. P. A. Mini, A. Balakrishnan, S. V. Nair, K. R. V. Subramanian, Highly super capacitive electrodes made of graphene/poly(pyrrole). *Chem. Commun.*, 2011, **47**(20), 5753–5.

38. S. Sahoo, G. Karthikeyan, G. C. Nayak, C. K. Das, Electrochemical characterization of *in situ* polypyrrole coated graphene nanocomposites. *Synth. Met.*, 2011, **161**(15–16), 1713–9.

39. C. H. Xu, J. Sun, L. Gao, Synthesis of novel hierarchical graphene/polypyrrole nanosheet composites and their superior electrochemical performance. *J. Mater. Chem.*, 2011, **21**(30), 11253–8.

40. D. C. Zhang, X. Zhang, Y. Chen, P. Yu, C. H. Wang, Y. W. Ma, Enhanced capacitance and rate capability of graphene/polypyrrole composite as electrode material for supercapacitors. *J. Power Sources*, 2011, **196**(14), 5990–6.

41. S. Bose, T. Kuila, M. E. Uddin, N. H. Kim, A. K. T Lau, J. H. Lee, *In situ* synthesis and characterization of electrically conductive polypyrrole/ graphene nanocomposites. *Polymer*, 2010, **51**(25), 5921–8.

42. J. Zou, X. H. Song, J. J. Ji, W. C. Xu, J. M. Chen, Y. Q. Jiang, *et al.*, Polypyrrole/graphene composite-coated fiber for the solid-phase micro-extraction of phenols. *J. Sep. Sci.*, 2011, **34**(19), 2765–72.

43. V. Chandra, K. S. Kim, Highly selective adsorption of Hg^{2+} by a polypyrrole-reduced graphene oxide composite. *Chem. Commun.*, 2011, **47**(13), 3942–4.

44. V. Chandra, S. U. Yu, S. H. Kim, Y. S. Yoon, D. Y. Kim, A. H. Kwon, *et al.*, Highly selective CO(2) capture on N-doped carbon produced by chemical activation of polypyrrole functionalized graphene sheets. *Chem. Commun.*, 2012, **48**(5), 735–7.

45. H. F. Yang, C. S. Shan, F. H. Li, Q. X. Zhang, D. X. Han, L. Niu, Convenient preparation of tunably loaded chemically converted graphene oxide/epoxy resin nanocomposites from graphene oxide sheets through two-phase extraction. *J. Mater. Chem.*, 2009, **19**(46), 8856–60.

46. M. Fang, Z. Zhang, J. F. Li, H. D. Zhang, H. B. Lu, Y. L. Yang, Constructing hierarchically structured interphases for strong and tough epoxy nanocomposites by amine-rich graphene surfaces. *J. Mater. Chem.*, 2010, **20**(43), 9635–43.

47. J. J. Qiu, S. R. Wang, Enhancing polymer performance through graphene sheets. *J. Appl. Polym. Sci.*, 2011, **119**(6), 3670–4.

48. K. S. Kim, I. Y. Jeon, S. N. Ahn, Y. D. Kwon, J. B. Baek, Edge-functionalized graphene-like platelets as a co-curing agent and a nanoscale additive to epoxy resin. *J. Mater. Chem.*, 2011, **21**(20), 7337–42.

49. D. R. Bortz, E. G. Heras, I. Martin-Gullon, Impressive fatigue life and fracture toughness improvements in graphene oxide/epoxy composites. *Macromolecules*, 2012, **45**(1), 238–45.

50. C. L. Bao, Y. Q. Guo, L. Song, Y. C. Kan, X. D. Qian, Y. Hu, *In situ* preparation of functionalized graphene oxide/epoxy nanocomposites with effective reinforcements. *J. Mater. Chem.*, 2011, **21**(35), 13290–8.

51. S. C. Kim, H. I. Lee, H. M. Jeong, B. K. Kim, J. H. Kim, C. M. Shin, Effect of pyrene treatment on the properties of graphene/epoxy nanocomposites. *Macromol. Res.*, 2010, ov **18**(11), 1125–8.

52. Y. Q. Guo, C. L. Bao, L. Song, B. H. Yuan, Y. Hu, *In Situ* polymerization of graphene, graphite oxide, and functionalized graphite oxide into epoxy resin and comparison study of on-the-flame behavior. *Indust. Eng. Chem. Res.*, 2011, **50**(13), 7772–83.

53. Z. Wang, X. Z. Tang, Z. Z. Yu, P. Guo, H. H. Song, X. S. Du, Dispersion of graphene oxide and its flame retardancy effect on epoxy nanocomposites. *Chin. J. Polym. Sci.*, 2011, **29**(3), 368–76.

54. I. Zaman, T. T. Phan, H. C. Kuan, Q. S. Meng, L. T. B. La, L. Luong, *et al.*, Epoxy/graphene platelets nanocomposites with two levels of interface strength. *Polymer*, 2011, **52**(7), 1603–11.

55. S. Y. Yang, W. N. Lin, Y. L. Huang, H. W. Tien, J. Y. Wang, C. C. M. Ma, *et al.*, Synergetic effects of graphene platelets and carbon nanotubes on the mechanical and thermal properties of epoxy composites. *Carbon*, 2011, **49**(3), 793–803.

56. M. Martin-Gallego, R. Verdejo, M. A. Lopez-Manchado, M. Sangermano, Epoxy-graphene UV-cured nanocomposites. *Polymer*, 2011, **52**(21), 4664–9.

57. S. L. Qiu, C. S. Wang, Y. T. Wang, C. G. Liu, X. Y. Chen, H. F. Xie, *et al.*, Effects of graphene oxides on the cure behaviors of a tetrafunctional epoxy resin. *Express Polym. Lett.*, 2011, **5**(9), 809–18.

58. C. C. Teng, C. C. M Ma, C. H. Lu, S. Y. Yang, S. H. Lee, M. C. Hsiao, *et al.*, Thermal conductivity and structure of non-covalent functionalized graphene/epoxy composites. *Carbon*, 2011, **49**(15), 5107–16.

59. Q. Liu, X. Yao, X. Zhou, Z. Qin, Z. Liu, Varistor effect in Ag-graphene/ epoxy resin nanocomposites. *Scr. Mater.*, 2012, **66**(2), 113–6.

60. R. K. Layek, S. Samanta, D. P. Chatterjce, A. K. Nandi, Physical and mechanical properties of poly(methyl methacrylate) -functionalized graphene/poly(vinylidine fluoride) nanocomposites. Piezoelectric beta polymorph formation. *Polymer*, 2010, **51**(24), 5846–56.

61. K. P. Pramoda, H. Hussain, H. M. Koh, H. R. Tan, C. B. He, Covalent bonded polymer-graphene nanocomposites. *J. Polym. Sci., Part A: Polym. Chem.*, 2010, **48**(19), 4262–7.

62. T. Kuila, S. Bose, P. Khanra, N. H. Kim, K. Y. Rhee, J. H. Lee, Characterization and properties of *in situ* emulsion polymerized poly(-methyl methacrylate)/graphene nanocomposites. *Composite, Part A: Appl. Sci. Manuf.*, 2011, **42**(11), 1856–61.

63. M. Kiran, K. Raidongia, U. Ramamurty, C. N.R Rao, Improved mechanical properties of polymer nanocomposites incorporating gra-phene-like BN: Dependence on the number of BN layers. *Scr. Mater.*, 2011, **64**(6), 592–5.

64. J. R. Potts, S. H. Lee, T. M. Alam, J. An, M. D. Stoller, R. D. Piner, *et al.*, Thermomechanical properties of chemically modified graphene/poly(-methyl methacrylate) composites made by *in situ* polymerization. *Carbon.*, 2011, **49**(8), 2615–23.

65. J. C. Wang, H. T. Hu, X. B. Wang, C. H. Xu, M. Zhang, X. P. Shang, Preparation and mechanical and electrical properties of graphene nanosheets-poly(methyl methacrylate) nanocomposites via *in situ* suspen-sion polymerization. *J. Appl. Polym. Sci.*, 2011, **122**(3), 1866–71.

66. X. Wang, J. Y. Li, W. D. Qu, G. Chen, Fabrication of graphene/ poly(methyl methacrylate) composite electrode for capillary electrophore-tic determination of bioactive constituents in Herba Geranii. *J Chromatogr. A*, 2011, **1218**(32), 5542–8.

67. M. Fang, K. G. Wang, H. B. Lu, Y. L. Yang, S. Nutt, Single-layer graphene nanosheets with controlled grafting of polymer chains. *J. Mater. Chem.*, 2010, **20**(10), 1982–92.

68. A. S. Patole, S. P. Patole, H. Kang, J. B. Yoo, T. H. Kim, J. H. Ahn, A facile approach to the fabrication of graphene/polystyrene nanocomposite by *in situ* microemulsion polymerization. *J. Colloid Interface Sci.*, 2010, **350**(2), 530–7.

69. X. L. Wu, P. Liu, Facile preparation and characterization of graphene nanosheets/polystyrene composites. *Macromol. Res.*, 2010, **18**(10), 1008–12.

70. Y. Li, Z. Q. Wang, L. Yang, H. Gu, G. Xue, Efficient coating of polystyrene microspheres with graphene nanosheets. *Chem. Commun.*, 2011, **47**(38), 10722–4.

71. H. T. Hu, X. B. Wang, J. C. Wang, L. Wan, F. M. Liu, H. Zheng, *et al.*, Preparation and properties of graphene nanosheets-polystyrene nanocomposites via *in situ* emulsion polymerization. *Chem. Phys. Lett.*, 2010, **484**(4–6), 247–53.

72. M. Z. Kassaee, E. Motamedi, M. Majdi, Magnetic Fe₃O₄-graphene oxide/polystyrene: Fabrication and characterization of a promising nanocomposite. *Chem. Eng. J.*, 2011, **172**(1), 540–9.

73. S. Das, A. S. Wajid, J. L. Shelburne, Y. C. Liao, M. J. Green, Localized *In situ* polymerization on graphene surfaces for stabilized graphene dispersions. *ACS Appl. Mater.Interfaces*, 2011, **3**(6), 1844–51.

74. Y. Lu, X. Wang, D. F. Chen, G. Chen, Polystyrene/graphene composite electrode fabricated by *in situ* polymerization for capillary electrophoretic determination of bioactive constituents in Herba Houttuyniae. *Electrophoresis*, 2011, **32**(14), 1906–12.

75. D. A. Nguyen, A. V. Raghu, J. T. Choi, H. M. Jeong, Properties of thermoplastic polyurethane/functionalised graphene sheet nanocomposites prepared by the *in situ* polymerisation method. *Polym. Polym. Compos.*, 2010, **18**(7), 351–8.

76. X. Wang, Y. A. Hu, L. Song, H. Y. Yang, W. Y. Xing, H. D. Lu, *In situ* polymerization of graphene nanosheets and polyurethane with enhanced mechanical and thermal properties. *J. Mater. Chem.*, 2011, **21**(12), 4222–7.

77. Y. J. Huang, Y. W. Qin, Y. Zhou, H. Niu, Z. Z. Yu, J. Y. Dong, Polypropylene/graphene oxide nanocomposites prepared by *in situ* Ziegler–Natta polymerization. *Chem. Mater.*, 2010, **22**(13), 4096–102.

78. Z. Xu, , C. Gao. *In situ* polymerization approach to graphene-reinforced nylon-6 composites. *Macromolecules*, 2010, **43**(16), 6716–23.

79. V. Alzari, D. Nuvoli, R. Sanna, S. Scognamillo, M. Piccinini, J. M. Kenny, *et al.*, *In situ* production of high filler content graphene-based polymer nanocomposites by reactive processing. *J. Mater. Chem.*, 2011, **21**(41), 16544–9.

80. V. Alzari, D. Nuvoli, S. Scognamillo, M. Piccinini, E. Gioffredi, G. Malucelli, *et al.*, Graphene-containing thermoresponsive nanocomposite hydrogels of poly(*N*-isopropylacrylamide) prepared by frontal polymerization. *J. Mater. Chem.*, 2011, **21**(24), 8727–33.

81. B. D. Wang, D. Yang, J. Z. Zhang, C. B. Xi, J. H. Hu, Stimuli-responsive polymer covalent functionalization of graphene oxide by Ce(IV)-induced redox polymerization. *J. Phys. Chem. C*, 2011, **115**(50), 24636–41.

82. Y. F. Yang, X. H. Song, L. Yuan, M. Li, J. C. Liu, R. Q. Ji, *et al.*, Synthesis of PNIPAM polymer brushes on reduced graphene oxide based on click chemistry and RAFT polymerization. *J. Polym. Sci., Part A: Polym. Chem.*, 2012, **50**(2), 329–37.

83. H. M. Etmimi, R. D. Sanderson, New approach to the synthesis of exfoliated polymer/graphite nanocomposites by miniemulsion polymerization using functionalized graphene. *Macromolecules*, 2011, **44**(21), 8504–15.

84. R. C. Feng, G. H. Guan, W. Zhou, C. C. Li, D. Zhang, Y. N. Xiao, *In situ* synthesis of poly(ethylene terephthalate)/graphene composites using a catalyst supported on graphite oxide. *J. Mater. Chem.*, 2011, **21**(11), 3931–9.

85. K. Liu, L. Chen, Y. Chen, J. L. Wu, W. Y. Zhang, F. Chen, *et al.*, Preparation of polyester/reduced graphene oxide composites via *in situ* melt polycondensation and simultaneous thermo-reduction of graphene oxide. *J. Mater. Chem.*, 2011, **21**(24), 8612–7.

86. D. R. Wang, G. Ye, X. L. Wang, X. G. Wang, Graphene functionalized with azo polymer brushes: surface-initiated polymerization and photo-responsive properties. *Adv. Mater.*, 2011, **23**(9), 1122–5.

87. J. Y. Wang, S. Y. Yang, Y. L. Huang, H. W. Tien, W. K. Chin, C. C. M. Ma, Preparation and properties of graphene oxide/polyimide composite films with low dielectric constant and ultrahigh strength via *in situ* polymerization. *J. Mater. Chem.*, 2011, **21**(35), 13569–75.

88. H. Liu, Y. Q. Li, T. M. Wang, Q. H. Wang, *In situ* synthesis and thermal, tribological properties of thermosetting polyimide/graphene oxide nanocomposites. *J. Mater. Sci.*, 2012, **47**(4), 1867–74.

89. Y. Wang, Z. X. Shi, J. H. Fang, H. J. Xu, X. D. Ma, J. Yin, Direct exfoliation of graphene in methanesulfonic acid and facile synthesis of graphene/polybenzimidazole nanocomposites. *J. Mater. Chem.*, 2011, **21**(2), 505–12.

90. B. Zhang, Y. Chen, L. Q. Xu, L. J. Zeng, Y. He, E. T. Kang, *et al.*, Growing poly(*N*-vinylcarbazole) from the surface of graphene oxide via RAFT polymerization. *J. Polym. Sci., Part A: Polym. Chem.*, 2011, **49**(9), 2043–50.

91. N. N. Zhang, R. Q. Li, L. Zhang, H. B. Chen, W. C. Wang, Y. Liu, *et al.*, Actuator materials based on graphene oxide/polyacrylamide composite hydrogels prepared by *in situ* polymerization.*Soft Matter*, 2011, **7**(16), 7231–9.

92. S. M. Oh, H. I. Lee, H. M. Jeong, B. K. Kim, The properties of functionalized graphene sheet/poly(ethyl methacrylate) nanocomposites: The effects of preparation method. *Macromol. Res.*, 2011, **19**(4), 379–84.

93. L. K. H Trang, T. T. Tung, T. Y. Kim, W. S. Yang, H. Kim, K. S. Suh, Preparation and characterization of graphene composites with conducting polymers. *Polym. Int.*, 2012, **61**(1), 93–8.
94. M. Sangermano, S. Marchi, L. Valentini, S. B. Bon, P. Fabbri, Transparent and conductive graphene oxide/poly(ethylene glycol) diacrylate coatings obtained by photopolymerization. *Macromol. Mater. Eng.*, 2011, **296**(5), 401–7.

CHAPTER 6

Microstructure and Properties of Compatibilized Polyethylene–Graphene Oxide Nanocomposites

A. U. CHAUDHRY[1], VIKAS MITTAL*[1] AND
N. B. MATSKO[2]

[1] Department of Chemical Engineering, The Petroleum Institute, Abu Dhabi,
UAE; [2] Graz Centre for Electron Microscopy (ZFE), Steyrergasse 17, A-8010
Graz, Austria
*E-mail: vmittal@pi.ac.ae

6.1   Introduction

Polyolefins are important among commodity plastics owing to their use in a variety of applications. Especially, high density polyethylene (HDPE) has wide range of properties including low cost, ease of recycling, good processability, non-toxicity, biocompatibility, and good chemical resistance. The demands of modern applications, however, are not met solely by neat polymers. Therefore, in addition to the many applications of unmodified HDPE, it is also necessary to improve the performance of the polymer in terms of properties such as stiffness and rigidity by forming composites.[1–2] In general, composites with polyolefin matrices are formed to fulfil various requirements for different applications where cost and weight reduction, dimensional stability, opacity, heat stability, and processability are required. The advances in nanofillers and nanofibers have made it possible to produce lightweight nanocomposites with better physical and mechanical properties at a low filler concentration than

RSC Nanoscience & Nanotechnology No. 26
Polymer–Graphene Nanocomposites
Edited by Vikas Mittal
© The Royal Society of Chemistry 2012
Published by the Royal Society of Chemistry, www.rsc.org

conventional composites. This is owing to the much higher number of interfacial contacts of inorganic particles with the polymer chains in such nanocomposites, which generates a completely different interfacial morphology as compared to the bulk polymer.[3–4]

Graphene, which consists of sheets of covalently sp^2-bonded carbon atoms, one atomic layer thick, in a hexagonal arrangement, has already received attention from researchers for the generation of polymer nanocomposites.[5] Its choice as a filler is due to its excellent electrical and mechanical properties, which are significantly better than those of other inorganic filler materials. A single defect-free graphene layer has Young's modulus of ≈ 1.0 TPa, intrinsic strength ≈ 42 N/m, thermal conductivity ≈ 4840–5300 W m^{-1} K^{-1}), electron mobility exceeding 25 000 cm^2 V^{-1} s^{-1}, excellent gas impermeability and specific surface area of ≈ 2630 m^2 g^{-1}.[5] All these properties make this material even superior to carbon nanotubes for use in polymer nanocomposites. A number of studies on polymer nanocomposites based on graphene have been published in the short period of time since its development.[5–13] The parent material for graphene is graphite, which is occurs widely in nature. Graphene (and graphene oxide) can be produced from graphite by various methods such as thermal expansion of chemically intercalated graphite, micromechanical exfoliation of graphite, chemical vapour decomposition and chemical reduction of graphene oxide.[7] The most commonly used method is the exfoliation of intercalated graphite oxide (GO) by introducing it suddenly to a higher temperature.[14]

The macroscopic properties of polymer nanocomposites are mainly dependent on thermodynamic factors such as interfacial compatibility of polymer and filler phases, or polarity match between the filler surface and the polymer chains. In addition, nanoscale dispersion and distribution of the filler also depends on the size, shape, dispersion techniques and equipment, time of mixing and applied shear, *etc.* The full advantage of nanofillers can be only achieved by considering these factors which could lead to uniform transfer of superior properties of nanofiller to host polymer matrix.[4,15–19]

Significant research effort has focused on attaining the full potential of nanofillers by using different mixing techniques, modification of polymer backbone or filler surface, use of compatibilizer (functional polymers) and coupling agents, *etc.* Graphene has very low surface energy as compared to GO, which is a precursor of graphene. The presence of fewer functional groups (such as carboxyl, epoxide or hydroxyl) on the surface of pristine graphene leads to lower compatibility with polar polymer matrices, thus resulting in poor dispersion and lower enhancement in polymer properties.[5] Similarly, the dispersion of polar graphene oxide in non-polar polymers is not optimal, owing to absence of positive interactions between them. One route to overcome this limitation is the functionalization of the filler surface, which results in significant enhancement of the mechanical and electrical properties of polymer nanocomposites. Bing *et al.* grafted amine-functionalized multi-walled carbon nanotubes (MWCNT) with polyethylene by reactive blending using maleic anhydride.[20] The improved stiffness, strength, ductility and toughness of the polymer were attributed to uniform dispersion of nanofiller and improved interfacial adhesion owing to polyethylene grafted on CNT. Similar results were achieved by adding functionalized CNTs to polypropylene matrix.[18] Ramanathan *et al.* prepared nanocomposites of functionalized sheets of graphene and poly(methyl methacrylate) (PMMA) by sonication and high-speed shearing of expanded graphene.[21] Partially oxygenated wrinkled sheets showed a shift of 30 °C in T_g of PMMA, which was superior to that obtained by using single-walled carbon nanotubes (SWCNT) and expanded graphite platelets.

The other method described in the literature is the use of a compatibilizer. In the case of polyethylene, the lack of polar groups in its backbone is a considerable hurdle in homogenous dispersion and exfoliation of nanofillers. The introduction of an amphiphilic compatibilizer having polar and non-polar groups acts as a bridge between filler and host polymer, thus resulting in improved filler dispersion. Valdes *et al.* reported that introduction of ethylene acid copolymer compatibilizer in linear low-density polyethylene (LLDPE) and clay nanocomposite improved exfoliation of clay particles, which resulted in better thermal properties.[22] A master-batch technique was used for the preparation of composites. Similarly, Kim *et al.* used maleic anhydride as a bridge for nanocomposites of low-density polyethylene (LDPE) and exfoliated graphite nanoplatelets.[14] Different dispersion techniques, *i.e.* solution and melt blending, along with different arrangements of screws, were used in the study. The better results in terms of filler dispersion were shown by solution mixing followed by counter-rotating screw arrangements.

Chlorinated polyethylene (CPE) has also been reported to be efficient as an adhesion promoter and compatibilizer between polymer blends and fillers in composites.[23] In poly(vinyl chloride) (PVC) and wood flour composites using CPE (chlorine content ranged from 25% to 42%) as a compatibilizer, improvements in processing, melt strength and elongation at break were observed. In a similar study, Simon *et al.* showed the effect of acid–base interaction between chlorine and hydroxyl groups on the adhesion of chlorinated polypropylene with polypropylene.[24] Significant changes in mechanical properties of blends of varying amounts of CPE with high-density polyethylene have also been reported by Maksimov *et al.*[25]

In the current study, two types of chlorinated polyethylene (25% and 35% chlorine content) have been used as compatibilizers in order to study the dispersion of graphene oxide in HDPE and the effect on resulting nanocomposite properties. A solution blending technique was used for the blending of compatibilizer and graphene oxide. The obtained master batches were then melt-mixed with HDPE. The effect of chlorine content in the compatibilizer as well as the amount of compatibilizer on the morphology, mechanical and rheological properties of the polyethylene graphene oxide nanocomposites was studied.

6.2   Experimental

6.2.1   Materials

Chlorinated polyethylene grades WeiprenR 6025 (25% chlorine content, CPE25) and CPE 135A (35% chorine content, CPE35) were obtained from Lianda Corporation, USA and Weifang Xuran Chemicals, China respectively. Matrix polymer, *i.e.* high-density polyethylene BB2581, was received from Abu Dhabi Polymers Company Limited (Borouge), UAE. The polymer materials were used as obtained. Graphite powder (325 mesh) was procured from Alfa Aesur GmbH and Co., Germany. Concentrated sulfuric acid (H_2SO_4, 95–98%), sodium nitrate ($NaNO_3$) and potassium permanganate ($KMnO_4$) were supplied by S. D. Fine Chemicals Ltd., India, Eurostar Scientific Ltd, UK and Fisher Scientific, UAE respectively.

6.2.2   Preparation of Graphite Oxide and Graphene Oxide

Graphene oxide was prepared by the thermal exfoliation of precursor GO,[26] using a modification of Hummer's method.[27] A short description for the preparation of GO and graphene oxide is as follows: 5 g of graphite powder was added to 125 mL concentrated H_2SO_4. Subsequently, 2.5 g of $NaNO_3$ was added to this mixture. The mixture was kept in an ice-bath (5 °C) under stirring. After 30 min, 15 g of $KMnO_4$ was added to the mixture and the temperature was allowed to rise gradually to 35 °C. The mixture was stirred for 2 h under these conditions. This was followed by the addition of deionized water till the temperature increased to 100 °C. After further stirring for 15 minutes while maintaining the same temperature, the mixture was quenched and diluted by pouring it into 1.5 L deionized water. 30% H_2O_2 was slowly added to the dilute solution until the evolution of bubbles of hydrogen stopped. The solution was then filtered using Buchner funnel to remove the non-GO waste. The residues were dispersed in 2 l de-ionized water and were added with dilute HCl (6%) (2 l) to remove the SO_42- ions. The dispersion was filtered and the filtrate was analyzed for SO_4^{2-} and Cl^- ions using $BaSO_4$ and $AgNO_3$ respectively (generation of white precipitates). The cleaning and filtration was continued until no SO_4^{2-} and Cl^- ions can be observed in the filtrate. The washed GO was dried under vacuum at 60 °C for 24 h.[27] Graphene oxide was generated via thermal exfoliation of dried GO. The process was carried out by placing 1 g GO in a long quartz tube with 25 mm internal diameter and sealed at one end. The other end of the quartz tube was closed using a rubber stopper. The sample was flushed with nitrogen, followed by insertion of the tube in a tube furnace preheated to 1050 °C. The tube was held in the furnace for 30 s.[26] The density of the obtained graphene oxide was measured by tapped density tester to be 0.0161 g/mL.

6.2.3 Nanocomposite Generation

The nanocomposites were prepared by either solution mixing followed by melt mixing or by direct melt mixing, as described earlier.[14] Table 6.1 also describes the compositions of different composites. For the solution mixing method, CPE was stirred in *p*-xylene (3% solid content) at 100 °C under reflux until the solution became clear. Graphene oxide (weighed according to the requirement for master batches) was suspended in few mL of *p*-xylene for 1 h at room temperature. In between, the suspension was sonicated for 10 min. The graphene oxide suspension was then added to polymer solution at 100 °C and further sonicated for 15 min. The mixture was stirred and gradually brought to room temperature. The solution was kept overnight at room temperature followed by 40 °C for 24 h in order to remove any solvent residues, which resulted in dried CPE/graphene oxide master batches.

To form nanocomposites, melt mixing of CPE/graphene oxide master batches with HDPE was carried out at 190 °C using a mini twin screw extruder (MiniLab HAAKE Rheomex CTW5, Germany). The screw length and screw diameter were 109.5 mm and 5/14 mm conical respectively. A batch size of 5 g was used and the shear mixing was performed for 5 min at 60 rpm. Direct melt mixing of HDPE and graphene oxide was also performed similarly. Pure HDPE was also processed by subjecting it to similar shear and thermal conditions. Disc- and dumbbell-shaped test specimens were injection moulded using a mini injection moulding machine (HAAKE MiniJet, Germany) at a processing temperature of 190 °C. The injection pressure was 700 bar for 6 s, whereas holding pressure was 400 bar for 3 s. The temperature of the mould was kept at 50 °C.

6.2.4 Material Characterization

The calorimetric properties of the nanocomposites were recorded on a Perkin-Elmer Pyris-1 differential scanning calorimeter under a nitrogen atmosphere. The scans were obtained from 50 to 190 °C at a heating rate of 20 °C/min. The

Table 6.1 Compositions of the composites (in wt%)

Code	Polymer/nanocomposite	HDPE, %	CPE25, %	CPE35, %	Graphene oxide, %
1	HDPE	100	-	-	-
2	HDPE/G	99.5	-	-	0.5
3	HDPE/1%CPE25/G	98.5	1	-	0.5
4	HDPE/2%CPE25/G	97.5	2	-	0.5
5	HDPE/5%CPE25/G	94.5	5	-	0.5
6	HDPE/10%CPE25/G	89.5	10	-	0.5
7	HDPE/5%CPE35/G	94.5	-	5	0.5
8	HDPE/10%CPE35/G	89.5	-	10	0.5

heat enthalpies (used to calculate the extent of crystallinity) were measured with an error of $\sim 0.1\%$ and were confirmed by repeating the runs.

Rheological properties such as storage modulus (G'), loss modulus (G''), viscosity and elasticity of the nanocomposites were measured using AR 2000 Rheometer (TA Instruments). The measuring temperature and gap opening were 185 °C and 1.6 mm respectively. Disc-shaped samples of diameter 25 mm and thickness 2 mm were used. Strain sweeps were recorded at $\omega = 1$ rad s^{-1} from 0.1% to 100% strain. The shear stability of the samples was observed up to 10% strain. Hence, as a safe approach, frequency sweeps (dynamic testing) were recorded at 4% strain from $\omega = 0.1$ to 100 rad s^{-1}.[28]

Tensile testing of composites was performed on universal testing machine (Testometric, UK). The sample dimensions for tensile testing were: sample length 73 mm, gauge length 30 mm, width 4 mm and thickness 2 mm. A loading rate of 4 mm min^{-1} was used and the tests were carried out at room temperature. Tensile modulus and yield stress were calculated using built-in software (Win Test Analysis). An average of three values is reported.

Transmission electron microscopy (TEM) of graphene oxide, master batch and the nanocomposite samples was performed using EM 912 Omega (Zeiss, Oberkochen, Germany) and Philips CM 20 (Philips/FEI, Eindhoven, The Netherlands) electron microscopes at 120 kV and 200 kV accelerating voltage, respectively. Thin sections of 70–90 nm thickness were microtomed from the block of the specimen and were subsequently supported on 100-mesh grids sputter coated with a 3 nm thick carbon layer.

6.3   Results and Discussion

In the current study, graphene oxide–polyethylene nanocomposites were generated by using solution mixing and melt mixing processes. Two chlorinated polyethylene compatibilizers differing in chlorination extent were used in order to study their effect on filler dispersion as well as resulting polymer properties. Solution mixing was used to generate master batches of chlorinated polyethylene with graphene oxide, which can help to better disperse the filler in the matrix polymer when the master batch is melt compounded with it. The EDX analysis of the graphene oxide surface revealed C/O ratio of 20, thus indicating the presence of polar surface groups (hydroxyl, epoxide, carboxyl, *etc.*) which can interact with the polar compatibilizers used in the study.

Table 6.2 and Figures 6.1–6.2 describe the calorimetric analysis of the pure polymer, compatibilizers as well as polymer nanocomposites. The melt enthalpy of pure crystalline HDPE was taken as 293 J/g and was used to determine the extent of crystallinity in the polymer.[28] CPE25 compatibilizer was semi-crystalline in nature as indicated by the crystalline melting peak in the DSC thermogram in Figure 6.1a. A peak melting temperature of 130 °C was observed. On the other hand, CPE35 was amorphous in nature as no melting transition was observed. Thus, the compatibilizers were different not

Table 6.2   Calorimetric analysis of the pure polymers and polymer nanocomposites

Code	Polymer/nanocomposite	ΔH J/g	Peak melting temp, °C	Crystallinity, %
1	HDPE	147	142	50
2	HDPE/G	142	144	49
3	HDPE/1%CPE25/G	153	143	52
4	HDPE/2%CPE25/G	149	143	51
5	HDPE/5%CPE25/G	151	147	52
6	HDPE/10%CPE25/G	150	145	51
7	HDPE/5%CPE35/G	147	143	48
8	HDPE/10%CPE35/G	150	144	46
9	CPE25	47	130	-
10	CPE35	-	-	-

only in the extent of chlorination, but also in morphology. The peak melting temperatures in the nanocomposites were always higher than in the pure polymer, indicating the impact of graphene oxide on polymer morphology. The impact was observed even on adding the amorphous compatibilizer to the system, though it was less in magnitude as compared to the system compatibilized with semi-crystalline compatibilizer. In the case of CPE35 composites, an increase of 1–2 °C in peak melting temperature as compared to pure HDPE was observed, whereas this increase for the same amount of CPE25 compatibilizer was 3–5 °C. The impact of graphene oxide (especially at higher compatibilizer content) was also confirmed by DSC analysis of blends of HDPE with CPE. In HDPE/CPE35 blends, decrease in peak melting temperatures was observed, whereas only a marginal increase was observed for the CPE25 compatibilized HDPE system. Also, the peak melting temperature increased on increasing the compatibilizer content (Figure 6.1b), indicating that the enhanced filler dispersion would have taken place which subsequently enhanced the thermal resistance of the crystals.[29] The degree of crystallinity of the polymer was also observed to be affected by graphene oxide as well as compatibilizer. The composite without any compatibilizer had a crystallinity of 49%, which was marginally lower than the pure polymer crystallinity of 50%. This indicated that the graphene oxide platelets slightly hindered the chain mobility and hence their packing into the crystal structure.[14] The addition of amorphous CPE35 also resulted in a further decrease in degree of crystallinity of the polymer, the magnitude of which increased on increasing the content of compatibilizer. Thus, increasing the number of amorphous chains in the matrix resulted in hindering the crystalline packing of HDPE chains. In the case of CPE25 compatibilizer, the extent of crystallinity was always higher than in the pure polymer irrespective of the compatibilizer content. However, changes in the melt transition curves were observed on increasing the compatibilizer content beyond 5 wt% (Figure 6.2), indicating changes in the crystallization

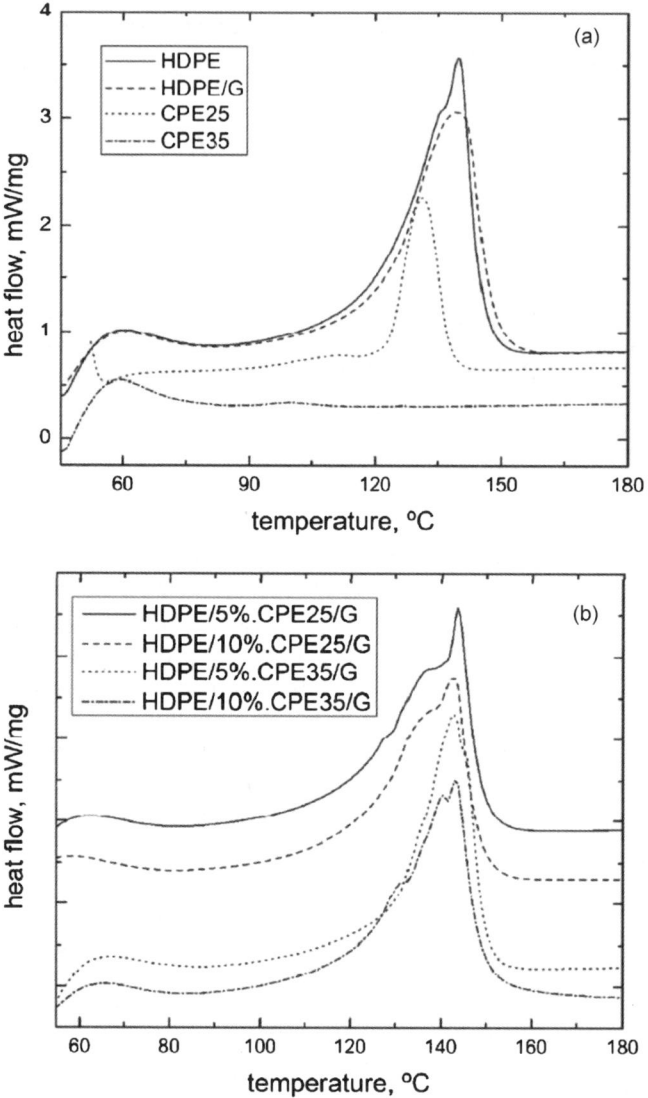

Figure 6.1   DSC thermograms of (a) HDPE, CPEs and HDPE/G composite and (b) HDPE/CPE/G composites with 5 and 10 wt% compatibilizer content. Reproduced from reference 35 with permission.

behaviour. There is a possibility that CPE crystallized separately from HDPE owing to either its incompatibility with HDPE or its interaction with graphene oxide surface which led to its separation from the matrix polymer. Figure 6.3 also show the microstructure of the matrix in HDPE/G composites with and without compatibilizers. An impact on the crystalline microstructure is evident with CPE35 composite the one most affected thus confirming the DSC findings.

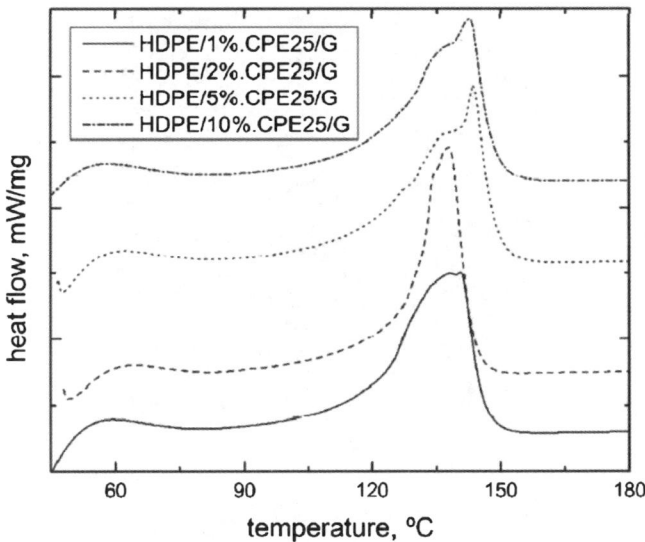

Figure 6.2   DSC thermograms of HDPE/CPE25/G composites with 1, 2, 5 and 10 wt% compatibilizer content. Reproduced from reference 35 with permission.

The network structure of the polymer nanocomposites was evaluated with shear rheology and the storage and loss moduli of the samples as a function of angular frequency are demonstrated in Figures 6.4–6.5. Strain sweep was conducted and samples were found to be safe up to 10% strain. Frequency sweep of the samples was performed with controlled shear strain at 4% using frequency range of 0.1 to 100 rad s^{-1}. As is evident in Figure 6.4a, the storage modulus of pure HDPE was lowest among all the samples at all frequencies. On addition of 0.5 wt% graphene oxide without compatibilizer, an order of magnitude increase in the storage modulus occurred. For example, at a frequency of 10 rad s^{-1}, the storage modulus for HDPE was 15 730 Pa, which was enhanced to 112 000 Pa with addition of only 0.5 wt% graphene oxide. The rate of increase in modulus decreased on increasing the angular frequency; however, both the samples became independent of frequency and showed sudden shear thinning at frequency of $\sim$20 rad s^{-1}. In the case of graphene oxide nanocomposite, such behaviour can be due to the rupture of the interface between the polymer and the graphene oxide surface at higher frequencies. Literature studies have suggested that this phenomenon may be a result of alignment of filler platelets in the direction of flow at high shear or due to slipping between the polymer and filler during high shear flow.[30] The compatibilized nanocomposites exhibited good low-frequency dependence followed by gradual decline in the modulus enhancement due to a shear thinning effect. CPE25 composites had higher storage moduli than the CPE35-containing nanocomposites. On increasing the amount of compatibilizer from 5 to 10 wt%, the storage modulus was observed to decrease owing to extensive

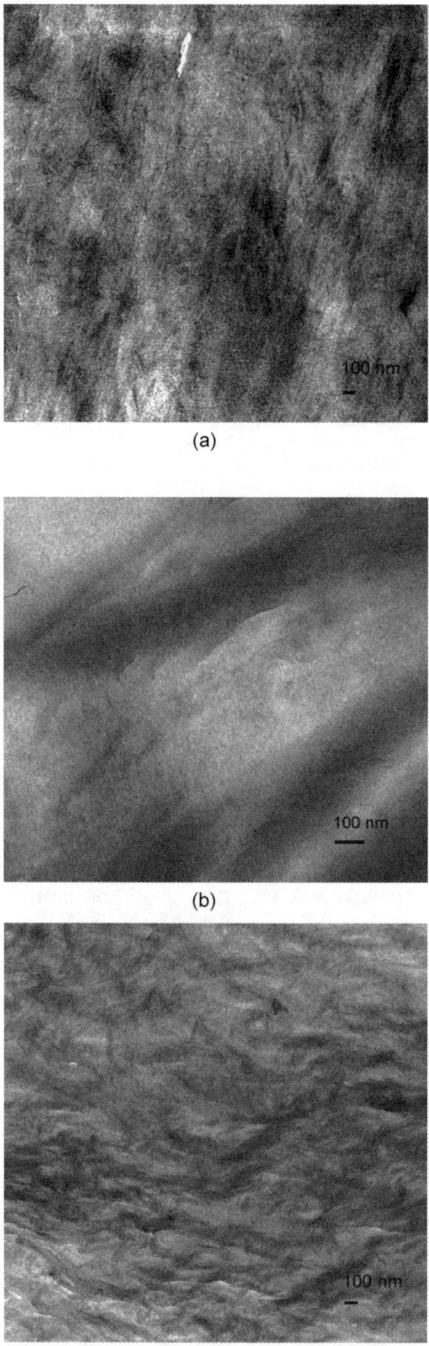

Figure 6.3   TEM micrographs depicting the matrix microstructure: (a) HDPE/G, (b) HDPE/5%CPE35/G and (c) HDPE/5%CPE25/G.

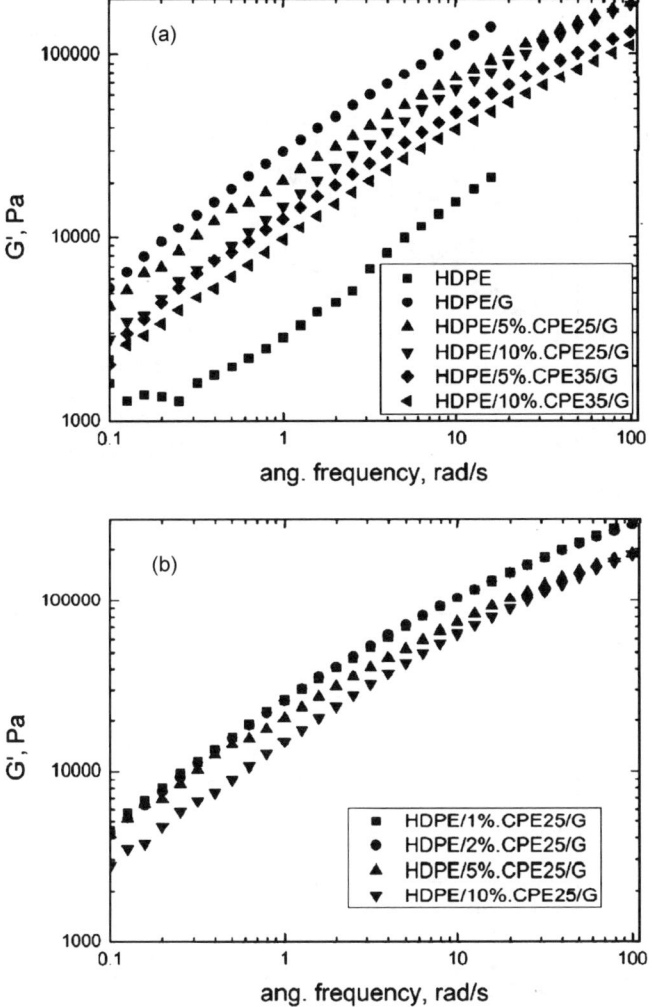

Figure 6.4   Storage modulus of HDPE and HDPE nanocomposites as a function of angular frequency. Reproduced from reference 35 with permission.

plasticization of the matrix (Figure 6.4a). Figure 6.4b also shows the effect of 1–10% CPE25 on the storage modulus of the nanocomposites. Composites with 1 and 2 wt% compatibilizer content were comparable in behaviour and had modulus value of 104 000 Pa at 10 rad s^{-1} frequency, which was similar to the value of 112 000 Pa for the nanocomposite without compatibilizer. Also, the storage modulus curves overlapped with each other for composites with 5 and 10 wt% compatibilizer at a higher frequency of 100 rad s^{-1}.

The loss modulus results of the pure polymer and nanocomposites are presented in Figure 6.5. On comparison with the storage moduli of the samples, it was observed that in pure HDPE, G'' was always higher than G' at

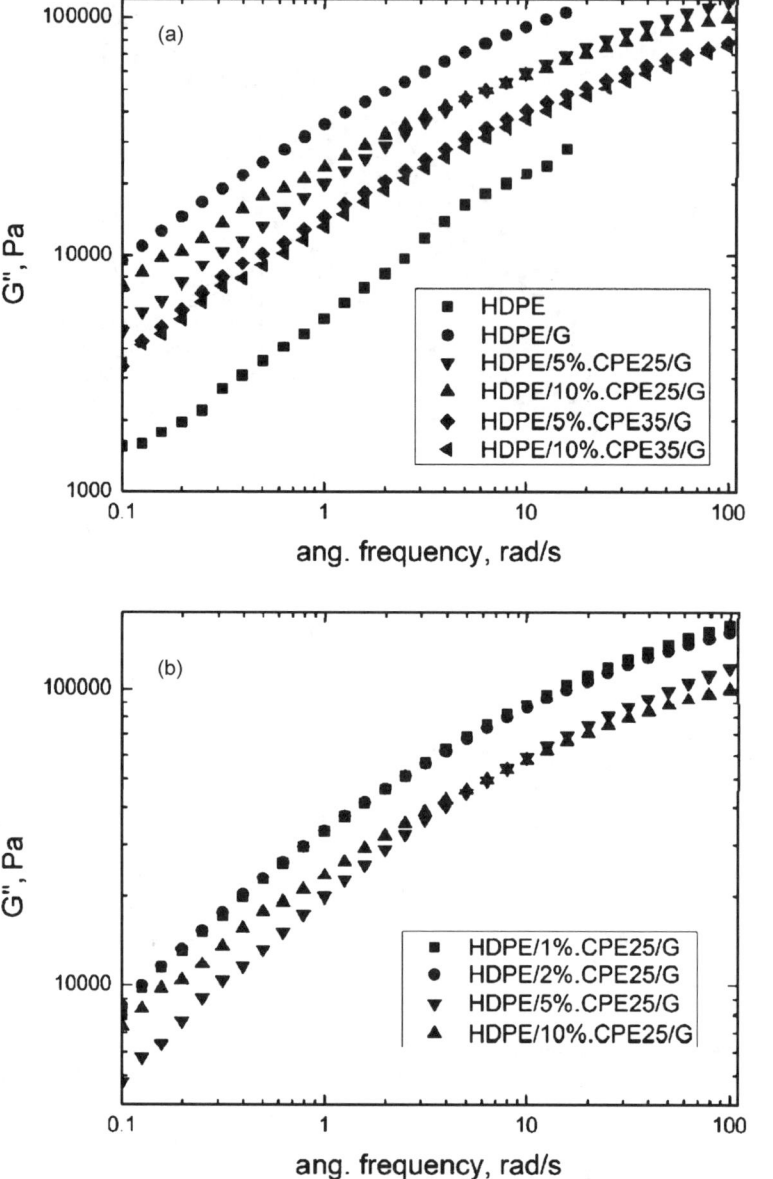

Figure 6.5   Loss modulus of HDPE and HDPE nanocomposites as a function of
                 angular frequency. Reproduced from reference 35 with permission.

any frequency. This indicated that the polymer chains had dominant viscous
behaviour with long relaxation times. In the case of nanocomposites, $G'' > G'$
at lower frequency, indicating the dominance of the viscous part. After that, a
transition was observed, after which $G' > G''$ for the whole range of frequency,

indicating strong elastic character of the material with shorter relaxation times. In case of polymer nanocomposite without compatibilizer, such transition was observed below 3 rad s^{-1} frequency. In the case of compatibilized systems, the transition frequency increased on increasing the compatibilizer content. In CPE 25 nanocomposites, the frequency increased from 2.5 to 6 rad s^{-1}, when the compatibilizer content was raised from 1 to 10 wt%. Similarly, in the case of CPE35-containing nanocomposites, the transition frequency increased from 3 to 8 rad s^{-1} on increasing the compatibilizer content from 5 to 10 wt%. This indicated that the material behaviour became more strongly viscous on increasing the compatibilizer content, with CPE35 having a stronger effect than CPE25. The loss modulus in the case of CPE25-containing composites was higher than the CPE35-containing composites. Also, the magnitude of loss modulus decreased on increasing the content of compatibilizer in the composite (Figure 6.5a). Figure 6.5b also shows the impact of CPE25 compatibilizer on the loss modulus when its content was enhanced from 1 to 10 wt%. The loss modulus at 10 rad s^{-1} was observed to be 88 590 Pa for 1% compatibilizer content, which was reduced to 59 010 Pa at a compatibilizer amount of 10 wt% in the composite. It is also worth noting that the improvements in the rheological properties on the addition of a small amount of graphene are very significant when compared to other filler systems. For example, only a slight increase in the storage and loss modulus of polypropylene nanotube nanocomposites was observed at 1 wt% nanotube content as compared to an order of magnitude increase at 0.5 wt% graphene oxide reported in the current study.[31]

Figures 6.6–6.7 show the viscosity and elasticity of the samples as a function of angular velocity. Contrary to the modulus, all of these quantities were observed to decrease on increasing angular frequency. Due to the increase in frequency, the polymer structure shows a temporary network of entanglements which leads to more flexibility (lower viscosity). In composites, this allowed more deformation energy to be stored, resulting in elastic dominance which leads to increased modulus. Thus, it simultaneously reduced the contribution

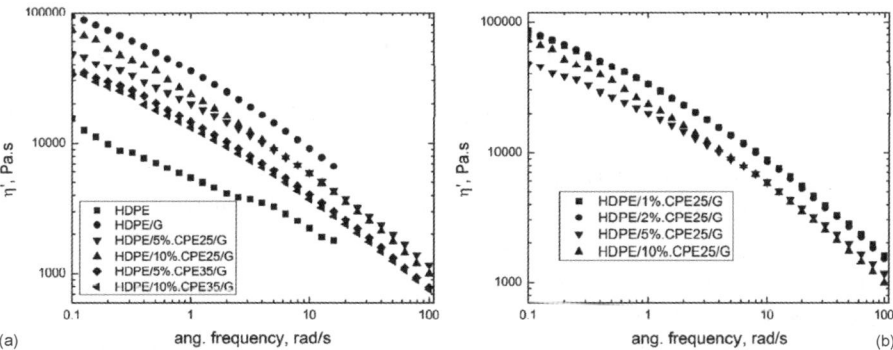

Figure 6.6   Viscosity curves of HDPE and HDPE nanocomposites as a function of angular frequency. Reproduced from reference 35 with permission.

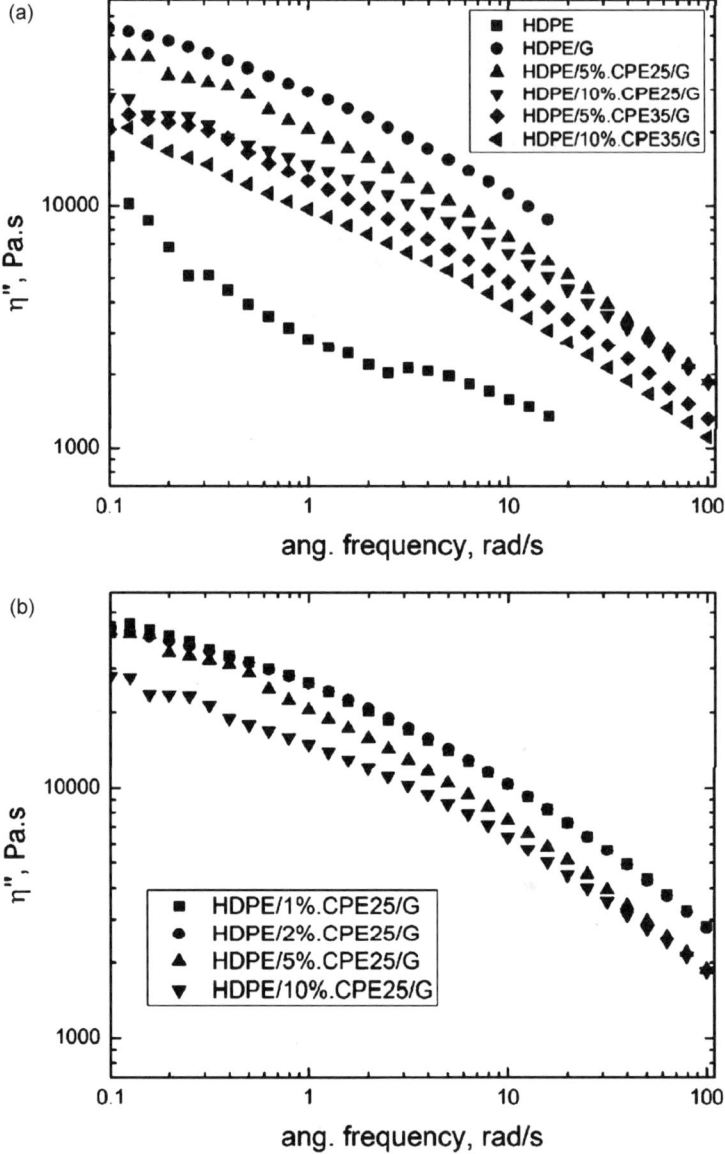

Figure 6.7   Elasticity of HDPE and HDPE nanocomposites as a function of angular
frequency. Reproduced from reference 35 with permission.

of lost deformation energy and hence viscous behaviour.[17] The lowest values
were observed in the case of HDPE, and a composite with graphene oxide
without compatibilizer exhibited the highest values. For example, at a
frequency of 10 rad s^{-1}, the viscosity of the pure HDPE was 2204 Pa s which
was enhanced to 9137 Pa s in the case of HDPE–graphene oxide

nanocomposite (Figure 6.6a). Similarly, the elasticity increased from 1573 Pa s for pure polymer to 11 230 Pa s for HDPE–graphene oxide nanocomposite (Figure 6.7a), indicating the strong impact on both viscosity and elasticity of adding only a small amount of graphene oxide. The slope of decrease in these quantities as a function of angular frequency also decreased after a frequency of ~ 10 rad s^{-1}. CPE25 composites had higher viscosity and elasticity than the CPE35-containing nanocomposites. On increasing the amount of compatibilizer from 5 to 10 wt%, the viscosity and elasticity were observed to decrease further (Figure 6.6a, 6.7a). Figures 6.6b and 6.7b also show the effect of increasing the CPE25 content from 1% to 10% on the viscosity and elasticity of the nanocomposites. Composites with 1 and 2 wt% compatibilizer content were comparable in behaviour and had viscosity and elasticity values of 8859 and 10 360 Pa s at 10 rad s^{-1} frequency, which was similar to the nanocomposite without compatibilizer. On the other hand, these values decreased to 5901 and 6402 Pa s in the case of 10 wt% compatibilizer system.

Similar to the shear moduli, n$'>$ n$''$ was true for HDPE at all angular frequency values, indicating that the viscous contribution dominated the effect of elasticity in the pure polymer (Figure 6.7a). In the case of HDPE/G composite, n$'>$ n$''$ was true only at lower angular frequency values. At ~ 3 rad s^{-1}, a transition in behavior was observed indicating the dominance of the elasticity component at higher shear frequencies. Similarly, for nanocomposites with compatibilizer, the transition between n$'$ and n$''$ occurred between 3 and 6 rad s^{-1} frequency.

The morphology of the master batches as well as nanocomposites was also analysed using microscopy, as shown in Figure 6.8 (for 5% compatibilizer content). The CPE35 master batch (Figure 6.8a) was observed to have better graphene oxide dispersion than the CPE25 master batch (Figure 6.8b). Though complete nanoscale delamination of the graphene oxide platelets was not observed in the composites, the composites with CPE35 compatibilizer also had much better filler dispersion than the corresponding CPE25 nanocomposites. Graphene oxide stacks of varying thicknesses (single layers to multiple layers) can be observed for CPE35 containing composites in Figures 6.8c and d, whereas, the stack thickness was much higher for CPE25 nanocomposite, as shown in Figure 6.8e. As the compatibilizers differ in their extent of chlorination, the resulting morphology can therefore be related to the interaction of polar chlorine atoms with the graphene oxide surface. More chlorination in the matrix resulted in a higher magnitude of interfacial interactions between the polymer and the filler surface, resulting in an increased extent of filler delamination. Thus, increased chlorination content decreased the polymer crystallinity, but also increased the susceptibility of filler platelets to delaminate in the polymer matrix. Figure 6.9 also shows AFM images of the HDPE/G, HDPE/5%CPE35/G and HDPE/5%CPE25/G composites respectively. The change in the crystalline microstructure on the addition of compatibilizer was confirmed again, as the microstructure was nearly lost in the case of amorphous compatibilizer.

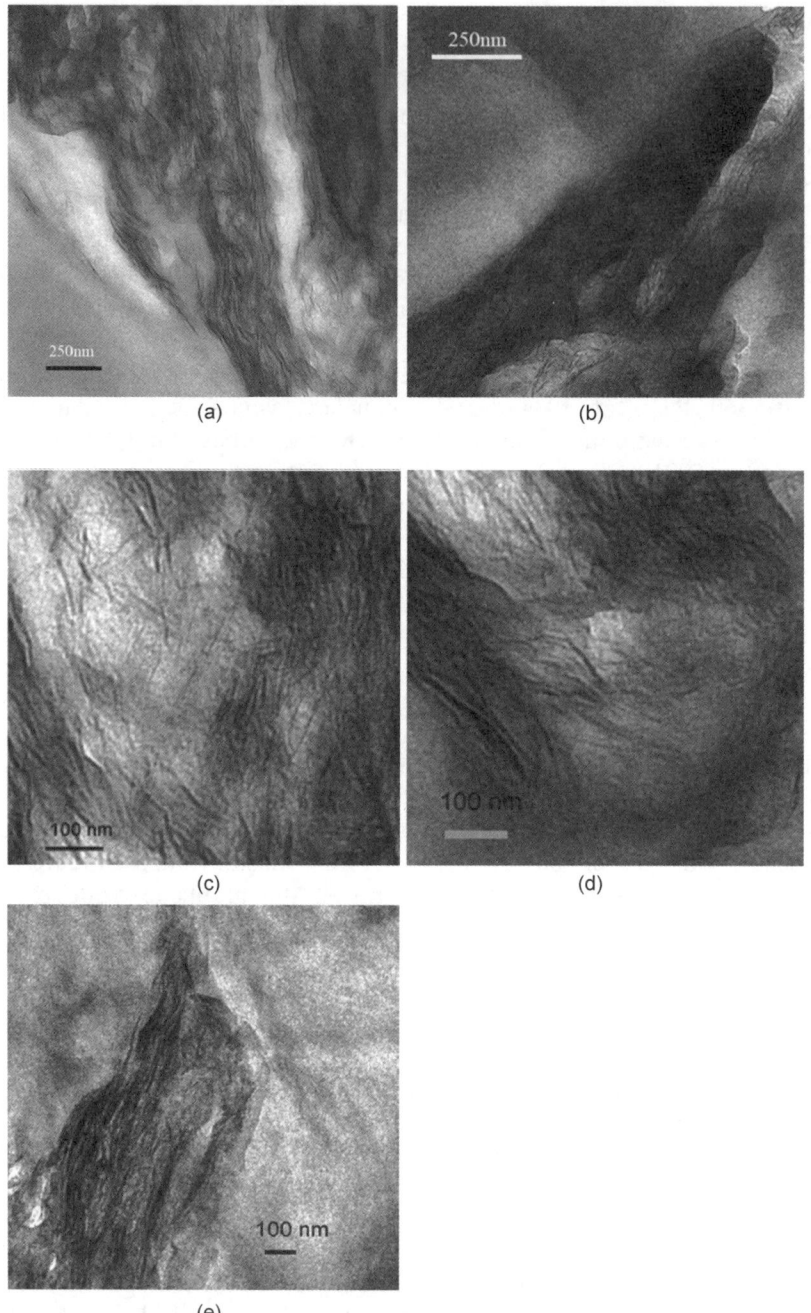

Figure 6.8 TEM micrographs of (a) master batch of CPE35/G, (b) master batch of CPE25/G, (c)(d) HDPE/5%CPE35/G and (e) HDPE/5%CPE25/G nano-composites. The black lines are the intersection of graphene oxide platelets. Reproduced from reference 35 with permission.

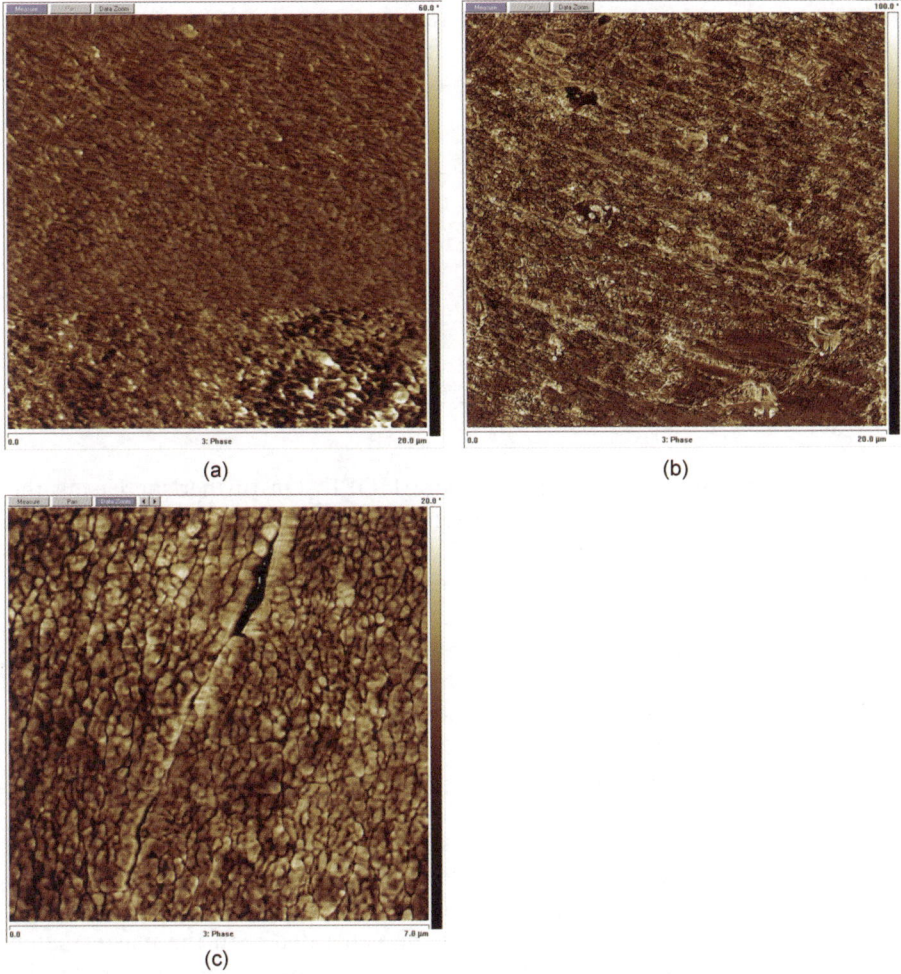

Figure 6.9   AFM images of (a) HDPE/G, (b) HDPE/5%CPE35/G and (c) HDPE/5%CPE25/G composites.

Table 6.3 also reports the tensile properties of HDPE and its graphene oxide composites. The tensile modulus for pure polymer was observed to be 1063 Mpa, which increased to 1150 Mpa in a graphene oxide–HDPE nanocomposite without compatibilizer. The addition of the semi-crystalline compatibilizer CPE25 increased the modulus gradually till 5 wt%, after which a reduction in the modulus was recorded. Similar observations have also been made earlier,[32–33] where the modulus was described to be affected by a balance between increase in modulus owing to enhanced filler dispersion by the compatibilizer and a simultaneous decrease in modulus due to plasticization of the matrix caused by it. Up to 5 wt% content of compatibilizer in the composite, the delamination affect dominated, resulting in an increment of

Table 6.3 Tensile properties of the pure polymer and polymer-graphene oxide nanocomposites

Sr. No.	Polymer/Nanocomposite	Young's modulus[a], MPa	Peak stress[b], MPa	Peak strain[c], %
1	HDPE	1063	62	8.4
2	HDPE/G	1150	58	7.8
3	HDPE/1%CPE25/G	974	56	8.5
4	HDPE/2%CPE25/G	1181	69	7.8
5	HDPE/5%CPE25/G	1228	75	7.6
6	HDPE/10%CPE25/G	1160	65	7.8
7	HDPE/5%CPE35/G	905	48	8.7
8	HDPE/10%CPE35/G	1105	68	8.0

[a]Relative probable error 2%. [b]Relative probable error 2%. [c]Relative probable error 5%.

16% in the modulus as compared to pure HDPE. On further increasing the compatibilizer content, the plasticization effect dominated the performance, resulting in a decrease in tensile modulus. The increased extent of interfacial interactions (hence of filler delamination) also resulted in different behaviour of CPE25 and CPE35 compatibilizers, as CPE35 composites exhibited an increase in modulus even at 10% compatibilizer content. However, the modulus of CPE35 composites was lower than that of the corresponding CPE25 composites owing to the amorphous nature of CPE35. It should also be noted that though the increments in the modulus are not very large, these enhancements are still significant owing to the very low amount of graphene oxide used to achieve them. The importance of the master batch approach was also confirmed by comparing the tensile modulus value of the HDPE/G/5%CPE25 with the similar composite generate by only melt mixing. A value of 969 MPa was obtained for such a melt-mixed composite, which was much lower than the 1228 MPa for the composite generated with the master batch approach. The peak stress also showed similar behaviour to the tensile modulus. The addition of graphene oxide to HDPE without compatibilizer resulted in a slight decrease of peak stress due to restriction in segmental mobility *via* mechanical interlocking with graphene oxide tactoids. On the other hand, addition of CPE25 gradually enhanced the strength up to 5 wt% content, resulting in an increment of 21% as compared to pure polymer. The CPE35 composites showed an increase of 10% in strength at a compatibilizer content of 10 wt%. The composites still remained rigid, as the peak strain was not significantly affected by the addition on compatibilizers. A comparison of the mechanical properties of the generated HDPE–graphene oxide nanocomposites with other systems like HDPE–clay nanocomposites also revealed their high potential. For example, HDPE–clay nanocomposites without compatibilizer showed an increase of 4% in Young's modulus at 2 wt% filler content,[34] which was much lower than the graphene oxide nanocomposites even with 0.5 wt% filler content.

6.4 Conclusions

Nanocomposites of HDPE, graphene oxide and two different chlorinated polyethylene compatibilizers were generated using master batch (by solution mixing of chlorinated polyethylene and graphene oxide) and melt mixing methods. Addition of even 0.5 wt% of graphene oxide and different amounts of compatibilizers significantly impacted the polymer morphology and properties. The addition of graphene oxide caused a slight reduction in the polymer crystallinity due to reduction in chain mobility and packing. The compatibilizer with less chlorination exhibited a semi-crystalline nature and did not decrease the overall crystallinity of polymer in the composites, whereas the compatibilizer with higher chlorination content was amorphous and led to a decrease in the polymer crystallinity. The rheological characterization concluded that the addition of CPE improved the processing of HDPE nanocomposites, rather than pure HDPE and HDPE/G nanocomposites, which showed sudden shear thinning at low frequency. The CPE25 composites were superior in performance to the corresponding CPE35 composites. The performance also reduced on increasing the amount of compatibilizer. The compatibilizer with higher chlorination content also resulted in better interfacial interactions with graphene oxide leading to a greater extent of filler delamination. Interplay of increased mechanical performance owing to filler delamination and decreased properties due to matrix plasticization affected the tensile response of the nanocomposites. The CPE25 composites had maximum improvement of 16% and 21% in modulus and strength at 5 wt% compatibilizer content. The CPE35 composites exhibited an increase in the properties even at 10 wt% compatibilizer content due to the greater magnitude of interfacial interactions. The mechanical properties in CPE35 composites, however, were lower than those of corresponding CPE25 composites due to the amorphous nature of the CPE35 compatibilizer.

Acknowledgements

A more detailed version of this chapter[35] is to be published in *Polymer Engineering and Science*.

References

1. S. Kanagaraj, in *Polymer Nanotube Nanocomposites: Synthesis, Properties, and Applications*, ed. V. Mittal, John Wiley and Scrivener Publishing, New York, 2010.
2. J. Jancar, in *Mineral Fillers in Thermoplastics I: Raw Materials and Processing*, ed. J. Jancar, Springer-Verlag, Berlin, 1999.
3. U. Hippi, in *Polyolefin Composites*, ed. D. Nwabunma and T. Kyu, John Wiley & Sons, New York, 2008.
4. H. Liu and L. C. Brinson, *Compos. Sci. Technol.*, 2008, **68**, 1502.

5. P. Mukhopadhyay and R. K. Gupta, *Plastics Eng.*, 2011, 32.
6. H. Kim, A. A. Abdala and C. W. Macosko, *Macromolecules*, 2010, **43**, 6515.
7. T. Kuilla, S. Bhadra, D. Yao, N. H. Kim, S. Bose and J. H. Lee, *Prog. Polym. Sci.*, 2010, **35**, 1350.
8. D. Cai and M. Song, *J. Mater. Chem.*, 2010, **20**, 7906.
9. M. A. Rafiee, J. Rafiee, Z. Wang, H. Song, Z. Z. Yu and N. Koratkar, *ACS Nano*, 2009, **3**, 3884.
10. P. Steurer, R. Wissert, R. Thomann and R. Muelhaupt, *Macromol. Rapid Commun.*, 2009, **30**, 316.
11. H. Kim, Y. Miura and C. W. Macosko, *Chem. Mater.*, 2010, **22**, 3441.
12. D. A. Nguyen, Y. R. Lee, A. V. Raghu, H. M. Jeong, C. M. Shin and B. K. Kim, *Polym. Int.*, 2009, **58**, 412.
13. M. Fang, K. Wang, H. Lu, Y. Yang and S. Nutt, *J. Mater. Chem.*, 2009, **19**, 7098.
14. H. C. Schniepp, J.-L. Li, M. J. McAllister, H. Sai, M. Herrera-Alonso, D. H. Adamson, R. K. Prud'homme, R. Car, D. A. Saville and I. A. Aksay, *J. Phys. Chem. B*, 2006, **110**, 8535.
15. L. L. Lebel and D. Therriault, in *Advances in Diverse Industrial Applications of Nanocomposites*, ed. B. Reddy, InTech, Rijeka, 2011.
16. L. Hui, R. C. Smith, X. Wang, J. K. Nelson and L. S. Schadler, *Electrical Insulation Dielectric Phenomena (CEIDP 2008)*, 2008, 317.
17. I. Manas-Zloczower, *Rheology Bull.*, 1997, **66**, 5.
18. S. H. Lee, E. N. R. Cho, S. H. Jeon and J. R. Youn, *Carbon*, 2007, **45**, 2810.
19. M. A. Serageldin and H. Wang, *Thermochim. Acta*, 1987, **117**, 157.
20. B. X. Yang, K. P. Pramoda, G. Q. Xu and S. H. Goh, *Adv. Func. Mater.*, 2007, **17**, 2062.
21. T. Ramanathan, A. A. Abdala, S. Stankovich, D. A. Dikin, M. Herrera-Alonso, R. D. Piner, D. H. Adamson, H. C. Schniepp, X. Chen, R. S. Ruoff, S. T. Nguyen, I. A. Aksay, R. K. Prud'Homme and L. C. Brinson, *Nat. Nanotechnol.*, 2008, **3**, 327.
22. S. Sanchez-Valdes and M. L. Lopez-Quintanilla, *Adv. Sci. Technol.*, 2006, **45**, 1399.
23. V. O. Guffey and A. B. Sabbagh, *J. Vinyl Additive Technol.*, 2002, **8**, 259.
24. S. Waddington and D. Briggs, *Polym. Commun.*, 1991, **32**, 506.
25. R. D. Maksimov, T. Ivanova and M. Kalnins, *Mechanics Compos. Mater.*, 2004, **40**, 331.
26. M. J. McAllister, J. L. Li, D. H. Adamson, H. C. Schniepp, A. A. Abdala, J. Liu, M. Herrera-Alonso, D. L. Milius, R. Car, R. K. Prud'homme and I. A. Aksay, *Chem. Mater.*, 2007, **19**, 4396.
27. W. S. Hummers and R. E. Offeman, *J. Am. Chem. Soc.*, 1958, **80**, 1339.
28. M. Joshi, B. S. Butola, G. Simon and N. Kukaleva, *Macromolecules*, 2006, **39**, 1839.
29. M. Trujillo, M. L. Arnal, A. J. Muller, S. Bredeau, D. Bonduel, P. Dubois, I. W. Hamley and V. Castelletto, *Macromolecules*, 2008, **41**, 2087.
30. J. W. Cho and D. R. Paul, *Polymer*, 2001, **42**, 1083.

31. K. Prashantha, J. Soulestin, M. F. Lacrampe, P. Krawczak, G. Dupin and M. Claes, *Compos. Sci. Technol.*, 2009, **69**, 1756.
32. V. Mittal, *J. Appl. Polym. Sci.*, 2008, **107**, 1350.
33. V. Mittal, *J. Thermoplastic Compos. Mater.*, 2007, **20**, 575.
34. M. A. Osman, J. E. P. Rupp and U. W. Suter, *Polymer*, 2005, **46**, 1653.
35. A. U. Chaudhry and V. Mittal, *Polym. Eng. Sci.*, in press.

CHAPTER 7

pH-Sensitive Graphene–Polymer Nanocomposites

JINGQUAN LIU[1] AND THOMAS P. DAVIS*[2]

[1] Qingdao University, College of Chemistry, Chemical and Environmental Engineering, Laboratory of Fiber Materials and Modern Textile, the Growing Base for State Key Laboratory, 266071 Qingdao, China; [2] The University of New South Wales, Centre for Advanced Macromolecular Design, School of Chemical Engineering, Sydney, NSW 2052, Australia
*E-mail: t.davis@unsw.edu.au

7.1   Introduction

The two-dimensional structure of graphene, with single-atom thickness and extensive conjugation, endow it with unique thermal, electric and mechanical properties[1–6] and numerous potential applications in optoelectronic devices,[7,8] electrode materials,[9] supercapacitors,[10] sensors[11] and biomedical[12] and catalytic materials.[13] A number of methods, *e.g.* oxidation–reduction, mechanical exfoliation,[14] thermal deposition,[15,16] and liquid-phase exfoliation of graphite,[17,18] have been explored to prepare graphene sheets. Presently, oxidative exfoliation of natural graphite by acid treatment and subsequent chemical reduction has been considered as one of the most efficient methods for the low-cost, large-scale production of graphene.[19,20] However, the reduction process gives relatively hydrophobic graphene sheets, with a tendency to aggregate irreversibly, hindering production, storage and processing. To tackle this problem a number of methods have been used to stabilize or functionalize graphene sheets, among which modification of

RSC Nanoscience & Nanotechnology No. 26
Polymer–Graphene Nanocomposites
Edited by Vikas Mittal
© The Royal Society of Chemistry 2012
Published by the Royal Society of Chemistry, www.rsc.org

Table 7.1 Structures of versitile pH sensitive polymers

PAA	PMAA	PMA	PDDA
PDMAEA	P4VP	PPY	PEDOT
PAH	PSS	poly(β-amino ester)	

graphene by various polymers has been reported. Graphene–polymer composites can be tailored with versatile properties and functionalities, offering potential applications in intelligent sensors, biology, medicine, nanoelectronics and other relevant areas.

pH-sensitive polymers are mainly polyelectrolytes. As shown in Table 7.1 the most common pH-sensitive polymers are poly(acrylic acid) (PAA), poly(diallyldimethyl ammoniumchloride) (PDDA), poly(methacrylic acid) (PMAA), poly(maleic anhydride) (PMA), poly(2-*N,N'*-(dimethyl amino ethyl acrylate) (PDMAEA), polypyrrole (PPY) and Nafion. Polymers containing phosphoric and sulfonate acid derivatives have also been reported.[21,22] pH-sensitive polymers rely on the protonation/deprotonation equilibrium, which depends on the pK_a of the acidic and/or basic moieties present in the polymer. Therefore, a pH-sensitive polymer can be charged (yielding a swollen state) or uncharged (yielding a hydrophobic/collapsed state), depending on the environmental pH.[23] pH sensitive graphene–polymer composites have been prepared by grafting graphene with pH-sensitive polymers. In some cases, such composites can also be fabricated using polymers without obvious pH sensitivity. For example, polyacrylamide itself is not pH sensitive, but its composite with graphene formed in water by hydrophobic interactions showed a reversible pH-responsive property.[24] Graphene–polymer composites can also exist in the form of hydrogels. Shi and coworkers[25] prepared a pH-sensitive hydrogel with graphene oxide (GO) and poly(vinyl alcohol) (PVA) where the GO worked like a 2D macromolecule and PVA as crosslinking agent. These hydrogels exhibited sol–gel transition behaviour in different pH environments.

Based on its superior electrical properties, 2D features, flexibility and chemical stability graphene has been used to prepare hybrid composites with polyelectrolytes for high-capacity intercalating cathode of lithium ion batteries,[26] conductive films,[27–30] biomolecular sensing devices [31] and drug delivery vehicles.[21,32] In this chapter we discuss the preparation, properties and future perspectives for the application of these composites.

7.2   Preparation of Graphene–Polymer Nanocomposites

pH-sensitive graphene–polymer composites can be prepared *via* either covalent or non-covalent methods. The covalent methods are based on strong bonds between the graphene sheets and polymer chains. However, covalent interactions are not always desirable as they can have a deleterious effect on the properties of the graphene substrate. Non-covalent synthetic methods can be advantageous, as they can preserve graphene's conjugated structure, while imparting stability and functionality.

7.2.1   Covalent Bonding

Covalent methods are desirable as stronger bonds are formed between the graphene and polymer. However, covalent bonding can be difficult to realize as graphene lacks functional groups for anchoring the polymer chains. In some cases,[33–35] when graphene sheets were prepared from graphene oxide, an incomplete reduction process left traces of oxygen-containing functionalities subsequently available for further modification. Other covalent modification strategies involve disruption of the conjugated structure of graphene. Although covalent modification of graphene will compromise some of its natural conductivity, this method is still valuable in some cases when other properties of graphene are desirable. Shen *et al.*[36] modified graphene surfaces with PAA and poly(acryl amide) by ester and amide covalent bondings, introducing negative and positive charges, respectively, on the surfaces of graphene. In many cases, where the conductivity was unimportant, polymer was directly covalently attached to graphene oxide; Wang *et al.*[37] grafted stimuli-responsive polymers, PAA and poly(*N*-isopropylacrylamide) (PNIPAM), onto GO through a facile redox polymerization initiated by cerium ammonium nitrate in aqueous solution at mild temperatures providing a convenient way to prepare polymers covalently grafted to GO, utilizing various water-soluble vinyl monomers (Figure 7.1).

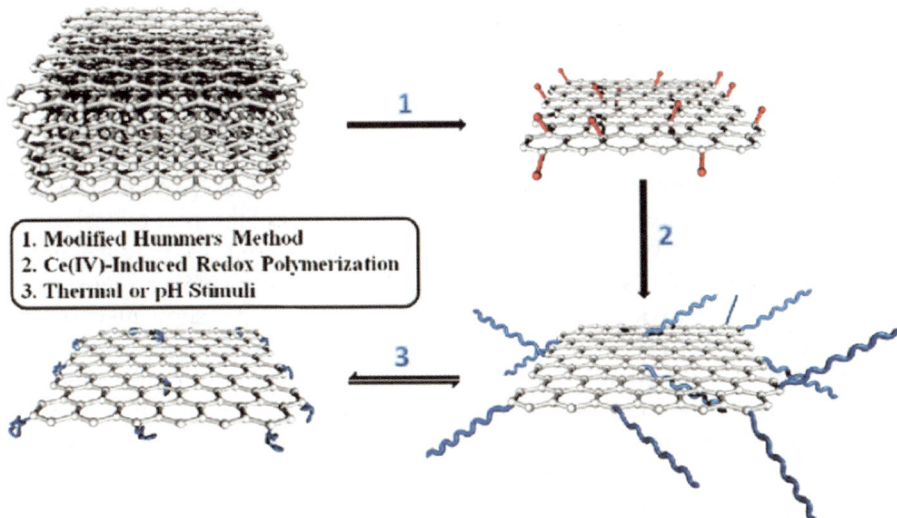

Figure 7.1   Schematic showing a synthetic route to polymer-grafted graphene oxide (GO).

7.2.2   Non-Covalent Interactions

7.2.2.1   Van der Waals Force

The preparation of graphene–polymer composites *via* van der Waals forces can be achieved by solvent processing, *in situ* polymerization or melt processing, as reviewed by Verdejo *et al.* [38] Each method has advantages and disadvantages. The solvent processing method is preferred for producing homogeneous graphene–polymer composites;[39,40] however, it is difficult to remove trace residual solvents.[41] The hydrophilicity of graphene can lead to irreversible aggregation induced by a re-stacking process. In order to avoid re-stacking-induced aggregation, the graphene–polymer composites can be prepared using the more stable GO and the required polymer in the presence of a reducing agent.[42] Generally speaking, *in situ* polymerization methods can used to prepare more homogeneous composites;[43] however, polymerization rates usually become very slow at the later stages of polymerization.[44] Compared with the other two methods, the melt processing route should be more attractive as it can be applied more extensively to polymers; however, the drawback with this approach is maintaining control over the feed of graphene filler, as bad control leads to phase separation and compromised mechanical and transport properties.[38]

As van der Waals forces are weak interactions, they can generate reversible surfaces. This reversibility was employed to prepare graphene-based sheets made highly visible under a fluorescence microscope by quenching the emission from a dye attachment. The weak van der Waals interaction allowed the dye to be conveniently removed by rinsing without disrupting the sheet structure.[45]

7.2.2.2 *Electrostatic Interaction*

Electrostatic interactions occur between two opposite charged materials and can be used as a convenient tool for the preparation of composite materials *via* self-assembly. Electrostatic interactions have been used to prepare novel hybrid materials of negatively charged sulfonated graphene and positively charged poly(3, 4-ethyldioxythiophene), with the composites showing excellent transparency, electrical conductivity, pH sensitivity and good flexibility, together with high thermal stability and ease of processing in both water and organic solvents.[7] Kim and coworkers[46] have used electrostatic attractions to prepare graphene–peptide hybrids for assembly into core-shell nanowires with peptide cores and graphene shells, which demonstrated electroconductivity *via* a continuous graphene shell. Thermal calcinations of the peptide cores led to the formation of highly entangled networks of the hollow graphene shell, with a demonstrated performance as a supercapacitor electrode. Similarly, Rani and Ren prepared graphene–polyelectrolyte composites.[47,48]

7.2.2.3 *π–π Stacking Interactions*

π–π stacking interactions can occur between two relatively large non-polar aromatic rings having overlapping π-orbitals. π–π interactions can be comparable in strength to covalent bonds and hence provide more stable alternatives to linkages based on hydrogen bonding, electrostatic bonding or coordination bonding. Furthermore, π–π stacking modification does not disrupt the conjugation of the graphene sheets, and hence preserves the electronic properties of graphene; such a modification strategy is envisaged to have enormous potential.

π–π stacking interactions have been widely used to modify graphene. Graphene has been modified with 1-pyrene sulfonic acid sodium salt stable and negative charged composite materials have been prepared for improving lithium ion charge–discharge properties and for increasing the power efficiency in heterojunction solar cells.[49,50] Using π–π stacking interactions, Liu *et al.* prepared pH sensitive graphene–polymer composites by modifying graphene with pyrene-terminated positive charged polymer (PDMAEA) and negatively charged polymer (PAA).[51] The pyrene-terminal PDMAEA and PAA were synthesized using a pyrene-terminated RAFT agent. These charged graphene–polymer composites were found to demonstrate phase transfer behaviour between aqueous and organic media at different pH values. These two oppositely charged graphene–polymer composites were self-assembled, to yield layer-by-layer (LbL) structures as evidenced by high-resolution scanning electron microscopy (SEM) analysis and quartz crystal microbalance (QCM) monitoring (Figure 7.2). Atom transfer radical polymerization (ATRP) has

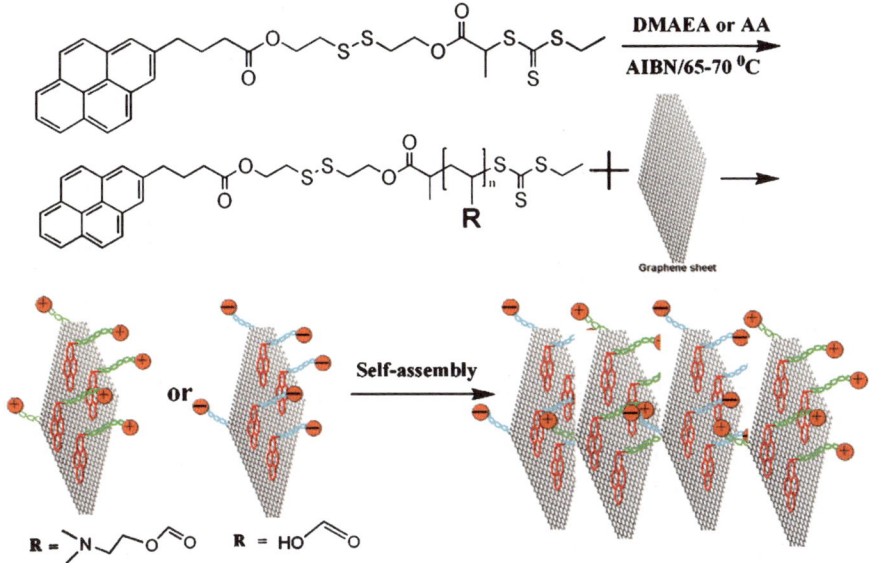

Figure 7.2 Synthesis of pH sensitive pyrene–polymer composites *via* π–π stacking interactions for the self-assembly of functionalized graphene into layered structures.

been used to synthesize π-orbital-rich polymers for modification of reduced graphene oxide (rGO) to afford fluorescent and water-soluble graphene composites *via* π–π stacking interactions.[52] Zhang *et al.*[53] designed a novel electrocatalytic biosensing platform by functionalizing rGO sheets with conducting polypyrrole graft copolymer, poly(styrenesulfonic acid-g-pyrrole) (PSSA-g-PPY), *via* π–π non-covalent interactions. The resulting nanocomposite was dispersed in water for at least 2 months with a solubility of 3.0 mg mL^{-1}. Qi[54] and Yang[55] have also successfully attached polyelectrolytes onto rGO *via* π–π interactions.

7.3 Applications of pH-Sensitive Graphene Polymer Nanocomposites

7.3.1 Sensors and Detection Devices

By judicious modification, pH-sensitive graphene polymer nanocomposites can be tailored for extensive applications in electrochemical sensors for DNA analysis of proteins, neurotransmitters, phytohormones, pollutants, metal ions, gases and hydrogen peroxide. The large specific surface area of graphene nanosheets (GNs) enables them to load metal nanoparticles (NPs) for a variety of applications such as catalysis, sensors and biomedicine. However, it remains challenging to achieve a uniform distribution of NPs on GNs due to a limited display of oxygen functional groups. Fang *et al.*[34] reported a method to

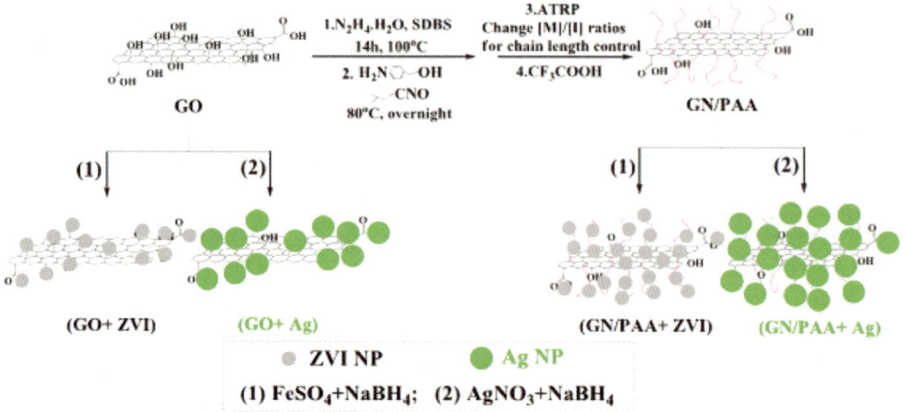

Figure 7.3   Covalent grafting of pH-sensitive polymer PAA on the GN surface and the deposition of the ZVI NPs.

improve the spatial and size distributions of ZVI and Ag NPs on GNs by PAA brushes. It was found that metallic NPs formed in PAA brushes have smaller sizes and a more uniform spatial distribution than metal NPs directly deposited on GO (Figure 7.3).

PDDA is a commonly used pH-sensitive polymer, and has been widely used in the functionalization of graphene as it can imbue the graphene surface with positive charge as precursors to composites materials with NPs, with potentially wide applications.[56–61] Li et al.[57] reported a novel strategy for the fabrication of PDDA-protected graphene–CdSe (P-GR-CdSe) composites for use as advanced electrogenerated chemiluminescence (ECL) immunosensors for the sensitive detection of human IgG (HIgG) by using P-GR-CdSe composites. Fang et al.[34,62] demonstrated the use of cationic PDDA-functionalized GNs as building blocks in the self-assembly of GN/Au–NP heterostructure to enhance electrochemical catalytic ability (Figure 7.4); the GN/Au–NP hybrids were employed as an electrochemical enhanced material for H_2O_2 sensing (as H_2O_2 is a by-product of several highly selective oxidases, it is important in enzyme-based biosensor applications). A nanohybrid of Au NPs and chemically reduced GNs (cr-Gs) has also been prepared by the in situ growth of Au NPs on the surface of GNs in the presence of PDDA to improve the dispersion of Au NPs and stabilize cholinesterase with high activity and loading efficiency.[63]

A single-step electropolymerization has been carried out by incorporating an anionic stacked graphene nanofibre dopant into polypyrrole films, at a disposable screen–printed electrode;[64] cyclic voltammetry of guanine and H_2O_2 had high selectivity at this electrode, allowing further applications of this modified electrode in the detection of biomedically important compounds and DNA sensing. Similarly, another novel polyelectrolyte, poly[N-(1-one-butyric acid) benzimidazole](PBI-BA), was also synthesized and used to fabricate an

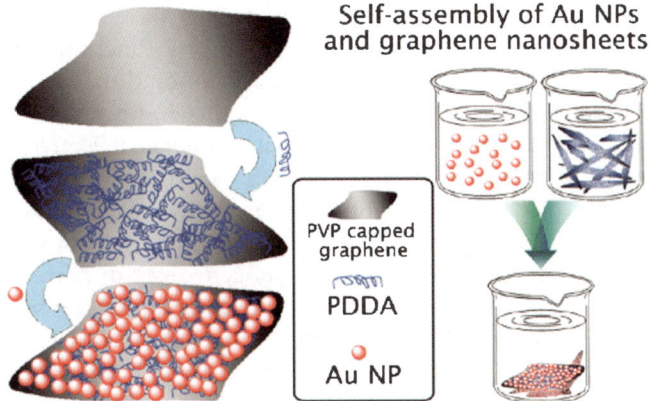

Figure 7.4 Procedure for the self-assembly of Au nanoparticles and PDDA-functionalized graphene nanosheets.

amperometric H_2O_2 sensor;[65] when graphene was modified with poly(sodium styrenesulfonate) the hybrid could be used as a biosensor to detect hydrazine.[22]

GNs functionalized with poly(diallyldimethylammonium chloride) (PDDA–graphene) were synthesized by Liu *et al.*[66] for combining with room temperature ionic liquids (RTILs). RTIL/PDDA–graphene composites were shown to display an enhanced capacity for immobilized haemoglobin to facilitate direct electrochemistry studies. In addition, RTIL/PDDA–graphene films had a strong electrocatalytic activity in nitric oxide reduction resulting from excellent conductivity and biocompatibility of RTIL/PDDA–graphene composite films.

In order to increase the surface/volume ratio, Alwarappan *et al.*[67] grew porous polypyrrole on glassy carbon electrodes (GCE) by controllable electrochemical polymerization, to accommodate graphene–GOx within its stable structure for applications as biosensors with a rapid, stable, and sensitive response toward glucose detection (improved performance over other reported carbon materials used for biosensors (Figure 7.5)).

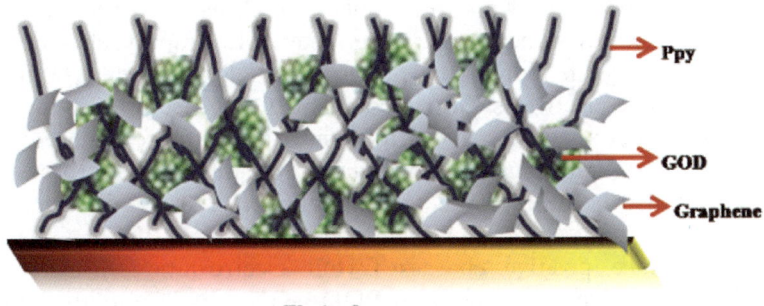

Figure 7.5 Representation of graphene glucose oxidase (GOD) entrapped within a porous PPY matrix.

pH-sensitive graphene–polymer nanocomposites have been tailored with opposite charges for LbL self-assembly to afford functional multilayered structures.[68] Zeng *et al.*[69] developed an unconventional method for the LbL assembly of pyrene-grafted PAA-modified graphene and poly(ethyleneimine) multilayer films on electrode surfaces, showing enhanced electron transfer for the redox reactions of $Fe(CN)_6^3$ and excellent electrocatalytic activity toward H_2O_2. Similarly, a novel positively charged polymeric ionic liquid functionalized graphene, poly(1–vinyl-3-butylimidazolium bromide)–graphene (poly(ViBuIm$^+$Br$^-$)–graphene) could be homogeneously dispersed in aqueous solutions for the immobilization of negatively charged GOx. The GCE modified with GOx/poly(ViBuIm$^+$Br$^-$)–graphene displayed excellent sensitivity, together with a wide linear range and excellent stability for the detection of glucose.[70]

Negatively charged Nafion has also been extensively explored as a component in graphene composite materials. Nafion has been used to prepare graphene polymer composites for immobilization of GOx.[71] Similarly, composites have been used to modify GCE to investigate the electrochemical behaviour of guanine and adenine on graphene by differential pulse voltammetry (DPV).[72] Li *et al.*[73] have also prepared an antitumour herbal drug aloe–emodin sensor based on graphene–Nafion modified GCE, displaying advantages of higher sensitivity and lower cost for aloe–emodin determination than electrodes modified with multiwalled carbon nanotubes.

7.3.2 Catalysis and Cells

The large surface area and unique electronic properties of graphene have stimulated research into applications in catalysis and cell biology. Shi *et al.*[74] prepared PDDA–RGO–PdPt nanocomposites using a sonoelectrochemical technique; a fast and simple synthetic strategy for the fabrication of nanomaterials. Cyclic voltammetric (CV) and chronoamperometric experiments were used to characterize catalytic activity and stability for the electro-oxidation of ethanol in alkaline media; a potentially promising use of electrocatalysts for direct alcohol fuel cells. PDDA has also been used as an electron acceptor for functionalizing graphene to impart electrocatalytic activity for the oxygen reduction reaction in (ORR) fuel cells.[75,76] PDDA-adsorbed graphene electrodes show remarkable ORR electrocatalytic activity with improved fuel selectivity, more tolerance to CO poisoning, and higher long-term stability than that of commercially available Pt/C electrodes.

PDDA has also been employed to functionalize GNs for direct methanol fuel cell applications.[77] PDDA-functionalized GNs (PDDA-GNs) with distributed surface positive charges favour electrostatic self-assembly of negatively charged $PtCl_6^{2-}$ ions onto PDDA-GNs. $PtCl^{6-}$-modified PDDA-GNs composites exhibited higher electrocatalytic activity in the electro-oxidation of methanol compared to the same reaction catalysed with $PtCl_6^{2-}$ supported on the GNs or PDDA. Therefore, PDDA–GNs can be an ideal

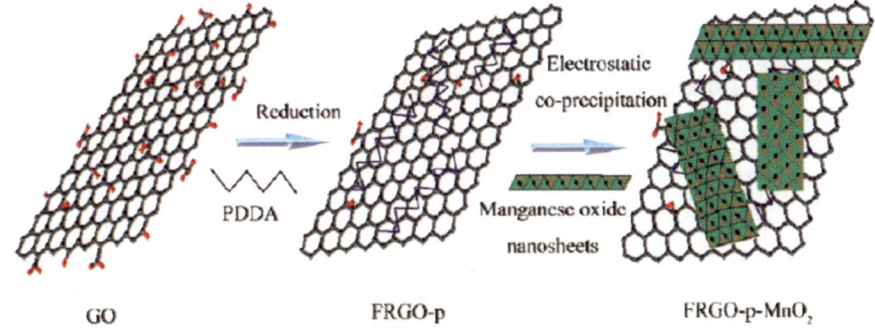

Figure 7.6   Preparation of functionalized rGO/MnO$_2$ composite materials.

support for Pt in direct methanol fuel cells. Zhang *et al.*[78] prepared functionalized graphene by reducing functionalized GO with PDDA, inducing a surface charge reversal of rGO from negative to positive, by dispersing negatively charged MnO$_2$ nanosheets on the functionalized rGO sheets *via* an electrostatic co-precipitation method. The composite material exhibited enhanced capacitive performances over pure functionalized rGO and Na-typed birnessite (Na/MnO$_2$) sheets, attributed to the synergic effect of both components (Figure 7.6). In order to reduce the cost of fabrication, Hasin *et al.*[79] used polyelectrolytes to tune the electrocatalytic activity of graphene films used as transparent conductive electrodes (TCEs) and counter-electrodes in dye-sensitized solar cells (DSCs).

7.3.3   Supercapacitors

Supercapacitors, also called ultracapacitors or electrochemical capacitors, store electrical charges on high-surface-area conducting materials. The performance of traditional supercapacitors is limited by their low energy storage density and relatively high effective series resistance. The excellent conductivity and large surface area of graphene make it a good alternative to replace traditional materials for the preparation of supercapacitors with high energy storage density and higher discharge rate. Li *et al.*[80] constructed a new class of multilayer film by electrostatic LbL self-assembly, using poly(sodium 4-styrenesulfonate) mediated graphene sheets (PSS–GS), manganese dioxide (MnO$_2$) sheets, and PDDA as building blocks for novel electrode materials for use as supercapacitors and other devices where controllable capacitance, good cycle stability, low cost and environmentally benign nature are important (Figure 7.7).

Wang *et al.*[81] fabricated a graphene–azo polyelectrolyte multilayer film through electrostatic LbL self-assembly, and investigated its performance as an electrochemical capacitor electrode, for use in electrical energy storage devices.

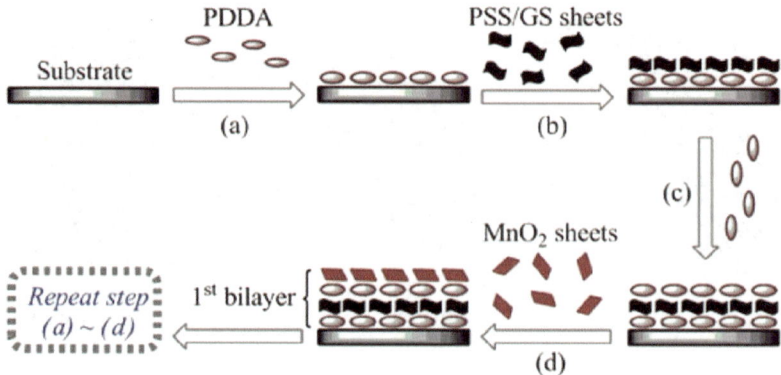

Figure 7.7   A schematic view for constructing multilayer films on substrate.

7.3.4   Drug Delivery

pH-sensitive polymers show a sharp change in properties upon a small or modest change in pH. This behaviour can be utilized for the preparation of so-called 'smart' drug delivery systems, mimicking biological response behaviour. Ionizable polymers with a pK_a value between 3 and 10 are candidates as building blocks in pH-responsive systems.[21,82] Polymers functionalized as weak acids and bases, *e.g.* carboxylic acids, phosphoric acid and amines,

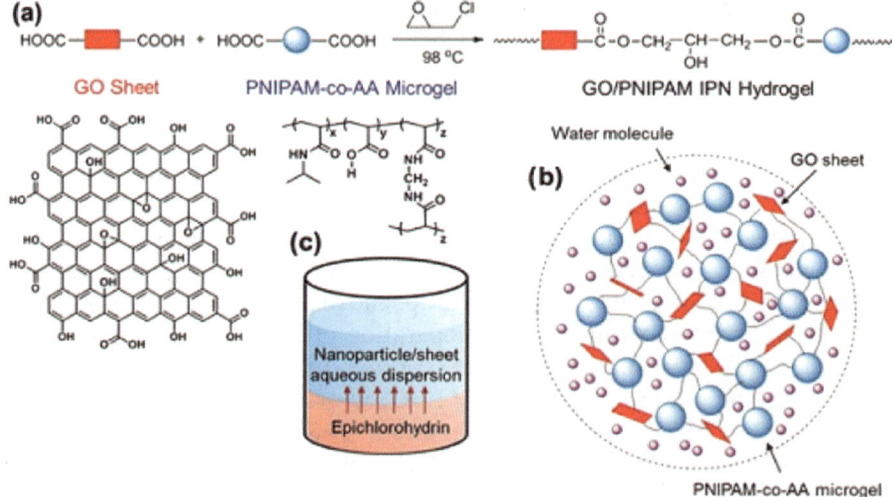

Figure 7.8   (a) Formation of GO–PNIPAM interpenetrating hydrogel networks *via* the reaction between epichlorohydrin (ECH) and carboxyl groups in GO sheets and PNIPAM-co-AA microgels. (b) Structural sketch of GO–PNIPAM IPN hydrogel. (c) The sealed reaction tube was placed at 98 °C for incubation while ECH would permeate into the aqueous phase to induce the cross-linking reaction.

exhibit a change in ionization state upon variation of pH, leading to a conformational change in solution or a change in the swelling behaviour of hydrogels. Thus pH-sensitive polymers can be used in drug delivery.

GO can be used to create barrier layers in multilayer thin films, trapping molecules of interest for controlled release. Protein-loaded polyelectrolyte multilayer films were fabricated by Hong *et al.*[83] using LbL assembly incorporating a hydrolytically degradable cationic poly(β-amino ester) with a model protein antigen, ovalbumin, in a bilayer architecture along with positively and negatively functionalized GO capping layers for the degradable protein films. Bai *et al.*[25] have demonstrated that GO sheets were able to form composite hydrogels with PVA. In these systems, GO sheets acted like 2D macromolecules. The formation of the hydrogels relied on the assembly of GO sheets and the cross-linking effect of PVA chains. The GO–PVA hydrogels exhibited pH-induced gel–sol transition and could be used for loading and the selective release of drugs at physiological pH. GO has also been used to prepare PNIPAM hydrogel networks by covalently bonding GO sheets and PNIPAM-co-AA microgels directly in water, producing hybrids with reversible dual thermal and pH responses. (Figure 7.8).

7.3.5 Others

pH-sensitive polymeric imidazolium molten salts can also be used to functionalize graphene membranes in the solution phase as a transfer medium to improve the electronic properties of graphene[84] (Figure 7.9). Tai *et al.*[85] synthesized silver nanocrystal/GN nanohybrids at room temperature, *via in*

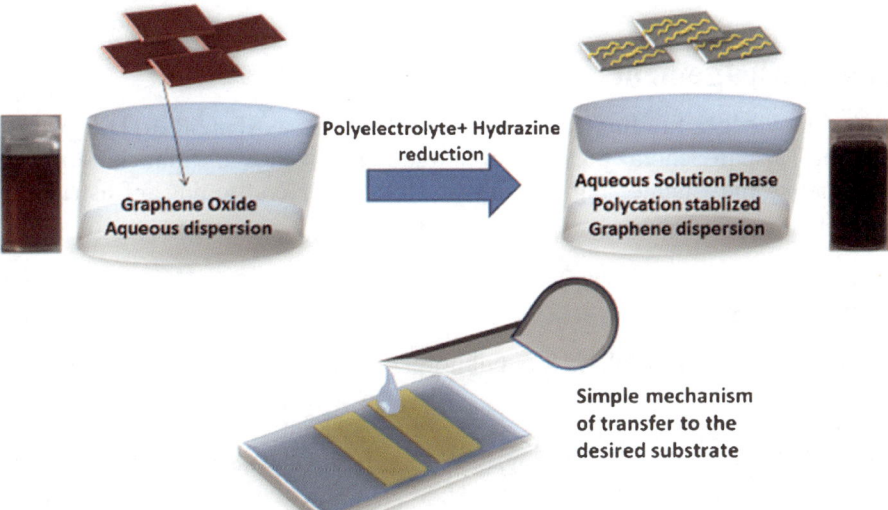

Figure 7.9 Aqueous solutions of GO and graphene after hydrazine reduction in the presence of polyelectrolyte and the schematic of the transfer mechanism.

situ PAA grafting followed by attachment of Ag nanocrystals; the hybrids exhibited good antimicrobial activity against Gram-positive *Staphylococcus aureus* and Gram-negative *Escherichia coli*.

7.4   Conclusions and Perspectives

In this chapter we have highlighted recent advances in the preparation and applications of pH-sensitive graphene–polymer nanocomposites. Composites can be prepared using either covalent or non-covalent approaches. Van der Waals forces, electrostatic interactions and π–π stacking interactions are the most widely used non-covalent methods for preparing pH-sensitive graphene–polymer nanocomposites. Composites have been explored as materials for sensors and detection devices, catalysis and cell studies, supercapacitors and drug delivery.

Since the application of pH-sensitive graphene–polymer nanocomposites mainly relies on graphene's large surface area and superconductivity, it is imperative to avoid irreversible aggregation of graphene during composite preparation. Since GO, the precursor to graphene, is much more stable, it is often used as a starting material to prepare more homogeneous composites by *in situ* reduction in the presence of polymers. Another alternative to preparing pH-sensitive graphene–polymer nanocomposites is direct exfoliation of graphene from graphite using pH-sensitive polymers with large aromatic pendant functionality. This may be the most attractive route in future work; it is more straightforward and versatile, as graphene sheets can be produced with large sizes, lower defects and higher conductivity compared to nanocomposites produced *via* the oxidation–reduction process.

Acknowledgement

JL acknowledges funding from NSFC (51173087), NSF (ZR2011EMM001) and the Taishan Scholar Program of Shandong Provence for financial support. TD thanks the Engineering Faculty at UNSW for continuing infrastructure support.

References

1. H. Chen, M. B. Muller, K. J. Gilmore, G. G. Wallace, D. Li, *Adv. Mater.*, 2008, **20**, 3557.
2. Y. B. Zhang, Y. W. Tan, H. L. Stormer, P. Kim, *Nature*, 2005, **438**, 201.
3. X. L. Li, X. R. Wang, L. Zhang, S. W. Lee, H. J. Dai, *Science*, 2008, **319**, 1229.
4. S. Stankovich, D. A. Dikin, G. H. B. Dommett, K. M. Kohlhaas, E. J. Zimney, E. A. Stach, R. D. Piner, S. T. Nguyen, R. S. Ruoff, *Nature*, 2006, **442**, 282.

5. Z. F. Liu, Q. Liu, Y. Huang, Y. F. Ma, S. G. Yin, X. Y. Zhang, W. Sun, Y. S. Chen, *Adv. Mater.*, 2008, **20**, 3924.
6. W. Yang, K. R. Ratinac, S. P. Ringer, P. Thordarson, J. J. Gooding, F. Braet, *Angew. Chem. Int. Ed.*, 2010, **49**, 2114.
7. Y. Xu, Y. Wang, J. Liang, Y. Huang, Y. Ma, X. Wan, Y. Chen, *Nano Res.*, 2009, **2**, 343.
8. Z. Yang, X. Shi, J. Yuan, H. Pu, Y. Liu, *Appl. Surf. Sci.*, 2010, **257**, 138.
9. X. Wang, J. Wang, H. Cheng, P. Yu, J. Ye, L. Mao, *Langmuir*, 2011, **27**, 11180.
10. F. Li, J. Song, H. Yang, S. Gan, Q. Zhang, D. Han, A. Ivaska, L. Niu, *Nanotechnology*, 2009, **20**, 455602.
11. W. Tu, J. Lei, S. Zhang, H. Ju, *Chemistry*, 2010, **16**, 10771.
12. G. Zeng, Y. Xing, J. Gao, Z. Wang, X. Zhang, *Langmuir*, 2010, **26**, 15022.
13. W. Hong, H. Bai, Y. Xu, Z. Yao, Z. Gu, G. Shi, *J. Phys. Chem. C*, 2010, **114**, 1822.
14. K. S. Novoselov, A. K. Geim, S. V. Morozov, D. Jiang, Y. Zhang, S. V. Dubonos, I. V. Grigorieva, A. A. Firsov, *Science*, 2004, **306**, 666.
15. C. Berger, Z. Song, X. Li, X. Wu, N. Brown, C. Naud, D. Mayou, T. Li, J. Hass, A. N. Marchenkov, E. H. Conrad, P. N. First, W. A. de Heer, *Science*, 2006, **312**, 1191.
16. D. Wei, Y. Liu, Y. Wang, H. Zhang, L. Huang, G. Yu, *Nano Lett.*, 2009, **9**, 1752.
17. X. Li, G. Zhang, X. Bai, X. Sun, X. Wang, E. Wang, H. Dai, *Nat. Nanotechnol.*, 2008, **3**, 538.
18. M. Choucair, P. Thordarson, J. A. Stride, *Nat. Nanotechnol.*, 2009, **4**, 30.
19. D. Li, M. B. Muller, S. Gilje, R. B. Kaner, G. G. Wallace, *Nat. Nanotechnol.*, 2008, **3**, 101.
20. S. Stankovich, D. A. Dikin, R. D. Piner, K. A. Kohlhaas, A. Kleinhammes, Y. Jia, Y. Wu, S. T. Nguyen, R. S. Ruoff, *Carbon*, 2007, **45**, 1558.
21. D. Schmaljohann, *Adv. Drug Delivery Rev.*, 2006, **58**, 1655.
22. C. Wang, L. Zhang, Z. Guo, J. Xu, H. Wang, K. Zhai, X. Zhuo, *Microchim. Acta*, 2010, **169**, 1.
23. T. Vermonden, R. Censi, W. E. Hennink, *Chem. Rev.*, 2012, DOI: 10.1021/cr200157d.
24. L. Ren, T. Liu, J. Guo, S. Guo, X. Wang, W. Wang, *Nanotechnology*, 2010, **21**, 335701.
25. H. Bai, C. Li, X. L. Wang, G. Q. Shi, *Chem. Commun.*, 2010, **46**, 2376.
26. T. Cassagneau, J. H. Fendler, *Adv. Mater.*, 1998, **10**, 877.
27. N. I. Kovtyukhova, P. J. Ollivier, B. R. Martin, T. E. Mallouk, S. A. Chizhik, E. V. Buzaneva, A. D. Gorchinskiy, *Chem. Mater.*, 1999, **11**, 771.
28. T. Szabó, A. Szeri, I. Dékány, *Carbon*, 2005, **43**, 87.
29. J. Wu, Q. Tang, H. Sun, J. Lin, H. Ao, M. Huang, Y. Huang, *Langmuir*, 2008, **24**, 4800.
30. B. S. Kong, H. W. Yoo, H. T. Jung, *Langmuir*, 2009, **25**, 11008.

31. N. Mohanty, V. Berry, *Nano Lett.*, 2008, **8**, 4469.
32. I. Y. Galaev, B. Mattiasson, *Trends Biotechnol.*, 1999, **17**, 335.
33. M. Fang, K. Wang, H. Lu, Y. Yang, S. Nutt, *J. Mater. Chem.*, 2009, **19**, 7098.
34. M. Fang, Z. X. Chen, S. Z. Wang, H. B. Lu, *Nanotechnology*, 2012, **23**, 08574.
35. M.-C. Hsiao, S.-H. Liao, M.-Y. Yen, P.-I. Liu, N.-W. Pu, C.-A. Wang, C.-C. M. Ma, *ACS Appl. Mater. Interfaces*, 2010, **2**, 3092.
36. J. F. Shen, Y. Z. Hu, C. Li, C. Qin, M. Shi, M. X. Ye, *Langmuir*, 2009, **25**, 6122.
37. B. Wang, D. Yang, J. Z. Zhang, C. Xi, J. Hu, *J. Phys. Chem. C*, 2011, **115**, 24636.
38. R. Verdejo, M. M. Bernal, L. J. Romasanta, M. A. Lopez-Manchado, *J. Mater. Chem.*, 2011, **21**, 3301.
39. H. Kim, Y. Miura, C. W. Macosko, *Chem. Mater.*, 2010, **22**, 3441.
40. H. Kim, S. Kobayashi, M. A. AbdurRahim, M. L. J. Zhang, A. Khusainova, M. A. Hillmyer, A. A. Abdala, C. W. Macosko, *Polymer*, 2011, **52**, 1837.
41. F. Barroso-Bujans, S. Cerveny, R. Verdejo, J. J. del Val, J. M. Alberdi, A. Alegra, J. Colmenero, *Carbon*, 2009, **48**, 1079.
42. T. Wei, G. Luo, Z. Fan, C. Zheng, J. Yan, C. Yao, W. Li, C. Zhang, *Carbon*, 2009, **47**, 2296.
43. Z. Gu, L. Zhang, C. Li, *J. Macromol. Sci., Part B: Phys.*, 2009, **48**, 1093.
44. R. Verdejo, F. J. Tapiador, L. Helfen, M. M. Bernal, N. Bitinis, M. A. Lopez-Manchado, *Phys. Chem. Chem. Phys.*, 2009, **11**, 10860.
45. J. Kim, L. J. Cote, F. Kim, J. Huang, *J. Am. Chem. Soc.*, 2009, **132**, 260.
46. T. H. Han, W. J. Lee, D. H. Lee, J. E. Kim, E.-Y. Choi, S. O. Kim, *Adv. Mater.*, 2010, **22**, 2060.
47. A. Rani, K. A. Oh, H. Koo, H. J. Lee, M. Park, *Appl. Surf. Sci.*, 2011, **257**, 4982.
48. L. L. Ren, S. Huang, W. Fan, T. X. Liu, *Appl. Surf. Sci.* 2011, **258**, 1132.
49. J.-H. Jang, D. Rangappa, Y.-U. Kwon, I. Honma, *J. Mater. Chem.*, 2011, **21**, 3462.
50. Q. Su, S. Pang, V. Alijani, C. Li, X. Feng, K. Mullen, *Adv. Mater.*, 2009, **21**, 3191.
51. J. Liu, L. Tao, W. Yang, D. Li, C. Boyer, R. Wuhrer, F. Braet, T. P. Davis, *Langmuir*, 2010, **26**, 10068.
52. L. Q. Xu, L. Wang, B. Zhang, C. H. Lim, Y. Chen, K.-G. Neoh, E.-T. Kang, G. D. Fu, *Polymer*, 2011, **52**, 2376.
53. J. Zhang, J. P. Lei, R. Pan, Y. D. Xue, H. X. Ju, *Biosens. Bioelectron.*, 2010, **26**, 371.
54. X. Y. Qi, K. Y. Pu, X. Z. Zhou, H. Li, B. Liu, F. Boey, W. Huang, H. Zhang, *Small*, 2010, **6**, 663.
55. H. F. Yang, Q. X. Zhang, C. S. Shan, F. H. Li, D. X. Han, L. Niu, *Langmuir*, 2010, **26**, 6708.

56. F. H. Li, J. Chai, H. F. Yang, D. X. Han, L. Niu, *Talanta*, 2010, **81**, 1063.
57. L. L. Li, K. P. Liu, G. H. Yang, C. M. Wang, J. R. Zhang, J. J. Zhu, *Adv. Func. Mater.*, 2011, **21**, 869.
58. X. X. Liu, H. Zhu, X. R. Yang, *Talanta*, **87**, 243.
59. K. P. Liu, J. J. Zhang, C. M. Wang, J. J. Zhu, *Biosens. Bioelectron.*, 2011, **26**, 3627.
60. S. Liu, J. Q. Tian, L. Wang, H. L. Li, Y. W. Zhang, X. P. Sun, *Macromolecules*, 2010, **43**, 10078.
61. D. B. Lu, Y. Zhang, S. X. Lin, L. T. Wang, C. M. Wang, *Analyst*, 2011, **136**, 4447.
62. Y. X. Fang, S. J. Guo, C. Z. Zhu, Y. M. Zhai, E. K. Wang, *Langmuir*, 2010, **26**, 11277.
63. Y. Wang, S. Zhang, D. Du, Y. Y. Shao, Z. H. Li, J. Wang, M. H. Engelhard, J. H. Li, Y. H. Lin, *J. Mater. Chem.*, 2011, **21**, 5319.
64. C. L. Scott, G. Zhao, M. Pumera, *Electrochem. Commun.*, 2010, **12**, 1788.
65. M. Y. Hua, H. C. Chen, R. Y. Tsai, Y. L. Leu, Y. C. Liu, J. T. Lai, *J. Mater. Chem.*, 2011, **21**, 7254.
66. K. P. Liu, J. J. Zhang, G. H. Yang, C. M. Wang, J. J. Zhu, *Electrochem. Commun.*, 2010, **12**, 402.
67. S. Alwarappan, C. Liu, A. Kumar, C. Z. Li, *J. Phys. Chem. C*, 2010, **114**, 12920.
68. T. Gan, S. Hu, *Microchim. Acta*, 2011, **175**, 1.
69. G. H. Zeng, Y. B. Xing, J. A. Gao, Z. Q. Wang, X. Zhang, *Langmuir*, 2010, **26**, 15022.
70. Q. Zhang, S. Wu, L. Zhang, J. Lu, F. Verproot, Y. Liu, Z. Xing, J. Li, X. M. Song, *Biosens. Bioelectron.* 2010.
71. X. Chen, H. Ye, W. Wang, B. Qiu, Z. Lin, G. Chen, *Electroanalysis*, **22**, 2347.
72. H. Yin, Y. Zhou, Q. Ma, S. Ai, P. Ju, L. Zhu, L. Lu, *Process Biochem.*, 2010, **45**, 1707.
73. J. Li, J. Chen, X. L. Zhang, C. H. Lu, H. H. Yang, *Talanta*, 2010, **83**, 553.
74. J. J. Shi, G. H. Yang, J. J. Zhu, *J. Mater. Chem.*, 2011, **21**, 7343.
75. S. Y. Wang, D. S. Yu, L. M. Dai, D. W. Chang, J. B. Baek, *ACS Nano*, 2011, **5**, 6202.
76. S. Y. Wang, X. Wang, S. P. Jiang, *Phys. Chem. Chem. Phys.*, 2011, **13**, 7187.
77. B. M. Luo, X. B. Yan, S. Xu, Q. J. Xue, *Electrochim. Acta*, 2011, **59**, 429.
78. J. T. Zhang, J. W. Jiang, X. S. Zhao, *J. Phys. Chem. C*, 2011, **115**, 6448.
79. P. Hasin, M. A. Alpuche-Aviles, Y. Y. Wu, *J. Phys. Chem. C*, 2010, **114**, 15857.
80. Z. P. Li, J. Q. Wang, X. H. Liu, S. Liu, J. F. Ou, S. R. Yang, *J. Mater. Chem.*, 2011, **21**, 3397.
81. D. G. Wang, X. G. Wang, *Langmuir*, 2011, **27**, 2007.
82. R. Siegel, K. Dusek, *Adv. Polym. Sci.*, 1993, **109**, 233.

83. J. Hong, N. J. Shah, A. C. Drake, P. C. DeMuth, J. B. Lee, J. Z. Chen, P. T. Hammond, *ACS Nano*, 2012, **6**, 81.

84. U. K. Hasan, M. O. Sandberg, O. Nur, M. Willander, *Nanoscale Res. Lett.*, 2011, **6**, 493.

85. Z. X. Tai, H. B. Ma, B. Liu, X. B. Yan, Q. J. Xue, *Colloids Surf., B*, 2011, **89**, 147.

CHAPTER 8

Dispersible Graphene Oxide–Polymer Nanocomposites

GANG LIU, KOON-GEE NEOH AND EN-TANG KANG

Department of Chemical & Biomolecular Engineering, National University of
Singapore, Kent Ridge, Singapore 119260
E-mail: cheket@nus.edu.sg

8.1 Introduction

As the thinnest material ever known, graphene has attracted a tremendous
amount of attention since its successful isolation a few years ago.[1–7] Because of
the two-dimensional (2D) π–π conjugation in the sp^2 hybridization state in an
atomic layer, graphene nanosheets are predicted to exhibit a range of unique
properties, *viz.*, high aspect ratio and large specific surface area,[8] superior
mechanical stiffness[9] and flexibility,[10] remarkable optical transmittance,[11]
extraordinary thermal response[12] and excellent electronic transport properties,[13] potentially useful for application in molecular electronics[14–19] and as
energy/hydrogen storage materials,[20,21] reinforcement and energy-dissipation
nanofiller for polymer and ceramic composites,[22,23] free-standing 'paper-like'
materials[24] and even biomedical materials.[25]

Graphene nanosheets do not occur naturally. Work is necessary to obtain a
stable dispersion of graphene or a derivative in order to exploit its fascinating
properties and applications. Among the various techniques of physical
separation or preparation of the nanosheets under specialized conditions,[26–52]
chemical exfoliation of graphite using Hummers methods has been considered as
the most efficient approach to the mass production of graphene oxide (GO)

RSC Nanoscience & Nanotechnology No. 26
Polymer–Graphene Nanocomposites
Edited by Vikas Mittal
© The Royal Society of Chemistry 2012
Published by the Royal Society of Chemistry, www.rsc.org

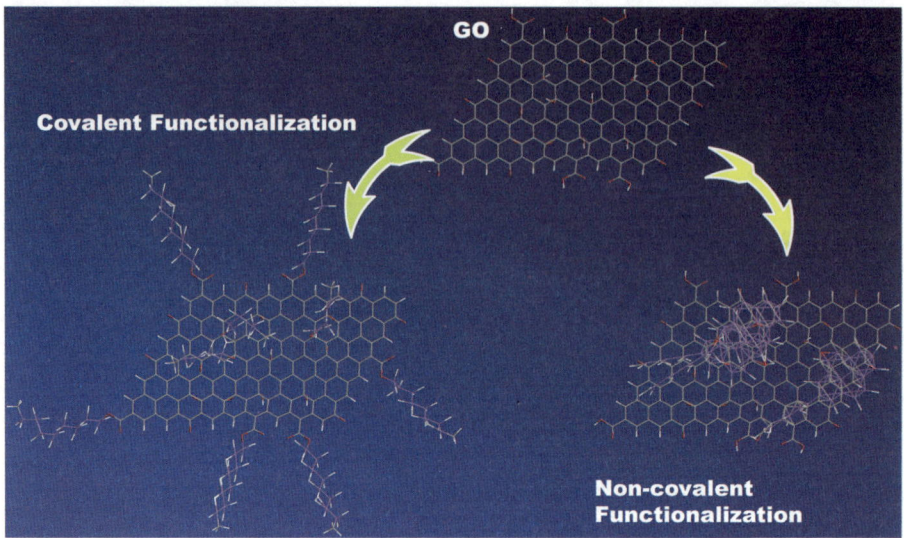

Figure 8.1 Functionalization of graphene oxide by polymers.

nanosheets that can be dispersed in aqueous media.[5,53–55] The high density of oxygen functionalities on the surface of basal planes and at the sheet edges, arising from the oxidation of the natural graphite powder with various oxidants in acidic media, offers the potential for solution processing and further functionalization, and makes GO an important precursor of graphene sheets. To date, by taking advantage of the well-developed technologies for dispersing carbon nanotubes (CNTs), many dispersible GO nanosheets have been prepared through both covalent modification and non-covalent functionalization (Figure 8.1).[4,56–60] For example, polymer materials have been covalently grafted from or grafted to the GO nanosheets, through oxygen-containing functional groups, *via* surface-initiated polymerization,[61–70] esterification,[71] amidation,[72–74] click chemistry [75] and other methods. The dispersion of GO nanosheets has also been facilitated by π–π, van der Waals and electrostatic interactions with organic and polymeric materials.[76–80] Surface functionalization of GO not only plays an important role in its dispersion behaviour and solution processability, but also fine-tunes the structural and functional properties of the nanosheets and extends their applications in many technological fields. In this chapter, we summarize recent advancements in the preparation, properties and potential applications of dispersible GO–polymer composite nanosheets that are of interest to both the academic and industrial communities.

8.2 Covalently Functionalized Graphene Oxide–Polymer Nanocomposites

As with the CNT allotrope, chemical functionalization is expected to play a key role in tailoring the structure, solubility, processability and properties of

GO nanosheets. The chemistry of GO deals mainly with the chemically reactive oxygen functionalities, including epoxy and hydroxyl groups on the basal planes, and carbonyl and carboxylic acid groups at the edges of the nanosheets.[4] As the density of oxygen-containing groups cannot be precisely controlled with the Hummers method of chemical exfoliation, surface modification based on the chemistry of oxygenated groups will inevitably lead to non-stoichiometric functionalization of the nanosheets. Nevertheless, the 'grafting to' and 'grafting from' approaches allow a variety of functionalized polymers to be covalently attached to the surfaces as well as the edges of GO nanosheets.

The 'grafting to' approach is among the most commonly used methods for tethering polymers to GO nanosheets. In this approach, the preformed polymer chains are attached directly to the oxygenated functionalities on the surface or at the edge of GO nanosheets through simple chemical reactions.[4,81] However, this approach may be limited by the steric hindrance of the polymer chains. Beyond a certain degree of functionalization, the reactive sites on the surface and at the edge of the nanosheet will be significantly screened by the bonded macromolecular species, and the convoluted polymers present in the

Table 8.1 Methods for covalent functionalization of graphene oxide by polymers

'Grafting from'			'Grafting to'		
Polymerization method	Polymer	Reference	Type of reaction	Polymer	Reference
ATRP	PS	61	Esterification	PVA	71
ATRP	PMMA/PBA	62	Esterification	PVA	96
ATRP	PtBA	63	Amidation	PEG	72
ATRP	PMMA	64	Amidation	TPAPAM	74
ATRP	PS	65	Nitrene cycloaddition	PEG/PS	104
ATRP	PEMA	66	Ring opening	Epoxy	107
ATRP	PS	67	Radical	PMMA	110
ATRP	PDMAEMA	68	ATNRC	PNIPAM	114
RAFT	PS	69	1,3-Dipolar cycloaddition	PFCF	120
RAFT	PVK	70	Click	PNIPAM	75
Polycondensation	PU	84	Click	SEBS	116
Ring opening	PCL	87	Diazonuim salts coupling	PFTB	125

ATNRC, atom transfer nitroxide radical coupling; ATRP, atomic transfer radical polymerization; PBA, poly(butyl acrylate); PCL, poly(ε-caprolactone); PDMAEMA, poly(2-(dimethylamino) ethyl methacrylate); PEG, poly(ethylene glycol); PEMA, poly(ethyl methacrylate); PFCF, fluorene-carbazole based copolymer; PFTB,fluorene-thiophene-benzothiadazole copolymer; PMMA, poly(methyl methacrylate); PNIPAM, poly(*N*-isopropylacrylamide); PS, polystyrene; PtBA, poly(*tert*-butyl acrylate); PU, polyurethane; PVA, poly(vinyl alcohol); PVK, poly(*N*-vinylcarbzole); RAFT, reversible addition fragmentation chain transfer; SEBS, poly(styrene-*b*-ethylene-*co*-butylene-*b*-styrene); TPAPAM, triphenylamine-based polyazomethine.

solution face increasing difficulty in diffusing to the coupling sites. As a result, an upper limit to the reactivity and degree of functionalization can be expected. The alternative 'grafting from' approach, or surface-initiated polymerization from the pre-immobilized initiators, is capable of achieving a higher density of grafted chains, although the type of polymers that can be grafted onto GO nanosheets may be limited by the surface-initiated polymerization technique. Benefiting from reduction in enthalpy contribution to the total free energy of the grafted system, GO–polymer nanosheets can be dispersed more readily in solvents and polymeric matrices (Table 8.1).

8.3   The 'Grafting from' Approach

This approach refers to the *in situ* synthesis of covalently attached polymer chains from the basal plane and edge of GO nanosheets. To achieve this object, polymerization initiators are first immobilized *via* chemical reactions with the oxygenated functionalities of GO, similar to those used in the sidewall modification of CNTs.[82,83] Due to the abundance of oxygenated functionalities with different chemical reactivities on nanosheets, GO is more versatile for a wider range of chemical modifications that can be undertaken.[4] Various polymerization methods, including condensation[84] and cationic,[85] anionic[86] and ring-opening polymerization,[87] radical polymerizations, such as ATRP[61–68] and reversible addition fragmentation chain transfer (RAFT) polymerization,[69,70] can be utilized to functionalize GO nanosheets with polymers. With the polymerization initiated from the GO nanosheets, the steric effect encountered by the growing chain is minimized. Despite the dimension of GO nanosheets, thick and dense polymer graft layers or brushes can be introduced on the carbon laminates.

The ATRP technique is the most widely used controlled/living radical polymerization method for preparing dispersible GO–polymer nanosheets.[61–68] The surface-initiated ATRP technique offers advantages including commercially available and inexpensive catalysts, ligands and initiators, a wide range of monomers, fast initiation process, dynamic equilibrium between dormant and growing radicals, controlled molecular weight and low polydispersibility, controlled composition of polymer chains grown from various substrate surfaces, possibility of producing block copolymers, and post-modification of the terminal alkyl halide for introducing new functionalities *via* conventional organic synthesis.[88]

Single-layer graphene nanosheets with grafted polystyrene chains of controlled length have been prepared using the ATRP technique.[61] Stabilized by sodium dodecylbenzene sulfone (SDBS, a surfactant), the single-layer graphene nanosheets were prepared *via* reduction of well-exfoliated graphite oxide by hydrazine hydrate in an aqueous medium. Diazonium moieties of controlled concentration were first immobilized on the edge of the reduced graphene oxide (rGO) nanosheets to modulate the bonding density of initiators and the subsequent graft density of polymer chains. The

ATRP initiator, carbonyl halide, was coupled to the hydroxyl groups on the rGO nanosheets to initiate the polymerization. By varying the monomer/initiator ratio, control of graft chain length of polystyrene from the single-layer graphene nanosheet surface was achieved during the subsequent ATRP reaction. The resultant rGO–polystyrene nanosheets can be readily dispersed in polar organic solvents. The single-layer structure of the polystyrene grafted rGO nanosheets was confirmed by atomic force microscopy (AFM) analysis. Pristine GO nanosheets exhibit a lateral dimension of 0.5–1.5 μm and an average thickness of ~1 nm. The thickness of initiator- and polymer-grafted rGO nanosheets has increased to ~3 nm and ~8 nm, respectively. The thermal stability of graphene nanosheets has also been improved upon covalent modification with the initiator and polymer chains. With the increase in density and molecular weight of the grafted polymer, the thermal stability of rGO–polystyrene nanosheets is further improved, indicating that the grafted polymer chains assist in inhibiting the decomposition of residual oxygenated functionalities. Benefiting from the grafted polystyrene chains, the rGO–polymer nanosheets exhibit good compatibility with polystyrene matrices and can form homogeneous polymer composites. The composite containing 2 wt% rGO exhibits a 161% increase in room temperature thermal conductivity, as compared to that of the pure polystyrene film. The thermal property is also superior to that of the single-walled CNT–polystyrene composites.

Besides styrene, methyl methacrylate and butyl acrylate polymers have also been covalently bonded to GO nanosheets by a surface-initiated ATRP technique (Figure 8.2).[62] Unlike the case of immobilizing the carbonyl bromide initiator through the pre-fixed diazonium compounds on rGO nanosheets, the halide initiator (isobutyryl bromide) is bonded to the GO laminates directly *via* an esterification reaction with the surface and periphery hydroxyl groups. With GO as a versatile substrate, covalent functionalization with polymer materials becomes easier and chemically more versatile as

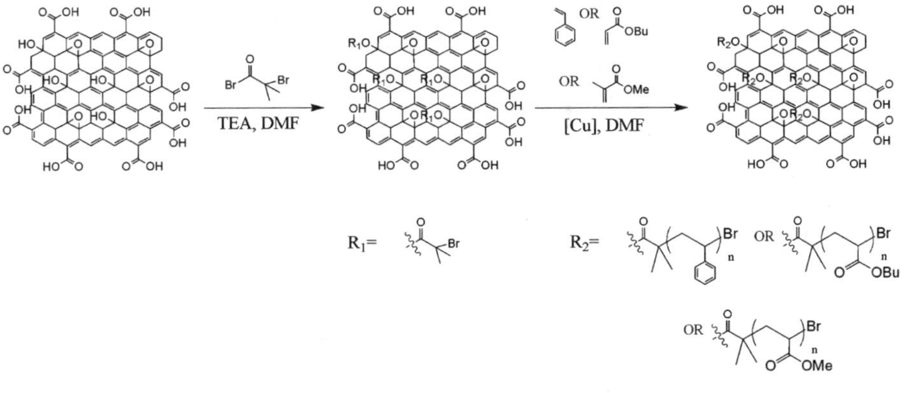

Figure 8.2 Schematic synthetic routes for controlled grafting of polystyrene, poly(butyl acrylate) and poly(methyl methacrylate) using ATRP technique onto single-layer graphene nanosheets. Reproduced with permission from reference.[62]

compared to the reduced form of GO nanosheets. Covalent bonding of the bromide-containing initiators is suggested by the appearance of C–Br bonds and reduction in concentration of the C–O bonds of the hydroxyl species in the X-ray photoelectron spectroscopy (XPS) C 1s core-level spectra of the initiator-coupled GO. Elemental analysis results indicate that 1.2 wt% of the initiator has been immobilized on the GO nanosheets, corresponding to 33 carbon atoms per initiator. Partial reduction of the oxygenated functionalities leads to two orders of magnitude improvement in electrical conductivity of the functionalized GO nanosheets. A series of monomers, including styrene, methyl methacrylate and butyl acrylate, can be polymerized from the active sites on the basal plane and at the edge of the GO nanosheets. The resultant GO–polymer nanosheets form stable solutions in organic solvents, significantly extending the processability of graphene-based materials.

Organo- and water-dispersible GO–poly(*tert*-butyl acrylate) composites have been prepared *via* surface-initiated ATRP of *tert*-butyl acrylate brushes from immobilized benzyl chloride initiator on the basal plane and laminate edge of the nanosheets (Figure 8.3).[63] With the presence of abundant hydroxyl groups on the surface and edge of the atomic carbon sheets, the ATRP initiator trichloro(4-chloromethylphenyl)silane can be readily coupled to the GO nanosheets through a simple hydrolysis reaction. The presence of the grafted polymer layer was revealed by Fourier-transform infrared (FT-IR) spectrum of the polymer-grafted GO nanosheets. The intense absorption at 1396/1365 and 1792 cm^{-1} are associated with the bending vibrations of the CH$_3$ groups and the C=O (ester) groups of the polymer brushes, respectively. These absorption bands disappears completely upon hydrolysis of the *tert*-butyl groups to produce the poly(acrylic acid) brushes. Thermal gravimetric analysis (TGA) results suggest that around 12.6 wt% of the polymer chains have been grafted onto the GO nanosheets, and the thickness of polymer coating layer is less than the XPS probing depth (~ 8 nm in organic matrices[89]). Due to improved compatibility between the hydrophobic *tert*-butyl moieties and non-polar solvents, the dispersibility of GO nanosheets in organic solvents has been substantially enhanced. Consequently, the GO–polymer nanosheets can be readily dispersed in toluene and incorporated into the poly(3-hexyl thiophene) matrices. The composite thin film of poly(3-hexyl thiophene) with GO–poly(*tert*-butyl acrylate) nanosheet exhibits bistable electrical conductivity switching and non-volatile memory effects, with an ON/OFF current ratio in excess of 10^3 and stability of up to 100 million read cycles. The memory-switching behaviour is ascribed to electric-field-induced charge transfer interaction between the electron-donating poly(3-hexyl thiophene) and electron-withdrawing GO–poly(*tert*-butyl acrylate) nanosheets. Water-dispersible GO–poly(acrylic acid) nanosheets can be obtained by hydrolysis of the poly(*tert*-butyl acrylate) brushes in the presence of trifluoroacetic acid.[63] Decoration of the GO–poly(acrylic acid) nanosheets with gold nanoparticles has been demonstrated, illustrating the potential of GO nanosheets as supports for immobilizing noble metal/metal oxide catalysts.

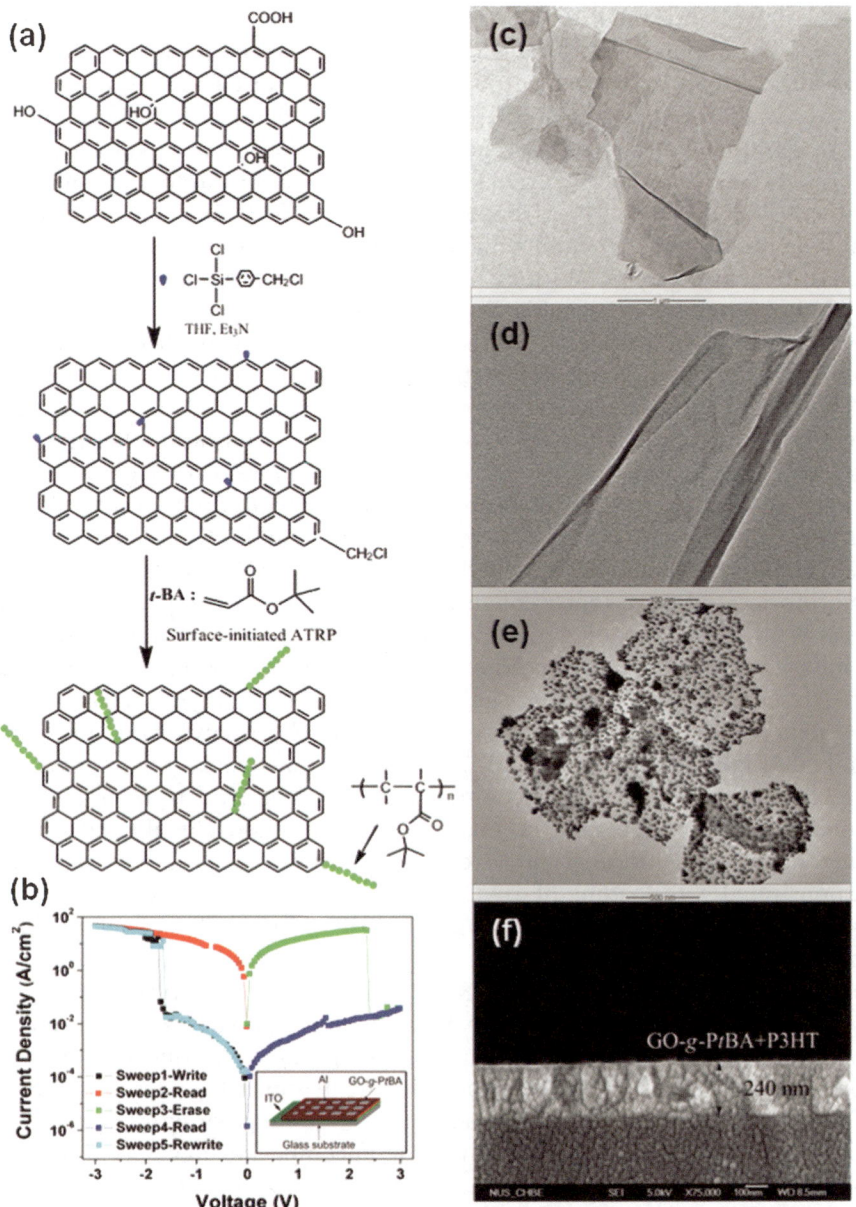

Figure 8.3 (a) Functionalization of graphene oxide (GO) with poly(*tert*-butyl acrylate) (P*t*BA) brushes *via* surfaces-initiated ATRP; (b) current density–voltage (J-V) characteristics of the Al/GO–*g*-P*t*BA+P3HT/ITO device; (c)–(e) transmission electron microscopy images of pristine GO, GO–*g*-P*t*BA and GO–*g*-PAAc decorated with gold nanoparticles; (f) field-emission scanning electron microscopy image (cross-section) of GO–*g*-P*t*BA+P3HT composite film with 5 wt% GO–*g*-P*t*BA. Reproduced with permission from reference 63.

GO–poly(methyl methacrylate) (GPMMA) nanosheets can also be prepared by surfaced-initiated ATRP and used as reinforcement fillers.[64] In order to enrich the density of hydroxyl groups on the surface of GO nanosheets, as well as to characterize the grafted poly(methyl methacrylate) from surface-initiated polymerization, an additional esterification reaction between the carboxylic groups of GO and ethylene glycol has been carried out prior to immobilization of the ATRP initiator. The carboxylic acid groups were first converted to aryl chloride, followed by reaction with ethylene glycol to yield the GO nanosheets rich in hydroxyl groups. With ester linkages between GO nanosheets and the grafted poly(methyl methacrylate) chains, post-hydrolysis can occur to cleave the grafted polymer chains from the GO nanosheets for further characterization. The grafted poly(methyl methacrylate) chains have a weight average molecular weight of 1280 with a relatively low polydispersity of 1.09, as measured by gel permission chromatography (GPC), indicating that the living radical polymerization is well controlled. The low molecular weight of grafted chains also suggests that surface modification with thionyl chloride and ethylene glycol seems to have introduced abundant ATRP initiator sites to compete for the available monomers. The GPMMA nanosheets are soluble in chloroform, and have much enhanced dispersibility as compared to pristine GO. With 1 wt% loading of GPMMA in the poly(methyl methacrylate) composite, homogeneous dispersion of the graphene nanosheets was observed. The presence of low molecular weight poly(methyl methacrylate) chains on the GO nanosheet surface increases its compatibility with poly(methyl methacrylate) through strong interfacial interactions and helps to improve the mechanical properties of the composite by efficient load transfer from the modified GO nanosheet to the matrices. A much more ductile and tougher material with significant improvement in elongation-at-break has been demonstrated with loading of 1 wt% of the GPMMA nanofillers.

Reversible addition–fragmentation chain transfer (RAFT) polymerization is another useful technique to prepare dispersible GO–polymer nanosheets *via* the 'grafting from' strategy.[90–92] RAFT polymerization utilizes dithioester compounds as chain transfer agents to mediate the polymerization through a chain transfer process, and has been shown to be advantageous over the other controlled polymerization processes, including its suitability over a wide range of reaction conditions and applicability to a wide range of monomers.

GO–polystyrene core-shell structured nanocomposites have been prepared *via* surface RAFT-mediated miniemulsion polymerization (Figure 8.4).[69] GO nanosheets prepared by the modified Hummers method were first modified with a RAFT agent, dodecyl isobutyric trithiocarbonate, through an esterification reaction of the hydroxyl groups on the basal surface of the laminates. The modified GO was dispersed in the styrene monomer, in the presence of a surfactant (sodium dodecylbenzene sulfonate) and a hydrophobe (hexadecane), for the synthesis of GO–polystyrene nanocomposites of core-shell morphology. As visualized by the transmission electron microscopy (TEM) images, the GO–polystyrene nanocomposites have a fairly narrow

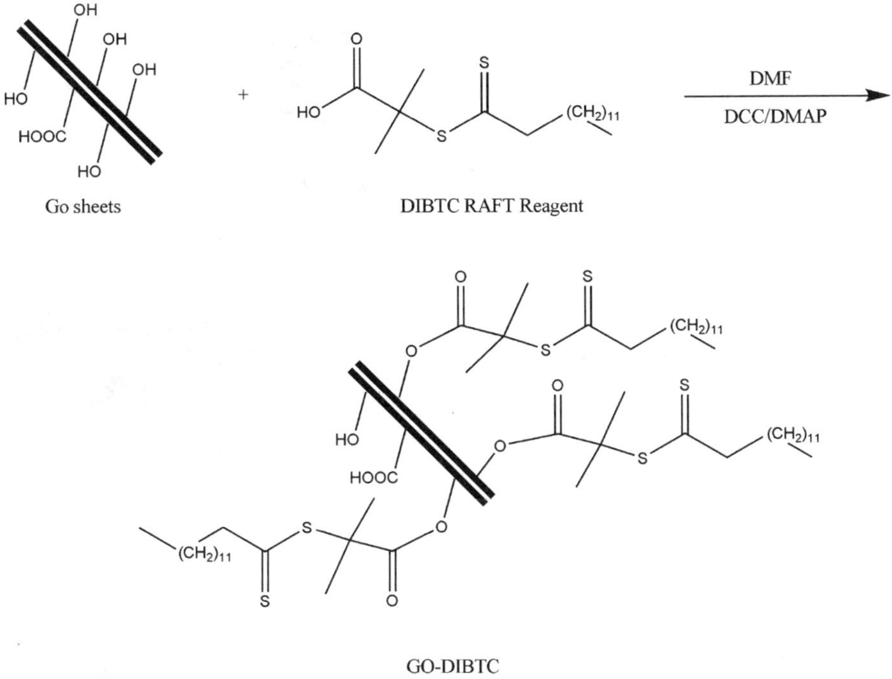

Go sheets   DIBTC RAFT Reagent

GO-DIBTC

Figure 8.4   Schematic synthetic routes for the preparation of RAFT-immobilized GO nanosheets. DCC, 1,3-dicyclohexyl carbodiimide; DMAP, 4-dimethyla-minopyridine; DMF, *N,N*-dimethylformamide. Reproduced with permission from reference.[69]

particle size distribution, suggesting that little or no secondary particle nucleation occurs during the polymerization process. The particles have a thin polymer shell with a solid core of GO laminates. Most of the GO nanosheets are of exfoliated morphology, which is in agreement with the X-ray diffraction patterns in which no diffraction peaks of graphite are discernible. The nanocomposites exhibit enhanced thermal stability in comparison to neat polystyrene. The onset decomposition temperature increases significantly when GO is present in the nanocomposite. The improved thermal stability of the nanocomposites can be ascribed to the intercalation of polymer chains into the lamellae of GO and restricted spatial chain movement at elevated temperatures. The GO nanosheets behave as an insulator to screen the heat source from the polymer, as well as to hinder the diffusion of volatile decomposition products within the nanocomposites by promoting char formation. The mechanical properties, especially the storage and loss modulus of the nanocomposites, also increase with increasing GO content in the samples. The enhancement can be attributed to the strong interfacial interaction between the polymer chains and the GO nanosheets of superior aspect ratio.

Poly(*N*-vinylcarbazole)-grafted GO nanosheets have also been prepared *via* surface-initiated RAFT polymerization (Figure 8.5).[70] Poly(*N*-vinylcarbazole)

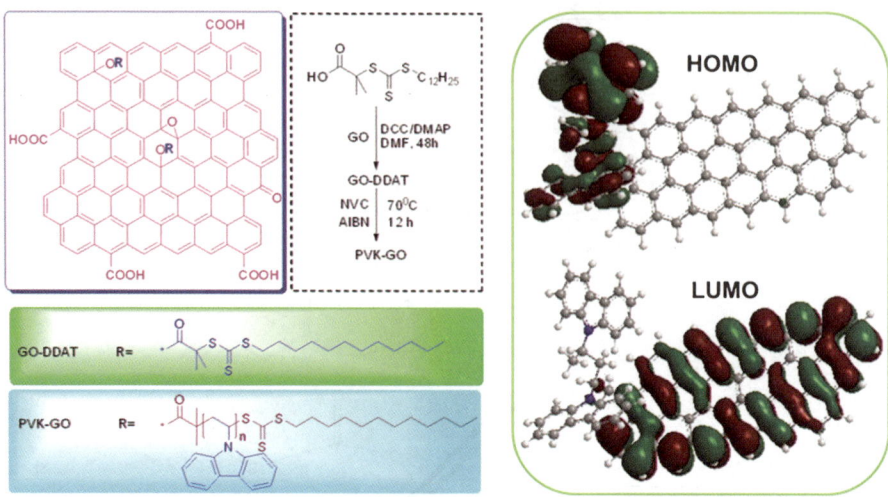

Figure 8.5 (Left) Schematic illustration for synthesis route of GO–PVK *via* surface-initiated RAFT polymerization. (Right) molecular orbitals of GO–PVK nanocomplex. Reproduced with permission from reference 70.

is a well-known hole-transporting/electron-donating material widely used for the fabrication of optoelectronic devices, such as electrophotographic photoreceptors, photoconductors, light-emitting diodes and organic solar cells,[93] while GO itself, with a large numbers of aromatic rings, can serve as the electron acceptor. Thus, it is expected that by covalent functionalization of GO with poly(*N*-vinylcarbazole) it should be possible to obtain not only solution-processable, but also electroactive, nanocomposites. *S*-1-dodecyl-*S*′-(α,α′-dimethyl-α″-acetic acid)trithiocarbonate (DDAT) modified GO was used as the RAFT agent for solution polymerization of *N*-vinylcarbazole monomer. Because the propagating *N*-vinylcarbazole radicals are relatively unstable, highly reactive, and with a tendency to undergo chain transfer and chain termination reactions, the synthesis of poly(*N*-vinylcarbazole) homopolymer with controlled molecular weight and narrow polydispersity is problematic. The poly(*N*-vinylcarbazole) chains grafted from GO have a relatively large polydispersity of 1.43, in comparison to those synthesized by use of free-DDAT as RAFT agent under the same condition with a polydispersity index of 1.16.

The electronic properties of GO–poly(*N*-vinylcarbazole) nanosheets have been studied by molecular simulation.[94] In contrast to the graphene nanosheet, which is a gapless material with ballistic charge transport characteristics, the π–π conjugation and charge carrier transport in GO–poly(*N*-vinylcarbazole) nanosheets are greatly disrupted by the oxygenated functionalities and the grafted chains. The nanocomplex exhibits an energy band gap of 2.15 eV, which is in good agreement with the experimental value obtained from cyclic voltammetry analysis. With the distinct asymmetric distribution of electron cloud over the molecular orbitals, electron transfer from the polymer chains to

GO laminates can give rise to charge transfer complex formation and bistable electrical conductivity switching behaviour.[70] Non-volatile rewritable memory performance with an ON/OFF current ratio of more than 10^3 has been demonstrated in the aluminium/GO–poly(*N*-vinylcarbazole)/indium tin oxide sandwich structures. Differing from the memory devices fabricated from physical blending of GO–poly(*tert*-butyl acrylate) nanosheets and poly(3-hexyl thiophene),[63] in which the electrical bistability arises from intermolecular charge transfer (CT) interaction between the poly(3-hexyl thiophene) donor and functionalized GO acceptor, covalent bonding of the polymer donor and GO acceptor prevents the problem of phase separation and promises greater stability in the electronic memory performance.[70,95]

Other methods, including polycondensation,[84] electrophilic polymerization[85,86] and ring-opening polymerization,[87] have also been investigated for the preparation of dispersible GO–polymer nanosheets. *In situ* polymerization of polyurethane (PU) on graphene nanosheets has resulted in enhanced mechanical and thermal properties.[84] The hydroxyl groups at the edge of chemically reduced GO nanosheets are utilized to form chemical bonds with the isocyanate groups of the 4,4′-diphenylmethane diisocyanate and to initiate the subsequent polycondensation of poly(tetramethylene glycol). Covalent functionalization leads to homogeneous dispersion of the rGO nanosheets in the PU matrices with large contact area and reduced interfacial electrical resistance. Due to strong interaction between graphene and the polymer matrices, increases in tensile strength and storage modules, both over 200%, have been observed in the rGO–PU nanocomposites with a low loading of 2 wt% graphene nanosheets.[84] Char formation of the integrated graphene nanosheets in rGO–PU nanocomposites delays the diffusion of volatile degradation products and leads to an increase of 40 °C in the onset decomposition temperature of PU.[84] Ring-opening polymerization has also been employed for the simultaneous reduction and surface functionalization of GO. Covalent bonding of poly(norepinephrine), or pNor, onto GO nanosheets, the occurrence of which is dependent on the pH of reaction medium, allows subsequent surface-initiated polymerization of ε-caprolactone from the pNor-modified GO nanosheets with spontaneous formation of metal nanoparticles.[87]

8.4 The 'Grafting to' Approach

The 'grafting to' approach allows direct covalent attachment of polymer chains to GO nanosheets. Generally, end-functionalized polymer materials react with the oxygenated functionalities located on the basal plane or at the laminate edges of the GO nanosheets to form tethered chains. With proper control of the grafting density, chain length and chemical composition of the polymer chains, 'polymer brushes' can be formed on the GO laminate. An alternative approach is to graft polymer chains to the GO nanosheets using the multiple functional groups in the side chains, similar to the cross-linking reactions. In this case, the polymer chains spread over several linkage sites on a single GO nanosheet, and the latter can be taken as part of the polymer chains. Benefiting from the versatile oxygen functionalities, various types of reactions, including esterification,[99–100] amidation,[72–74,101–103] nitrene cycloaddition,[104,105] ring-opening reaction,[106–109] radical grafting/coupling,[110–112] diazonium salts coupling,[113] atomic transfer nitroxide radical coupling reactions[114] and 'click' chemistry,[75,115–117] can be employed to prepare dispersible GO nanosheets *via* the 'grafting to' strategy. When immobilization of specific initiators for the 'grafting from' methods are not feasible, 'grafting to' approach *via* covalent linkages between the preformed polymers and GO nanosheets is an effective alternative to expand the types of polymers that can be bound to the laminate to form composite materials. However, due to the steric effects and activation barrier arising from the polymer chain attached earlier, the amount of polymer chains that can be tethered onto the GO nanosheets can become limited.

Following the procedures for modifying carbon nanotubes, esterification of the carboxylic groups of GO nanosheets with hydroxyl groups at the polymer side chains has been widely investigated. GO–poly(vinyl alcohol) composites have been prepared by direct esterification reaction, in the presence of a *N,N*-dicyclohexylcarbodiimide (DCC) and 4-(dimethylamino)-pyridine (DMAP) catalytic system (Figure 8.6).[71] The poly(vinyl alcohol)-modified GO so obtained is soluble in water and polar organic solvents, such as dimethyl sulfoxide, at elevated temperatures. Due to the presence of huge GO laminates and the covalent linkage between the polymer side chains and the carboxylic

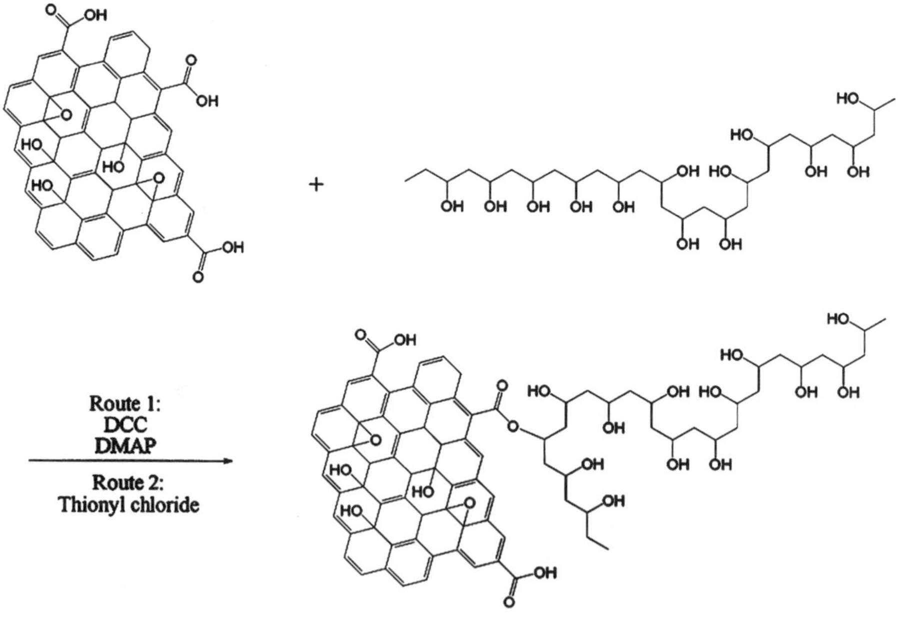

Figure 8.6   Schematic illustration for the esterification of GO with PVA. Reproduced with permission from reference 71.

groups at GO edges, segmental motions of the polymer chains are markedly restricted. As a result, the melting behaviour of neat poly(vinyl alcohol) has disappeared completely while the glass transition temperature of esterified polymer composite exhibits a dramatic shift of 35 °C. Hydrogen bonding between oxygenated moieties in the GO laminates and poly(vinyl alcohol) has also contributed to the changes in crystalline parameters of the composite material. Because of the water solubility of the covalently attached polymers, highly transparent films can be prepared *via* solution processing. Furthermore, the solubility also allows subsequent modification of the GO–poly(vinyl alcohol) composite in a homogeneous aqueous phase to give rise to rGO nanosheets with partially restored sp^2 hybridization in the composite material.

Conjugated polymers have also been grafted on GO nanosheets through esterification reaction at the chain ends (Figure 8.7).[98] Thionyl chloride was first attached to graphene oxide to prepare the aryl chloride functionalized GO. The hydroxyl-terminated regioregular poly(3-hexyl thiophene) can then be covalently grafted onto the modified GO nanosheets. The resultant poly(3-hexyl thiophene)–modified GO dissolves readily in chloroform and can be spin-cast into thin films. The GO–poly(3-hexyl thiophene) nanosheets exhibit a red shift in the absorption band associated with the π–π* transition of the polymer, arising from the graphene-induced enhancement in electron delocalization along the polymer chains. The red shift in adsorption is not observed in the physical mixture of poly(3-hexyl thiophene) and GO in the same solvent. The enhanced electron delocalization effect through CT from the electron-donating polymer to the electron-accepting GO nanosheets will effectively reduce the energy band gap, and is advantageous for photovoltaic applications. A photovoltaic device with GO–poly(3-hexyl thiophene)/C$_{60}$ as the active layer exhibits a 200% increase in the power conversion efficiency over its poly(3-hexyl thiophene)/C$_{60}$ counterpart.

An amidation reaction has been employed to functionalize GO nanosheets with branched and biocompatible poly(ethylene glycol) to render the complex soluble in aqueous media and stable in physiological solutions (Figure 8.8).[72] Prior to functionalization, the GO nanosheets are dispersible in water.

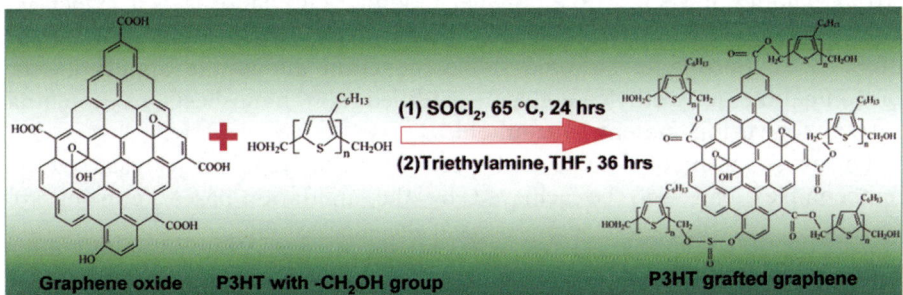

Figure 8.7 Synthetic procedure for chemical grafting of CH$_2$-OH terminated P3HT chains onto GO nanosheet. Reproduced with permission from reference 98.

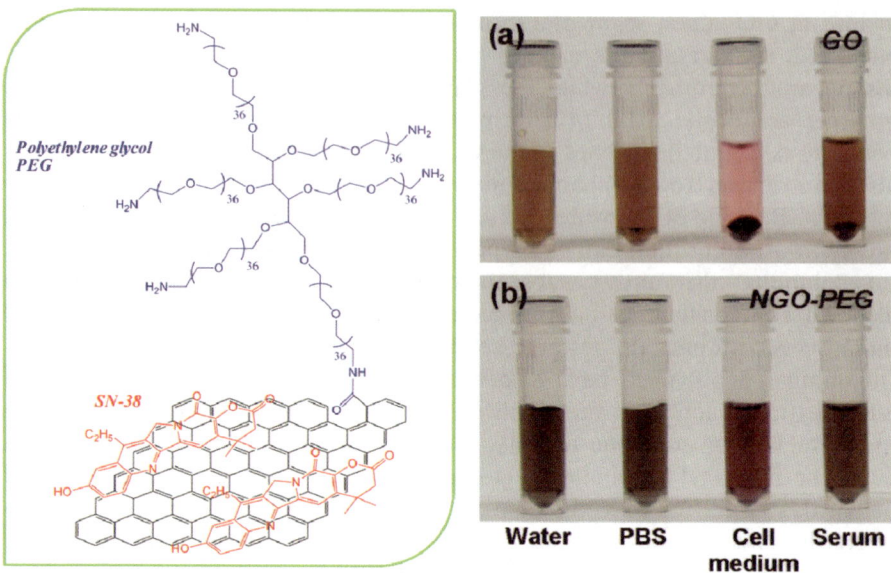

Figure 8.8   (Left) Schematic illustration of SN-38 loaded NGO–PEG. (Right) Photographs of GO (a) and NGO–PEG (b) in different solutions recorded after centrifugation at 10000 *g* for 5 min. Reproduced with permission from reference 72.

However, they aggregate readily when salts or proteins are present in the solution, due to the screening of the electrostatic charges and binding of proteins. Upon sonicating to dislodge the GO aggregates into small sizes (<50 nm), poly(ethylene glycol) modified by a six-armed amine can be bound to the carboxylic groups of GO *via* a carbodiimide-promoted amidation reaction. The resultant poly(ethylene glycol)-grafted GO nanosheets are stable in biological solutions, including serum, without aggregation. A camptothecin analogue of water-insoluble aromatic SN38 is loaded with the GO–poly(ethylene glycol) nanosheets *via* non-covalent van der Waals interaction to form water-soluble complexes at concentrations up to 1 mg mL^{-1}. The strong binding of SN38 to the polymer-modified GO nanosheets is evidenced by the dramatic fluorescence quenching of the drug. The slow but finite release of SN38 in phosphate buffered saline or serum allows the application of biocompatible poly(ethylene glycol)-modified GO nanosheets as potential carriers for various potent hydrophobic drugs.

The preparation of soluble and electroactive GO–poly(*N*-vinylcarbazole) nanocomplex can also be achieved by the amidation-based 'grafting to' method.[102] The carboxylic acid groups at the edges of GO nanosheets are first reacted with tolylene-2,5-diisocyanate (TDI) to produce the reactive GO–TDI. Then the carboxyl-terminated poly(*N*-vinylcarbazole), prepared *a priori* by RAFT polymerization of *N*-vinylcarbazole in the presence of 2-(dodecylthio-carbonothioylthio)-2-methylpropanoic acid (DDAT, a chain transfer agent), is coupled to GO–TDI to produce the GO–PVK nanosheets with significantly enhanced solubility of up to 10 mg mL^{-1} in non-polar organic solvents, such as toluene and tetrahydrofuran. The TGA results suggest the presence of about 4 wt% or 2.2 vol% (exceeding the percolation threshold of 0.1 vol%) of GO in the nanocomplex. GO with covalently grafted poly(azomethine) and con-jugated copolymer of fluorene and carbazole have also been prepared using the same strategy.[73,103]

Nitrene cycloaddition provides possibilities of modifying GO with various functionalities and promises graphene nanosheets with enhanced solubility and dispersion in polymer matrices. A series of azide-containing compounds can be utilized to introduce various functional groups onto graphene nanosheets (Figure 8.9).[104] The [2+1] cycloaddition of nitrenes, which are highly reactive intermediates generated upon either thermolysis or electromagnetic irradiation to the π-electron system, allows the preparation of functionalized graphene nanosheets with various end groups, such as $-OH$, $-COOH$, $-NH_2$ and $-Br$, as well as covalently anchored alkyl and polymer chains. The simultaneous occurrence of thermal reduction of GO nanosheets in the nitrene cycloaddition process gives rise to electrically conductive hybrid materials, contradicting the widely accepted idea that chemical functionalization always destroys the π-conjugation network of graphene to produce insulating species. The functional groups introduced onto the graphene nanosheets maintain their chemical reactivity, and can be used for further functionalization *via* ring-opening polymerization, amidation, surface-initiated ATRP and growth of metal

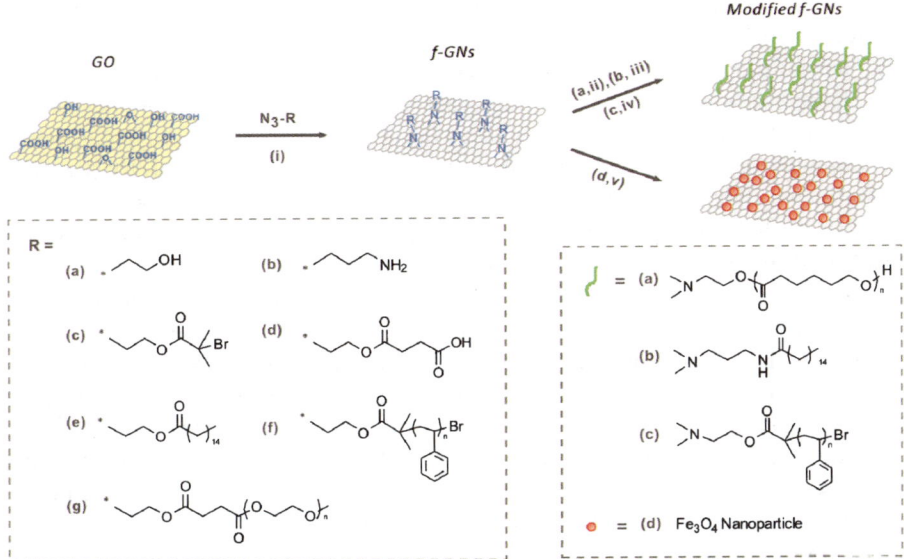

Figure 8.9 Schematic illustration of nitrene chemistry and further chemical modifications for the preparation of functionalized graphene nanosheets. Reproduced with permission from reference 104.

nanoparticles. Arising from the introduction of functional moieties, the thickness of modified graphene nanosheets is in the range of 1–3 nm and greater than that of the pristine graphene or GO. Single-layer dispersion of the functionalized graphene nanosheets can be directly visualized by microscopy, promising facile production of nanosheets with good solubility (6–8 mg mL^{-1} in DMF) and processability in desired solvents.[104] Homogeneous and flexible films can be formed by using 10% of functionalized graphene nanosheets as nanofiller in PU matrices during conventional solution processing. Moreover, the spherical protuberances present on the surfaces of graphene confirm the covalent anchoring of the polymer chains.[104] The relative simple one-step 'grafting to' approach is an alternative to the 'grafting from' method which usually involves multi-step processes to prepare the dispersible graphene–polymer nanosheets. The resultant two-dimensional macromolecular brushes constructed from the graphene–polymer hybrid can be used as building blocks or coatings for applications in water-based lubrication and surface wetting.

Unlike GO, which has an abundance of chemically reactive oxygen-containing functionalities and is electrically insulating due to disruption of the π-conjugation network, the electrical conductivity of reduced graphene oxide (rGO) can be recovered to a large extent by restoring the sp^2 hybridization state of the carbon crystalline lattice.[118] Partially due to the fewer number of reactive functionalities present on the rGO nanosheets, in particular the carboxylic acid groups at the edges and epoxy groups on the basal planes, less work has been done on the covalent modification of rGO, in comparison to that on the functionalization of GO. The 1,3-dipolar cycloaddition reaction of azomethine ylides provides a versatile approach to the functionalization of virtually defect-free graphene, including rGO, liquid-phase exfoliated graphene and its multilayers. Electroactive rGO–polymer nanosheets, with conjugated copolymers containing fluorene and carbazole groups in the polymer backbone and triphenylamine moieties in the pendant chains (poly[{4,4′-(9*H*-fluorene-9,9-diyl)bis(*N*,*N*-diphenylbenzenamine)}{4-(9*H*-carbazol-9-yl)benzaldehyde}(9,9-dihexyl-9*H*-fluorene)], or PFCF, and poly{[4,4′-(4-(9-phenyl-9*H*-fluoren-9-yl)phenylazanediyl)dibenzaldehyde]-[4,4′-(9*H*-fluorene-9,9-diyl)bis(*N*,*N*-diphenylbenzenamine)]-(9,9-dihexyl-9*H*-fluorene)}, or PFTPA), have been prepared (Figure 8.10).[119–121] *N*-Methylglycine is used as the agent to attach conjugated polymer chains onto the surface of rGO nanosheets. PFCF–rGO and PFPTPA–rGO are soluble in many common organic solvents, and exhibit reversible redox behaviour as characterized by cyclic voltammetry. Due to the regained π-conjugation network and improved electronic properties, electronic memory devices fabricated from PFCF–rGO and PFPTPA–rGO nanosheets exhibit better device performance with reduced switch-on and off voltages and increased ON/OFF current ratio, as compared to devices constructed from GO–polymer nanosheets.

In addition to the nitrene cycloaddition of azides and 1,3-dipolar cycloaddition of azomethine ylides, click chemistry provides approaches for linking polymers and GO nanosheets by Cu(I)-catalysed reactions between the

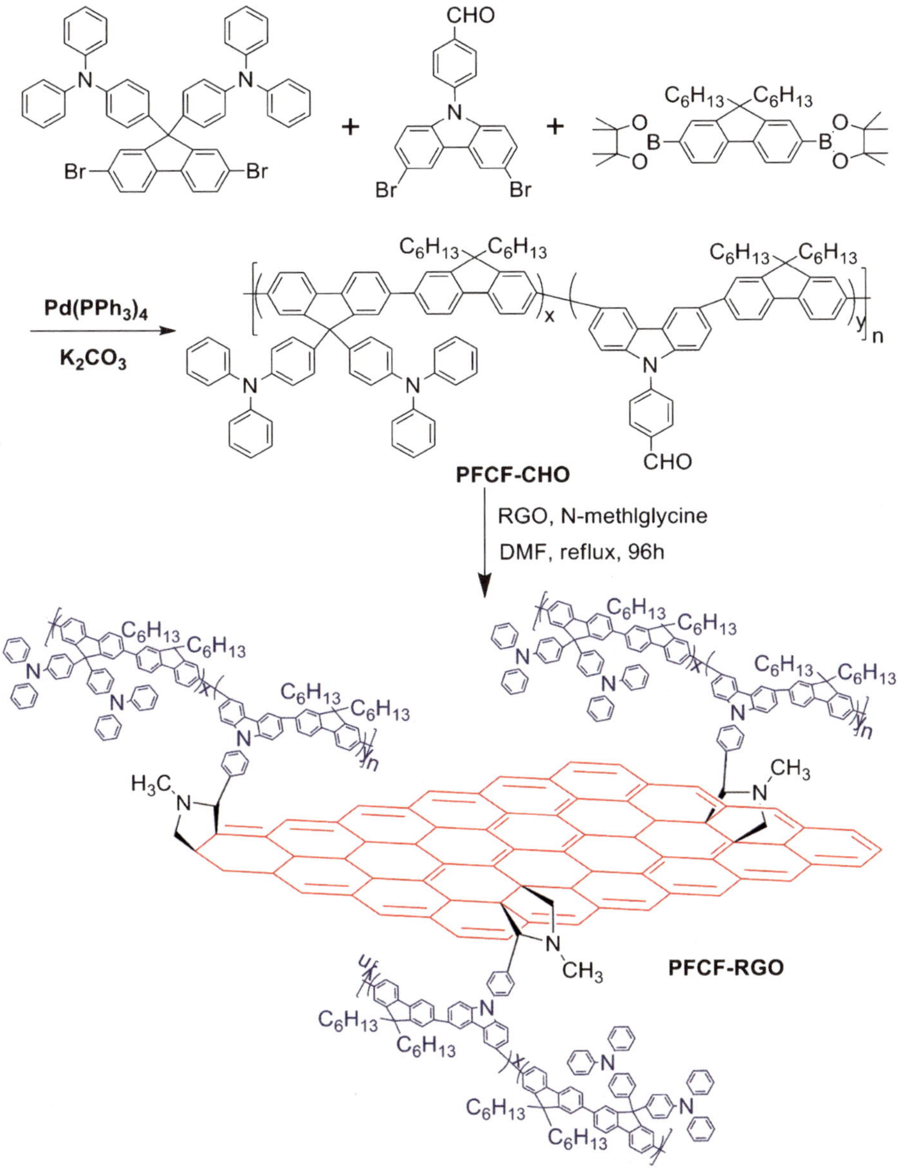

Figure 8.10 Synthetic procedure for PFCF–rGO. Reproduced with permission from reference 120.

azide end-group functionalized polymers and alkyne moieties–modified GO nanosheets.[122,123] Well-defined polymeric components can be grafted onto GO nanosheets with high specificity and nearly quantitative yield. Water-soluble and biocompatible GO–poly(*N*-isopropylacrylamide) nanosheets (GS-PNIPAM) have been prepared *via* click chemistry (Figure 8.11).[75] The alkyne

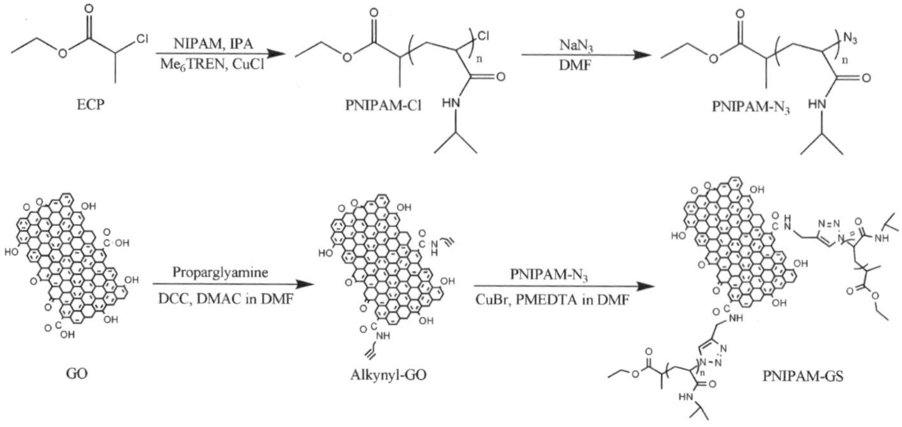

Figure 8.11 Synthesis route of PNIPAM-GS by click chemistry. Reproduced with permission from reference 75.

groups are first introduced onto the GO surface by the amidation reaction. The azide-terminated thermoresponsive PNIPAM homopolymer with a narrow distribution of molecular weight, prepared by ATRP, is then clicked onto the alkyne-functionalized GO nanosheets. GS-PNIPAM, containing about 50% polymer, can be readily dispersed in aqueous medium and various polar organic solvents, with a good solubility of up to 0.02 wt% and stability of more than a week under physiological conditions. The non-cytotoxic property of GS-PNIPAM even at high concentration makes it an ideal carrier in drug delivery. Through π–π stacking and hydrophobic interaction, an aromatic and hydrophobic anticancer drug (camptothecin, or CPT) can be loaded onto both sides of the GS-PNIPAM, to increase the solubility of the potent topoisomerase I inhibitor in an aqueous system.[75] The superior loading capacity of 15.6 wt%, ability to kill cancer cells, and stable and continuous release of drug in *in vitro* tests suggest GS-PNIPAM as an effective vehicle for anticancer drug delivery.

Besides the above-mentioned 'grafting to' methods, other approaches, including functionalization of GO nanosheets by PNIPAM *via* the atom transfer nitroxide radical coupling (ATNRC) chemistry,[114] amine-induced ring-opening reaction and subsequent cross-linking of epoxy resin,[106] grafting of linear *para*-poly(ether-ketone) (*p*PEK) *via* direct electrophilic substitution[124] and diazonium salts-aided coupling of donor–spacer–acceptor polymer triad[125] have also been reported recently.

8.5   Non-Covalent Functionalization of Graphene Oxide Nanosheets

Benefiting from the versatile chemistry of GO, a wide range of polymers can be covalently attached to GO nanosheets, resulting in highly dispersible and solution-processable GO–polymer composites. As an integral part of the polymers, the intrinsic physicochemical properties of the GO nanosheets can be effectively tuned upon covalent functionalization. However, the drawbacks of these methods are also obvious: covalent modification of the π-conjugation network disrupts the conjugated structure significantly. Thus, non-covalent functionalization, based on the π–π, electrostatic or hydrogen bonding interactions between the graphene nanosheets and stabilizers, has been considered as a non-destructive alternative having less negative impact on the chemical structure, yet still providing opportunities to fine-tune the electronic properties of graphene.[76–80,126–137] Accomplished by chemical reduction, polymer-stabilized dispersion of rGO nanosheets with restored aromaticity and electrical conductivity can be readily prepared (Table 8.2).

In practice, GO produced by the oxidative treatment of graphite powder forms a stable colloidal dispersion in water. Due to the removal of the oxygen functionalities, rGO will lose its hydrophilicity, and will agglomerate and precipitate from the aqueous dispersion irreversibly. However, when chemical reduction by hydrazine hydrate is conducted in the presence of an anionic polymer, poly(sodium 4-styrenesulfonate) (PSS), a stable black dispersion can be obtained.[76] The darker colour of the resulting rGO–PSS dispersion, as compared to that of the pristine GO, suggests the restoration of the aromatic graphitic network and electronic conjugation upon deoxygenation. Due to strong self-association between the rGO platelets, a high concentration of PSS

Table 8.2 Methods for non-covalent functionalization of graphene oxide by polymers.

Functionalization strategy	Polymer	Reference
Stabilization by polymer and reduction by hydrazine hydrate	PSS	76
Stabilization by polymer and reduction by hydrazine	SPANI	77
Stabilization by polymer and reduction by hydrazine	ssDNA	78
Stabilization by polymer and reduction by sodium borohydride	Cellulose	80
Stabilization by polyelectrolyte and reduction by hydrazine hydrate	PFVSO$_3$	127
Reduction by hydrazine and dispesion by end-functioanlized polymers	PS-COOH/PMMA-OH	130
Reduction and decoration by protein	BSA	135
Reduction and stabilization by saccharides	Glucose/ᴅ-fructose/sucrose	136
Reduction and stabilization by polymer	PDA	137

BSA, bovine serum albumin; PDA, polydopamine; PFVSO$_3$, conjugated polyclectrolyte with a planer backbond and charged sulfonate and oligo(ethylene glycol) side chains; PMMA-OH, hydroxyl-terminated poly(methyl methacrylate); PS-COOH, carbolyxic acid terminated polystyrene; PSS, poly(sodium 4-styrenesulfonate); SPANI, sulfonated polyaniline; ssDNA, single-stranded DNA.

(~40% by weight as measured by elemental analysis) is required to compete for the hydrophobic interaction and prevent the agglomeration of the rGO nanosheets. Once PSS has been absorbed onto the platelet surface, further agglomeration of the rGO nanosheets can be prevented. XPS analysis indicates that the reduction of GO leads to partial removal of the oxygen functionalities. Theoretical calculation shows that the edge-to-face interactions between the aromatic rings of the PSS polymer and graphitic surface of the rGO nanosheets, or the π–π interactions of the overlapping orbitals, are responsible for the partial coverage and stabilizing of the rGO nanosheets by the polymer chains in aqueous solutions.

Electrochemical functionalization can also be achieved simultaneously with the dispersion of GO nanosheets by electroactive sulfonated polyaniline (SPANI).[77] By reducing the exfoliated GO in the presence of SPANI, π–π stacking of the rGO basal planes and SPANI backbones results in a homogeneous black dispersion of graphene nanosheets in an aqueous medium. The dispersion is stable against gravity-induced sedimentation for weeks and does not precipitate when being centrifuged at ultra-high speed (8000 rpm over 5 min). The electrostatic repulsion between the charged rGO–SPANI nanosheets also leads to dispersion stability in highly acidic media. In addition to XPS analysis, which reveals the presence of predominantly C–C or C–H species in the rGO nanosheets and the incorporation of C–S and C–N species of SPANI, electrical conductivity measurement also confirms the successful reduction of GO and restoration of the electron conjugation network. Composite films prepared by vacuum-aided processing exhibit a conductivity of 0.3 S cm^{-1} which is close to that reported for rGO. The electrical conductivity of the drop-cast film can be further enhanced by a factor of ~10, when treated with ammonium persulfate to regain the electrochemically active state of the SPANI component. Unlike polyaniline, the electroactivity of which is strongly dependent on the pH of the solution, the self-doping effect enables sulfonated polyaniline to maintain its electroactivity in neutral or basic media. The cyclic voltammograms of the rGO–SPANI-modified electrode exhibit reversible redox behaviour in both acidic and neutral solution with good reproducibility and higher current densities, as compared to those of the SPANI-modified electrodes. Therefore, rGO–SPANI composites are potentially useful for biological applications where the media are usually in a neutral state.

Inspired by the growing interest in nanoelectronics and bionanotechnology, adsorption of biomolecules onto the surface of CNTs has been well studied. Similar work has been performed for the aqueous stabilization of chemically reduced rGO nanosheets by single-stranded (ss) DNA (Figure 8.12).[78] Instead of π–π interaction of the overlapping orbitals, adsorption of ssDNA onto the graphitic surface is achieved through weak electrostatic interaction or hydrogen bonding between the purine and pyrimidine bases of the DNA molecules with the carboxylic and phenolic groups of the rGO nanosheets. Consequently, the biopolymer is oriented in such a manner that the charged

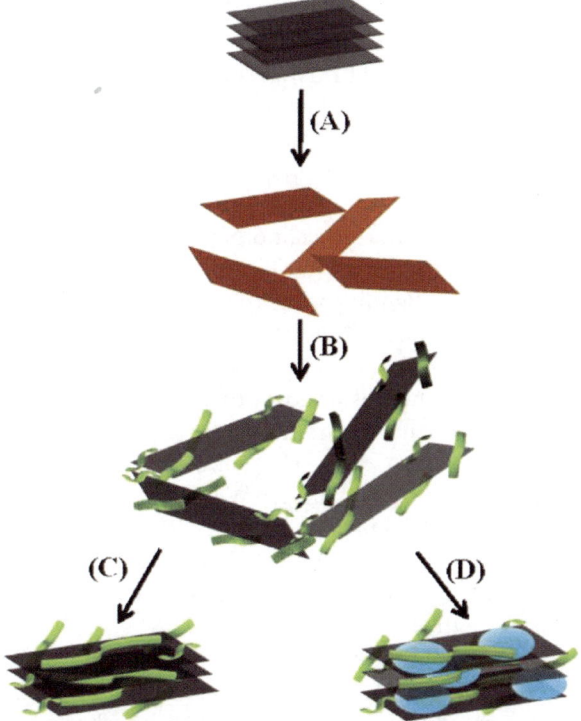

Figure 8.12 Schematic illustration for the preparation of DNA-stabilized graphene sheets and fabrication of lamellar multifunctional nanocomposites, *via* (A) oxidative exfoliation of graphite (grey-black) yielding graphite oxide (GO) nanosheets (brown); (B) chemical reduction of GO with hydrazine in the presence of ssDNA (green); (C) evaporation-induced deposition and self-assembly of lamellar nanocomposite films with intercalated *ss*DNA molecules; and (D) co-assembly of negatively charged ssDNA-G sheets and positively charged cytochrome *c*. Reproduced with permission from reference 78.

and polar moieties of the sugar-phosphate backbone are exposed to the solvent to facilitate dispersion of the wrapped rGO nanosheets in aqueous media. Restoration of electron conjugation in the chemically reduced rGO–ssDNA dispersion is consistent with changes in the lineshapes of the UV-visible absorption spectra, which exhibit the typical graphene absorption band at 266 nm, instead of the GO band at 231 nm, after hydrazine treatment. The dispersed rGO–ssDNA nanosheets can self-assemble into thin films of ordered lamellar nanostructures, *via* evaporation-induced or vacuum-aided processing, with a low angle of reflection at $2\theta = 5.8°$, corresponding to an expanded interlayer (d_{001}) spacing of 1.51 nm and an ordered lamellar nanostructure with d_{002} spacing of 0.71 nm ($2\theta = 12.3°$). The stacking arrangement of rGO nanosheets, with intercalated monolayer of adsorbed ssDNA molecules, provides a negatively charged platform for the fabrication of novel multifunctional bionanocomposite materials. Positively charged biomolecules, including redox metalloprotein and cytochrome *c*, can further facilitate the assembly of rGO–ssDNA nanohybrids *via* the charge neutralization effect. The entrapped biomolecules are located within the gallery regions of a stacked array of graphene nanosheets, being accessible to foreign molecules through diffusion from the external solution into the interlayer spaces of the DNA-stabilized graphene bionanocomposites and providing potential applications in biosensing, nanoelectronics and biotechnology.

Other environmentally friendly, biologically compatible and biodegradable natural polymers, such as lignin and cellulose, have also been used for the stabilization of rGO nanosheets in aqueous dispersions.[80] Due to the unique chemical structure and amphiphilic nature of sodium lignosulfonate (SLS), the alkyl chains of which can be adsorbed onto the graphene surface by strong hydrophobic interaction, the aromatic rings can also interact with the graphene surface by π–π stacking. The sulfonic groups provide sufficient electrostatic repulsion against the van der Waals attraction between graphene nanosheets, and excellent stabilizing effect of SLS towards rGO has been observed. A low SLS/rGO weight ratio of 3:1 is capable of stabilizing the rGO nanosheets in water at a concentration of 2 mg mL^{-1} for up to 4 months without conspicuous aggregation.

Water-soluble conjugated polyelectrolytes with highly efficient electron-delocalized backbones and ionic side chains are another family of polymeric stabilizers that are capable of dispersing rGO nanosheets *in situ* in polar solvents during hydrazine reduction (Figure 8.13).[127] The combined electronic properties of the conjugated backbone of a sulfonated polyfluorene derivative (PFVSO$_3$), which endow PFVSO$_3$ with a preferred coplanar geometry matching the flat geometry of rGO and enhancing the π–π interaction, together with the electrostatic behaviour of electrolytes and steric repulsion of ethylene glycol side chains, provide the stabilizing force to prevent the resultant PFVSO$_3$–rGO composites from agglomeration and precipitation. The PFVSO$_3$–rGO composites are soluble in a number of polar solvents and stable for at least 3 months. Amphiphilic rGO composites have also been prepared using an amphiphilic coil–rod–coil conjugated triblock copolymer of ethylene glycol and phenylene–ethynylene (PEG–OPE) as the stabilizer.[128] The conjugated rigid-rod polymer backbone can attach to the surface of rGO strongly *via* π–π interaction, while the lipophilic side chains and the end hydrophilic coils serve as the amphiphilic outlayer to facilitate the dispersion of the resultant PEG–OPE–rGO composites in both non-polar and polar solvents. Comprising 31 wt% of rGO, the solubility limit of the PEG–OPE–rGO composites in a series of solvents can reach as high as ~5 mg mL^{-1}. When using an ionic liquid polymer (PIL) as the stabilizer, poly(1-vinyl-3-ethylimidazolium) for instance, the dispersing behaviour of the rGO–polymer composites can be switched reversibly between water- and organo-soluble, by controlling the type of counter ion of the rGO–polymer composites.[129] The

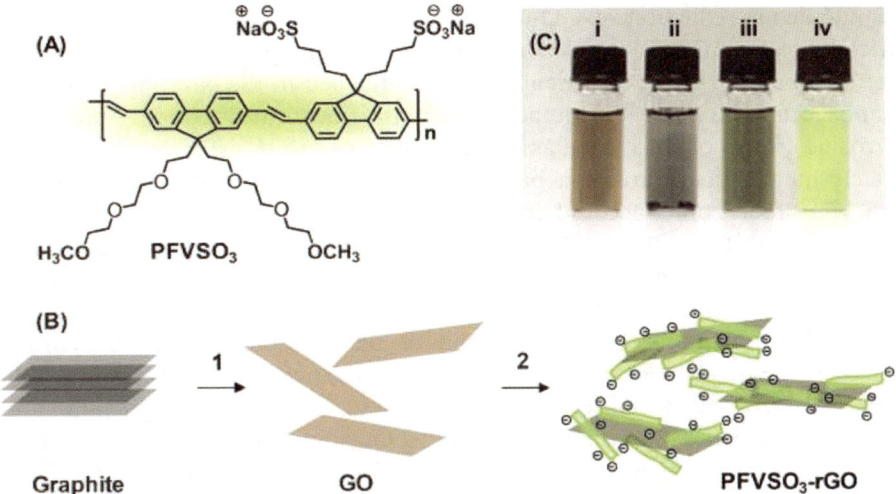

Figure 8.13   (A) Chemical structure of the conjugated polyelectrolyte PFVSO₃. (B) Schematic illustration for the preparation of PFVSO₃-stabilized rGO in H₂O *via* (1) oxidative exfoliation of graphite (grey-black) yielding single-layer GO sheets (brown) and (2) chemical reduction of GO with hydrazine in the presence of PFVSO₃. (C) Photograph of aqueous dispersions of (i) GO, (ii) rGO, (iii) PFVSO₃-rGO, and (iv) PFVSO₃. Reproduced with permission from reference.[127]

rGO–PIL dispersion, either in an aqueous or organic phase, is stable for over 6 months.

The dispersion of rGO nanosheets can also be achieved by polymer stabilizers in post-reduction functionalization. Because of the presence of unremoved oxygen functionalities, in particular the hydroxyl and carboxylic groups at the laminate edges, rGO nanosheets can be dispersed in water and polar solvents, including 1-methyl-2-pyrrilidone, 1,3-dimethyl-2-imidazolidinone, 1-propanol, ethanol, dimethylformamide, among others, at a concentration of 0.009 wt%.[130] When using the carboxylic groups as the modification sites for non-covalent (ionic) interaction with the protonated terminal amine groups of the end-functional polymer, polystyrene–NH₂ (PS–NH₂), the functionalized rGO nanosheets are converted from water-soluble to organosoluble. Owing to the chemical diversity of polymeric stabilizers, comparable dispersibility of the rGO nanosheets in a broad range of organic solvents can be anticipated upon post-reduction functionalization. Biocompatible chitosan bearing amine groups, water-soluble polymers containing pyrene and perylene groups, as well as conjugated polyelectrolytes, are capable of dispersing rGO nanosheets into aqueous solutions to form thermoresponsive, fluorescent and conductive rGO–polymer composites.[131]

Because of the hazardous nature of hydrazine-based reducing agents, the methods developed above to reduce GO may have practical limitations. On the other hand, biologically active polymeric dispersants are capable of providing simultaneous reduction of GO (Figure 8.14).[135] Through an environmentally friendly one-step reduction/decoration process, an amphiphilic protein, bovine serum albumin (BSA), has been successful attached to the basal planes of GO nanosheets which are simultaneously transformed to a reduced state. The tyrosine groups of BSA acted as the reducing agent to transfer electrons to GO at elevated temperature in a base medium while the hydrophobic and π–π stacking interactions account for the adsorption of the biopolymer and the stabilization of the BSA–rGO conjugates. In BSA–GO composites, hydrogen bonding between the oxygen functionalities of GO and nitrogen/oxygen-containing groups of BSA may also have participated in the polymer decoration process. Due to ionization of the phenolic groups of tyrosine residues, pH-dependent water solubility and enhanced stabilities against high ionic strength can be observed. Assembly and co-assembly of metallic and latex nanoparticles onto the BSA-decorated GO and rGO nanosheets are made possible *via* chemical bondings between the noble metal particles with the thiol, amine and imidazole groups on BSA, and strong interactions between the surface of pristine polystyrene latex with the hydrophobic patches on the protein molecules, respectively. This green chemistry route has provided a tool for the reduction and stabilization of GO aqueous dispersion, as well as a self-assembled platform of graphene-based nanohybrid materials for catalysis, battery electrodes, optoelectronics and sensing applications.

Saccharides, including glucose, fructose and sucrose, have also been used as reductants and capping reagents to prepare dispersible rGO nanosheets, due to their mild reductive ability and non-toxic property.[136] Upon oxidation to aldonic acid and further conversion to lactone in the presence of an ammonia

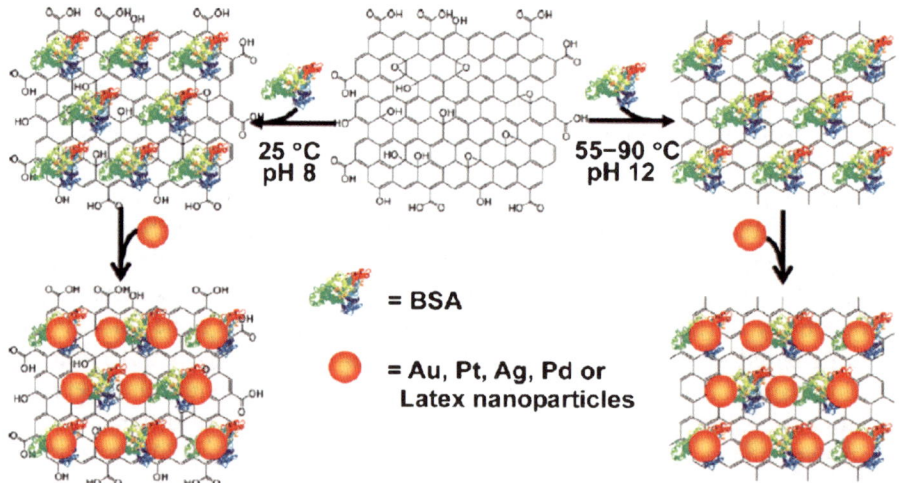

Figure 8.14   Schematic illustration of one-step reduction/decoration of GO nanosheets by amphiphilic protein bovine serum albumin for nanoparticle assembly. Reproduced with permission from reference 135.

solution, the oxidized products of glucose bearing a large number of hydroxyl and carboxyl groups interact with the rGO nanosheets strongly by hydrogen bonding between the oxygen functionalities on both components. The electrocatalytic response of the electrode modified by the resultant rGO nanosheets has been utilized for the detection of catecholamine compounds (dopamine, epinephrine [adrenaline] and norepinephrine [noradrenaline]). The oxygen functionalities on the oxidized products of saccharides may provide additional functional sites for further modification of the dispersible rGO nanosheets. With their reduction, self-polymerization and adhesion properties, dopamine has also been used to prepare polydopamine reduced graphene oxide (PDA–rGO) nanosheets.[137] The catechol groups of the PDA adlayer make further modification and functionalization of rGO with organic layers possible. Poly(ethylene glycol)-grafted rGO (rGO–*g*-PEG) nanosheets, synthesized *via* Michael addition of thiol-terminated PEG or Michael addition/Schiff base reaction of amino-terminated PEG, can be dispersed in both aqueous and organic solvents with a maxim solubility of 6.5 mg mL^{-1}.

8.6 Summary and Future Challenges

Combining a broad range of fascinating electronic, electrical, mechanical, thermal and optical features that have never been observed in a single material, graphene has been identified as one of the most important material in nanoscience and nanotechnology in the 21st century.[138] To fully realize these properties, the utilization of single-layer graphene nanosheets having good compatibility with either hydrophilic or hydrophobic media remains a challenge in graphene synthesis and modification. Owing to the ease of its large-scale and cost-effective preparation, GO has been considered as an excellent precursor to various graphene-based materials.[5,7] With their processability and functionalities, polymer materials have been widely used as stabilizers and molecular grafts in modifying the dispersibility, solution processability and intrinsic physicochemical properties of GO nanosheets. Ever-increasing efforts have been made to obtain dispersible nanosheets of GO–polymer composites in the past few years.

The strategies employed for the preparation of dispersible GO–polymer nanosheets, through either covalent or non-covalent functionalization, are similar to those well established for the CNT analogues, given the fact that both graphene and CNTs are constructed from sp^2-bonded carbon atoms. The oxygen functionalities, such as hydroxyl, epoxy, carbonyl and carboxylic groups, on the basal plane and edge of GO, provide versatile and effective sites for chemical modifications by various types of polymers. Regardless of whether the 'grafting from' or 'grafting to' method is used, the advances in carbon chemistry allow almost any kind of polymers to be tethered to GO nanosheets. With the 'grafting from' approach, polymers of high molecular weight and low polydispersity can be grown directly from GO nanosheets with immobilized initiators to form a thick brush coating. The 'grafting to' methodology allows the anchoring of a wide variety of functional polymer chains, including those that cannot be grown *via* surfaced-initiated polymerization, to the GO nanosheets. By using the covalent functionalization strategy, stable aqueous and organo-dispersion of GO–polymer nanosheets with adequately controlled nanostructures, enhanced compatibility with the polymer matrices, and finely tuned electronic properties useful for a wide range of applications can be obtained. However, the oxygen-containing functional groups and the covalently attached polymer chains may disrupt the graphitic π-conjugation network, with severe scattering of the ballistic charge transport. As a result the GO–polymer nanosheets prepared by covalent functionalization are usually insulating in nature. On the other hand, the use of non-covalent interactions, such as intermolecular π–π stacking interactions, hydrogen bonding or electrostatic forces, is a promising alternative to attach polymer chains to the extended aromatic surfaces of GO nanosheets. With minimal modification of the aromaticity of the graphite structure, this non-destructive approach is still capable of dispersing and stabilizing GO nanosheets in various aqueous and organic media. Though more difficult to control and quantify, the non-covalent functionalization is effective in stabilizing the dispersion of rGO nanosheets, which has been obtained by hydrazine reduction to restore the aromaticity and electrical conductivity of the supramolecular structure.

Graphene science has just begun in the past decade, neither is covalent nor non-covalent functionalization an ideal methodology to produce dispersible GO–polymer nanosheets. Based on the current understanding of structure–property relationships at the polymer–GO interface, further work, including molecular design, synthesis and simulation, is essential to develop new and improved functionalization strategies for the enhancement of both the dispersibility and physicochemical properties of the graphene nanosheets. After making their debut, dispersible nanosheets of GO–polymer composites are already an attractive candidate for diverse applications.

References

1. K. S. Novoselov, A. K. Geim, S. V. Morozov, D. Jiang, Y. Zhang, S. V. Dubonos, I. V. Grigorieva and A. A. Firsov, *Science*, 2004, **306**, 666.
2. J. Wu, W. Pisula and K. Müllen, *Chem. Rev.*, 2007, **107**, 718.
3. F. Chen and N. J. Tao, *Acc. Chem. Res.*, 2009, **42**, 429.
4. D. R. Dreyer, S. Park, C. W. Bielawski and R. S. Ruoff, *Chem. Soc. Rev.*, 2010, **39**, 228.
5. S. Park and R. S. Ruoff, *Nat. Nanotechnol.*, 2009, **4**, 217.
6. C. N. R. Rao, A. K. Sood, K. S. Subrahmanyam and A. Govindaraj, *Angew. Chem. Int. Ed.*, 2009, **48**, 7752.
7. A. K. Geim and K. S. Novoselov, *Nat. Mater.*, 2007, **6**, 183.
8. S. Stankovich, D. A. Dikin, G. H. B. Dommett, K. M. Kohlhaas, E. J. Zimney, E. A. Stach, R. D. Piner, S. T. Nguyen and R. S. Ruoff, *Nature*, 2006, **442**, 282.

9. C. Lee, X. Wei, J. W. Kysar and J. Hone, *Science*, 2008, **321**, 385.
10. D. A. Dikin, S. Stankovich, E. J. Zimney, R. D. Piner, G. H. B. Dommett, G. Evmenenko, S. T. Nguyen and R. S. Ruoff, *Nature*, 2007, **448**, 457.
11. R. R. Nair, P. Blake, A. N. Grigorenko, K. S. Novoselov, T. J. Booth, T. Stauber, N. M. R. Peres and A. K. Geim, *Science*, 2008, **320**, 1308.
12. A. A. Balandin, S. Ghosh, W. Bao, I. Calizo, D. Teweldebrhan, F. Miao and C. N. Lau, *Nano Lett.*, 2008, **8**, 902.
13. H. B. Heersche, P. Jarillo-Herrero, J. B. Oostinga, L. M. K. Vandersypen and A. F. Morpurgo, *Nature*, 2007, **446**, 56.
14. T. J. Echtermeyer, M. C. Lemme, M. Baus, B. N. Szafranek, A. K. Geim and H. Kurz, *IEEE Electr. Device Lett.*, 2008, **29**,952.
15. J. Y. Son, Y. H. Shin, H. J. Kim and H. M. Jang, *ACS Nano*, 2010, **4**, 2655.
16. F. Schedin, A. K. Geim, S. V. Morozov, E. W. Hill, P. Blake, M. I. Katsnelson and K. S. Novoselov, *Nat. Mater.*, 2007, **6**, 652.
17. J. Balapanuru, J. X. Yang, S. Xiao, Q. L. Bao, M. Jahan, L. Polavarapu, J. Wei, Q. H. Xu and K. P. Loh, *Angew. Chem. Int. Ed.*, 2010, **49**, 6549.
18. Z. F. Liu, Q. Liu, Y. Huang, Y. F. Ma, S. G. Yin, X. Y. Zhang, W. Sun and Y. S. Chen, *Adv. Mater.*, 2008, **20**, 3924.
19. S. P. Pang, H. N. Tsao, X. L. Feng and K. Müllen, *Adv. Mater.*, 2009, **21**, 3488.
20. M. D. Stoller, S. Park, Y. W. Zhu, J. H. An and R. S. Ruoff, *Nano Lett.*, 2008, **8**, 3498.
21. G. Srinivas, Y. W. Zhu, R. Piner, N. Skipper, M. Ellerby and R. S. Ruoff, *Carbon*, 2010, **48**, 630.
22. M. Terrones, O. Martín, M. González, J. Pozuelo, B. Serrano, J. C. Cabanelas, S. M. Vega-Díaz and J. Baselga, *Adv. Mater.*, 2011, **23**, 5302.
23. L. S. Walker, V. R. Marotto, M. A. Rafiee, N. Koratkar and E. L. Corral, *ACS Nano*, 2011, **5**, 3182.
24. C. M. Chen, Q. H. Yang, Y. G. Yang, W. Lv, Y. F. Wen, P. X. Hou, M. Z. Wang and H. M. Cheng, *Adv. Mater.*, 2009, **21**, 3007.
25. L. M. Zhang, J. G. Xia, Q. H. Zhao, L. W. Liu and Z. J. Zhang, *Small*, 2010, **6**, 537.
26. K. S. Novoselov, D. Jiang, F. Schedin, T. J. Booth, V. V. Khotkevich, S. V. Morozov and A. K. Geim, *Proc. Natl. Acad. Sci. U S A*, 2005, **102**, 10451.
27. X. L. Li, X. R. Wang, L. Zhang, S. W. Lee and H. J. Dai, *Science*, 2008, **319**, 1229.
28. X. Li, G. Zhang, X. Bai, X. Sun, X. Wang, E. Wang and H. Dai, *Nat. Nanotechnol.*, 2008, **3**, 538.
29. J. H. Lee, D. W. Shin, V. G. Makotchenko, A. S. Nazarov, V. E. Fedorov, Y. H. Kim, J. Y. Choi, J. M. Kim and J. B. Yoo, *Adv. Mater.*, 2009, **21**, 4383.
30. P. K. Ang, S. Wang, Q. Bao, J. T. L. Thong and K. P. Loh, *ACS Nano*, 2009, **3**, 3587.

31. P. Blake, P. D. Brimicombe, R. R. Nair, T. J. Booth, D. Jiang, F. Schedin, L. A. Ponomarenko, S. V. Morozov, H. F. Gleeson, E. W. Hill, A. K. Geim and K. S. Novoselov, *Nano Lett.*, 2008, **8**, 1704.

32. Y. Hernandez, V. Nicolosi, M. Lotya, F. M. Blighe, Z. Sun, S. De, I. T. McGovern, B. Holland, M. Byrne and Y. K. Gun'Ko, *Nat. Nanotechnol.*, 2008, **3**, 563.

33. W. T. Gu, W. Zhang, X. M. Li, H. W. Zhu, J. Q. Wei, Z. Li, Q. K. Shu, C. Wang, K. L. Wang, W. C. Shen, F. Y. Kang and D. H. Wu, *J. Mater. Chem.*, 2009, **19**, 3367.

34. S. Biswas and L. T. Drzal, *Nano Lett.*, 2009, **9**, 167.

35. L. Gomez De Arco, Y. Zhang, C. W. Schlenker, K. Ryu, M. E. Thompson and C. Zhou, *ACS Nano*, 2010, **4**, 2865.

36. M. Kalbac, A. Reina-Cecco, H. Farhat, J. Kong, L. Kavan and M. S. Dresselhaus, *ACS Nano*, 2010, **4**, 6055.

37. A. Michon, S. Vezian, A. Ouerghi, M. Zielinski, T. Chassagne and M. Portail, *Appl. Phys. Lett.*, 2010, 97, Art. No. 171909..

38. S. J. Chae, F. Gunes, K. K. Kim, E. S. Kim, G. H. Han, S. M. Kim, H. J. Shin, S. M. Yoon, J. Y. Choi, M. H. Park, C. W. Yang, D. Pribat and Y. H. Lee, *Adv. Mater.*, 2009, **21**, 2328.

39. V. Lopez, R. S. Sundaram, C. Gomez-Navarro, D. Olea, M. Burghard, J. Gomez-Herrero, F. Zamora and K. Kern, *Adv. Mater.*, 2009, **21**, 4683.

40. X. S. Li, W. W. Cai, J. H. An, S. Kim, J. Nah, D. X. Yang, R. Piner, A. Velamakanni, I. Jung, E. Tutuc, S. K. Banerjee, L. Colombo and R. S. Ruoff, *Science*, 2009, **324**, 1312.

41. K. S. Kim, Y. Zhao, H. Jang, S. Y. Lee, J. M. Kim, K. S. Kim, J. H. Ahn, P. Kim, J. Y. Choi and B. H. Hong, *Nature*, 2009, **457**, 706.

42. P. W. Sutter, J. I. Flege and E. A. Sutter, *Nat.Mater.*, 2008, **7**, 406 .

43. Y. S. Guo, X. P. Jia and S. S. Zhang, *Chem. Commun.*, 2011, **47**, 725.

44. M. Acik, C. Mattevi, C. Gong, G. Lee, K. Cho, M. Chhowalla and Y. J. Chabal, *ACS Nano*, 2010, **4**, 5861.

45. Y. Gao, H. L. Yip, S. K. Hau, K. M. O'Malley, N. C. Cho, H. Z. Chen and A. K. Y. Jen, *Appl. Phys. Lett.*, 2010, **97**, 203306.

46. C. Berger, Z. Song, T. Li, X. Li, A. Y. Ogbazghi, R. Feng, Z. Dai, A. N. Marchenkov, E. H. Conrad, P. N. First and W. A. de Heer, *J. Phys.Chem. B*, 2004, **108**, 19912.

47. K. P. Loh, Q. L. Bao, G. Eda and M. Chhowalla, *Nat. Chem.*, 2010, **2**, 1015.

48. K. V. Emtsev, A. Bostwick, K. Horn, J. Jobst, G. L. Kellogg, L. Ley, J. L. McChesney, T. Ohta, S. A. Reshanov, J. Rohrl, E. Rotenberg, A. K. Schmid, D. Waldmann, H. B. Weber and T. Seyller, *Nat. Mater.*, 2009, **8**, 203.

49. L. J. Cote, J. Kim, Z. Zhang, C. Sun and J. X. Huang, *Soft Matter*, 2010, **6**, 6096.

50. J. Sloan, Z. Liu, K. Suenaga, N. R. Wilson, P. A. Pandey, L. M. Perkins, J. P. Rourke and I. J. Shannon, *Nano Lett.*, 2010, **10**, 4600.

51. B. Gulbakan, E. Yasun, M. I. Shukoor, Z. Zhu, M. X. You, X. H. Tan, H. Sanchez, D. H. Powell, H. J. Dai and W. H. Tan, *J. Am. Chem. Soc.*, 2010, **132**, 17408.

52. M. Mueller, C. Kubel and K. Müllen, *Chem. Eur. J.*, 1998, **4**, 2099.

53. W. S. Hummers and R. E. Offeman, *J. Am. Chem. Soc.*, 1958, **80**, 1339.

54. J. Yue, Z. H. Wang, K. R. Cromack, A. J. Epstein and A. G. MacDiarmid, *J. Am. Chem. Soc.*, 1991, **113**, 2665.

55. D. Li, M. B. Müller, S. Gilje, R. B. Karner and G. G. Wallace, *Nat. Nanotechnol.*, 2008, 3, 101.

56. X. Huang, Z. Y. Yin, S. X. Wu, X. Y. Qi, Q. Y. He, Q. C. Zhang, Q. Y. Yan, F. Boey and H. Zhang, *Small*, 2011, **14**, 1876.

57. V. Singh, D. Joung, L. Zhai, S. Das, S. I. Khondaker, S. Seal, *Prog. Mater. Sci.*, 2011, **56**, 1178.

58. H. Bai, C. Li and G. Q. Shi, *Adv. Mater.*, 2011, **23**, 1089.

59. H. J. Salavagione, G. Martínez and G. Ellis, *Macromol. Rapid Commun.*, 2011, 1771.

60. D. Y. Cai and M. Song, *J. Mater. Chem.*, 2010, **20**, 7906.

61. M. Fang, K. Wang, H. Lu, Y. Yang and S. Nutt, *J. Mater. Chem.*, 2010, **20**, 1982.

62. S. H. Lee, D. R. Dreyer, J. An, A. Velamakanni, R. D. Piner, S. Park, Y. Zhu, S. O. Ki, C. W. Bielawski and R. S. Ruoff, *Macromol. Rapid Commun.*, 2010, **31**, 281.

63. G. L. Li, G. Liu, M. Li, D. Wan, K. G. Neoh and E. T. Kang, *J. Phys. Chem. C*, 2010, **114**, 12742.

64. G. Gonçalves, P. A. A. P. Marques, A. Barros-Timmons, I. Bdkin, M. K. Singh, N. Emami and J. Grácio, *J. Mater. Chem.*, 2010, **20**, 9927.

65. S. H. Lee, H. W. Kim, J. O. Hwang, W. J. Lee, J. Kwon, C. W. Bielawski, R. S. Ruoff and S. O. Kim, *Angew. Chem. Int. Ed.*, 2010, **49**, 10084.

66. D. Wang, G. Ye, X. Wang and X. Wang, *Adv. Mater.*, 2011, **23**, 1122.

67. M. Fang, K. Wang, H. Lu, Y. Yang and S. Nutt, *J. Mater. Chem.*, 2009, **19**, 7098 .

68. Y. Yang, J. Wang, J. Zhang, J. Liu, X. Yang and H. Zhao, *Langmuir*, 2009, **25**, 11808.

69. H. M. Etmimi, M. P. Tonge and R. D. Sanderson, *J. Polym. Sci., Part A: Polym. Chem.*, 2011, **49**, 1621.

70. B. Zhang, Y. Chen, L. Q. Xu, L. J. Zeng, Y. He, E. T. Kang and J. J. Zhang, *J. Polym. Sci., Part A: Polym. Chem.*, 2011, **49**, 2043.

71. H. J. Salavagione, M. A. Gómez and G. Martínez, *Macromolecules*, 2009, **42**, 6331.

72. Z. Liu, J. T. Robinson, X. Sun and H. Dai, *J. Am. Chem. Soc.*, 2008, **130**, 10876.

73. D. Yu and L. Dai, *J. Phys. Chem. Lett.*, 2010, **1**, 467.

74. X. D. Zhuang, Y. Chen, G. Liu, P. P. Li, C. X. Zhu, E. T. Kang, K. G. Neoh, B. Zhang, J. H. Zhu and Y. X. Li, *Adv. Mater.*, 2010, **22**,1731.

75. Y. Pan, H. Bao, N. G. Sahoo, T. Wu and L. Li, *Adv. Funct. Mater.*, 2011, **21**, 2754.

76. S. Stankovch, R. D. Piner, X. Q. Chen, N. Q. Wu, S. T. Nguyen and R. S. Ruoff, *J. Mater. Chem.*, 2006, **16**, 155.

77. H. Bai, Y. X. Xu, L. Zhao, C. Li and G. Q. Shi, *Chem. Commun.*, 2009, 1667.

78. A. J. Patil, J. L. Vickery, T. B. Scott and S. Mann, *Adv. Mater.*, 2009, **21**, 3159.

79. J. Q. Liu, L. Tao, W. R. Yang, D. Li, C. Boyer, R. Wuher, F. Braet and T. P. Davis, *Langmuir*, 2010, **26**, 10068.

80. Q. Yang, X. Pan, F. Hunag and K. C. Li, *J. Phys. Chem. C*, 2010, **114**, 3811.

81. S. Minko, in *Polymer Surfaces and Interfaces: Characterization, Modification and Applications*, ed. M. Stamm, Springer Verlag, Berlin, 2008, p 215.

82. Y. P. Sun, K. F. Fu, Y. Lin and W. J. Huang, *Acc. Chem. Res.*, 2002, **35**, 1096.

83. A. Hirsch and O. Vostrowsky, *Top. Curr. Chem.*, 2005, **245**, 193.

84. X. Wang, Y. Hu, L. Song, H. Yang, W. Xing and H. Lu, *J. Mater. Chem.*, 2011, **21**, 4222.

85. W. R. Collins, E. Schmois and T. M. Swager, *Chem. Commun.*, 2011, **47**, 8790.

86. J. Li, H. Lin, W. Zhao and G. Chen, *J. Appl. Polym. Sci.*, 2008, **109**, 1377.

87. S. M. Kang, S. Park, D. Kim, S. Y. Park, R. S. Ruoff and H. Lee, *Adv. Funct. Mater.*, 2011, **21**, 108.

88. V. Coessens, T. Pintauer and K. Matyjaszewsky, *Prog. Polym. Sci.*, 2001, **26**, 337.

89. K. L. Tan, L. L. Woon, H. K. Wong, E. T. Kang and K. G. Neoh, *Macromolecules*, 1993, **26**, 2832.

90. C. B. Kwollik, *Handbook of RAFT Polymerization*, Wiley-VCH, Weinheim, 2008.

91. A. B. Lowe and C. L. McCormick, *Prog. Polym. Sci.*, 2007, **32**, 283.

92. Y. K. Chong, T. P. T. Le, G. Moad, E. Rizzardo and S. H. Thang, *Macromolecules*, 1999, **32**, 2071.

93. J. V. Grazuleviciusa, P. Strohrieglb, J. Pielichowskic and K. Pielichowskic, *Prog. Polym. Sci.*, 2003, **28**, 1297.

94. M. J. Frisch, G. W. Trucks, H. B. Schlegel, G. E. Scuseria, M. A. Robb, J. R. Cheeseman, *et al.*, *Gaussian 09* (Revision A.02), Gaussian, Inc., Wallingford, CT, 2009.

95. Q. D. Ling, D. J. Liaw, C. X. Zhu, D. S. H. Chan, E. T. Kang and K. G. Neoh, *Prog. Polym. Sci.*, 2008, **33**, 917.

96. L. M. Veca, F. Lu, M. J. Meziani, L. Cao, P. Zhang, G. Qi, L. Qu. M. Shrestha and Y. P. Sun. *Chem. Commun.*, 2009, **45**, 2565.

97. H. J. Salavagione and G. Martínez, *Macromolecules*, 2011, **44**, 2685.

98. D. Yu, Y. Yang, M. Durstock, J. B. Baek and L. Dai, *ACS Nano*, 2010, **4**, 5633.
99. S. P. Zhang, P. Xiong, X. J. Yang and X. Wang, *Nanoscale*, 2011, **3**, 2169.
100. T. A. Pham, N. A. Kumar and Y. T. Jeong, *Synth. Met.*, 2010, **160**, 2028.
101. G. Liu, X. D. Zhuang, Y. Chen, B. Zhang, J. H. Zhu, C. X. Zhu, K. G. Neoh and E. T. Kang, *Appl. Phys. Lett.*, 2009, **95**, 253301.
102. B. Zhang, Y. Chen, X. D. Zhuang, G. Liu, B. Yu, E. T. Kang, J. H. Zhu and Y. X. Li, *J. Polym. Sci., Part A: Polym. Chem.*, 2010, **48**, 2642.
103. B. Zhang, Y. L. Liu, Y. Chen, K. G. Neoh, Y. X. Li, C. X. Zhu, E. S. Tok and E. T. Kang, *Chem. Eur. J.*, 2011, **17**, 10304.
104. H. He and C. Gao, *Chem. Mater.*, 2010, **22**, 5054.
105. X. Xu, Q. Luo, W. Lv, Y. Dong, Y. Lin, Q. Yang, A. Shen, D. Pang, J. Hu, J. Qin and Z. Li, *Macromol. Chem. Phys.*, 2011, **212**, 768.
106. M. Fang, Z. Zhang, J. Li, H. Zhang, H. Lu and Y. Yang, *J. Mater. Chem.*, 2010, **20**, 9635.
107. Y. Fu and W. H. Zhong, *Thermochin. Acta*, 2010, **516**, 58.
108. Y. Lin, J. Jin and M. Song, *J. Mater. Chem.*, 2011, **21**, 3455.
109. C. Shan, H. Yang, D. Han, Q. Zhang, A. Ivaska and L. Niu, *Langmuir*, 2009, **25**, 12030.
110. D. Vugula, J. M. Thomassin. I. Moenberg, I. Huynen, B. Gilbert, C. Jerome, M. Alexandre and C. Detremlbeur, *Chem. Commun.*, 2011, **47**, 2544.
111. L. Kan, Z. Xu and C. Gao, *Macromoleules*, 2011, **44**, 444.
112. J. Shen, Y. Hu, C. Li, C. Qin and M. Ye, *Small*, 2009, **5**, 82.
113. J. R. Lomeda, C. D. Doyle, M. V. Kosynkin, W. F. Hwang and J. M. Tour, *J. Am. Chem. Soc.*, 2008, **130**, 16201.
114. Y. Deng, Y. Li, J. Dai, M. Lang and X. Huang, *J. Polym. Sci., Part A: Polym. Chem.*, 2011, **49**, 1582.
115. S. T. Sun, Y. W. Cao, J. C. Feng and P. Y. Wu, *J. Mater. Chem.*, 2010, **20**, 5605.
116. Y. W. Cao, Z. L. Lai, J. C. Feng and P. Y. Wu, *J. Mater. Chem.*, 2011, **21**, 9271.
117. X. M. Yang, L. J. Ma, S. Wang, Y. W. Li, Y. F. Tu and X. L. Zhu, *Polymer*, 2011, **52**, 3046.
118. I. Jung, D. A. Dikin, R. D. Piner and R. S. Ruoff, *Nano Lett.*, 2008, **8**, 4283.
119. V. Georgakilas, A. B. Bourlinos, R. Zboril, T. A. Steriotis, P. Dallas, A. S. Stuboscd and C. Trapalis, *Chem. Commun.*, 2010, **46**, 1766.
120. B. Zhang, G. Liu, Y. Chen, L. J. Zeng, C. X. Zhu, K. G. Neoh. C. Wang and E. T. Kang, *Chem. Eur. J.*, 2011, **17**, 13646.
121. B. Zhang, Y. Chen, G. Liu, L. Q. Xu, J. N. Chen, C. X. Zhu, K. G. Neoh and E. T. Kang, *J. Polym. Sci. Part A*, 2012, **50**, 378.
122. V. V. Rostovtsev, L. G. Green, V. V. Fokin and K. B. Sharpless, *Angew. Chem. Int. Ed.*, 2002, **41**, 2596.

123. C. W. Tornoe, C. Christensen and M. Meldal, *J. Org. Chem.*, 2002, **67**, 3057.
124. E. K. Choi, I. Y. Jeon, S. J. Oh and J. B. Baek, *J. Mater. Chem.*, 2010, **20**, 10936.
125. A. Midya, V. Mamidala, J. X. Yang, P. K. L. Ang, Z. K. Chen and K. P. Loh, *Small*, 2010, **6**, 2292.
126. J. Q. Liu, W. R. Yang, L. Tao, D. Li, C. Boyer and T. P. Davis, *J. Polym. Sci., Part A: Polym. Chem.*, 2010, **48**, 425.
127. X. Y. Qi, K. Y. Pu, X. Z. Zhou, H. Li, B. Liu, F. Boey, W. Huang and H. Zhang, *Small*, 2010, **6**, 663.
128. X. Y. Qi, K. Y. Pu, H. Li, X. Z. Zhou, S. X. Wu, Q. L. Fan, B. Liu, F. Boey, W. Huang and H. Zhang, *Angew. Chem. Int. Ed.*, 2010, **49**, 9426.
129. T. Y. Kim, H. W. Lee, J. E. Kim and K. S. Suh, *ACS Nano*, 2010, **4**, 1612.
130. E. Y. Choi, T. H. Han, J. H. Hong, J. E. Kim, S. H. Lee, H. W. Kim and S. O. Kim, *J. Mater. Chem.*, 2010, **20**, 1907.
131. X. M. Yang, Y. F. Tu, L. Li, S. M. Shang and X. M. Tao, *ACS Appl. Mater. Interface*, 2010, **2**, 1707.
132. L. Q. Xu, L. Wang, B. Zhang, C. H. Lim, Y. Chen, K. G. Neoh, E. T. Kang and G. D. Fu, *Polymer*, 2011, **52**, 2376.
133. H. F. Yang, Q. X. Zhang, C. S. Shan, F. H. Li, D. X. Han and L. Niu, *Langmuir*, 2010, **26**, 6708.
134. S. Liu, J. Q. Tian, L. Wang, H. L. Li, Y. W. Zhang and X. P. Sun, *Macromolecules*, 2010, **43**, 10078.
135. J. B. Liu, S. H. Fu. B. Yuan, Y. L. Li and Z. X. Deng, *J. Am. Chem. Soc.*, 2010, **132**, 7279.
136. C. Z. Zhu, S. J. Guo, Y. X. Fang and S. J. Dong, *ACS Nano*, 2010, **4**, 2429.
137. L. Q. Xu, W. J. Yang, K. G. Neoh, E. T. Kang and G. D. Fu, *Macromolecules*, 2010, **43**, 8336.
138. M. J. Allen, V. C. Tung and R. B. Kaner, *Chem. Rev.*, 2010, **110**, 132.

CHAPTER 9

Graphene–Conducting Polymer Nanocomposites Prepared by Interfacial Polymerization

SERGIO H. DOMINGUES, RODRIGO V. SALVATIERRA AND ALDO J. G. ZARBIN*

Grupo de Química de Materiais, Departamento de Química, Universidade Federal do Paraná (UFPR), CP 19081, CEP 81531-990, Curitiba–PR, Brazil
*E-mail: aldozarbin@ufpr.br

9.1 Introduction

Nanocomposites of conducting polymers (*e.g.* polyaniline [PANI], polypyrrole and polythiophene) and carbon nanostructures (mainly carbon nanotubes (CNTs) and recently graphene) are important advanced materials with novel and improved properties compared to the isolated components.[1] Similar to other polymeric nanocomposites, the chemical and physical properties of the conducting polymer–carbon nanostructures are strongly dependent both on the dispersibility of the carbonaceous nanomaterial throughout the polymer and on the nature of interactions between the components, which are modulated by the chemical or physical routes employed to prepare the nanocomposites.[2] The carbon nanomaterial must be uniformly dispersed and coated with the polymer, which minimizes the presence of stress concentration centres. The type of interaction governs the transfer of properties between the components (*i.e.* a good interaction leads to a good transfer of properties, which consequently results in a good nanocomposite). However, in contrast to

RSC Nanoscience & Nanotechnology No. 26
Polymer–Graphene Nanocomposites
Edited by Vikas Mittal

other polymeric nanocomposites, the nature of the polymer also affects the properties of the material. Each conducting polymer can adopt different chemical structures as the oxidation state and the overall charge of the polymeric chain are varied, which leads to the presence of counter-ions (dopants) to balance the charge if the net charge of the polymeric chain is non-zero. In addition to the important experimental variables that can be controlled during the preparation of polymeric nanocomposites as a whole, the synthesis of conducting polymer-based nanocomposites requires full control of the parameters that govern the numerous combinations of structures and dopants that each polymer can adopt.

In this study, a novel concept for the preparation of graphene–conducting polymer nanocomposites based on interfacial polymerization is presented. The main advantages of this technique are that the majority of the experimental variables can be finely controlled and that the final product is directly obtained as a transparent and self-assembled film that can be easily transferred to ordinary substrates.

In the following sections, a short introduction to conducting polymers is presented, with an emphasis on PANI (one of the most highly studied and widely used conducting polymers), followed by a discussion regarding common synthetic routes and potential applications of graphene–PANI nanocomposites. After factors regarding the process of interfacial polymerization are considered, the synthesis of graphene–PANI nanocomposites is discussed on the basis of this approach.

9.2 Conducting Polymers

Intrinsically conductive polymers (CPs) have been intensively studied in recent years due to potential use in several applications such as electronic, electrochromic and photoelectrochemical devices;[3–4] rechargeable batteries;[5] smart windows;[6] sensors;[7] light-emitting diodes;[8] and non-linear optical devices.[9] Generally, CP structures have a long π-conjugated system of alternating single and double bonds along the backbone of the chain. Neutral CPs become conductive after charge injection or charge withdrawal (*i.e.* a partial reduction or oxidation of the polymer chain, respectively), which is a process known as doping (analogous to the doping process in inorganic semiconductors). The chemical species responsible for the doping is the dopant, which remains incorporated in the polymer chain as a counter-ion to neutralize the charge. The properties of CPs depend strongly on the microstructure and morphology, which are determined by the synthetic method, counter-ions (dopants), and several other variables.[10] This structure–property relationship is not trivial to understand. In addition, the poly-dispersity and complex structure of CPs complicates the unequivocal interpretation of some experimental data.[11–12] Among CPs, PANI has one of the most fascinating and complex structure–property relationships, as discussed in the following section.

9.2.1 Polyaniline (PANI)

Among the CPs, PANI has a unique collection of properties such as high stability under ambient conditions, high conductivity, facile synthesis, a low-cost and highly available monomer, and a high electrochromic effect. The electrochromism of PANI is due to different redox states: the most reduced form is yellow (leucoemeraldine base), the half-oxidized form is blue (emeraldine base), and the most oxidized form is purple (pernigraniline base).[13] Additionally, PANI can be protonated by acids, which results in the so-called salt structure.[14] The protonation of the intermediate oxidized form of PANI (emeraldine base) occurs via electron donation from the imine-like nitrogens present in the polymer structure, which results in a green material (emeraldine salt) that is the most conductive form of PANI. The doping process of PANI produces two types of charge carriers (each with a different associated band structure): polarons (a structure with benzene-like rings that is paramagnetic and a spin = 1/2) and bipolarons (a dication structure and quinoid-like rings that is diamagnetic and spinless). The type and amount of carriers are key factors that affect the conductivity, optical properties, reactivity and microstructure of PANI. A schematic representation of all the possible structures and carriers for PANI is shown in Figure 9.1.[15]

PANI is obtained through the oxidation of aniline, which can be done chemically *via* common oxidizing compounds (*e.g.* hydrogen peroxide, ammonium persulfate or potassium permanganate) or electrochemical processes. In addition to a dependence on the chemical structure, the properties of PANI are also strongly connected to the morphology, porosity and chain organization, which are closely related to the synthetic route.

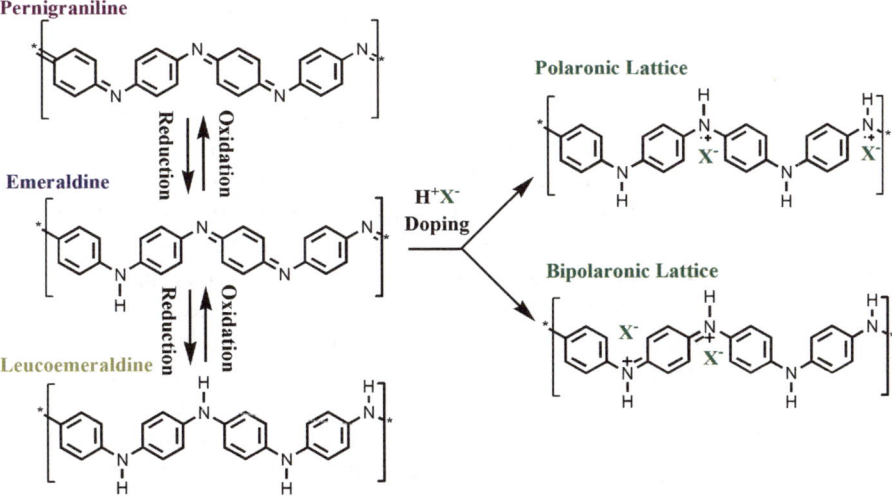

Figure 9.1 Schematic representation of the various structures and charge carriers adopted by PANI.

Generally, chemical polymerization is known to produce PANI samples with granular morphology, while electrochemical methods produce fibrous materials. Other morphologies, such as tubular, plate-like and dendrimer-like morphologies can also be obtained simply by varying the synthetic conditions.[16–17]

9.3 Graphene–Polyaniline Nanocomposites

Following the trend observed for CNT-based PANI nanocomposites, interest in the preparation, characterization and application of graphene–PANI nanocomposites has recently increased. The impressive improvement in the electrical, thermal, mechanical, optical and redox properties observed for both materials relative to the individual components has resulted in the use of the material in many possible applications for different devices, mainly in supercapacitors,[18] electrochemical sensors,[19] thermoelectric materials[20] and electrocatalysts.[21]

The synthesis of graphene–PANI nanocomposites is usually performed following one of five main approaches: (1) the solid state mixing of powders of pre-synthesized materials;[18] (2) the mixture of dispersions of each component in suitable solvents, followed by controlled evaporation of the solvent;[22] (3) electrochemical polymerization of aniline over a pre-constructed graphene-based electrode;[23] (4) *in situ* chemical polymerization of aniline in a dispersion of graphene;[24–25] or (5) a combination of two or more approaches described above.[24] Stable graphene dispersions are required during all polymerization steps, and except for route (3), all of the routes require the synthesis of graphene in bulk quantities.

For the synthesis of nanocomposites, the method most commonly used to obtain sufficient quantities of graphene to form the dispersions is based on the oxidation of graphite to graphite oxide, which is exfoliated to a single layer or a few layers of graphene oxide (GO), and then the GO is reduced to graphene (*i.e.* reduced graphene oxide, rGO).[26] The reduction can be done using hydrazine, sodium borohydride, heat treatment or solvothermal methods.[27–28] Due to their high water solubility, the nanocomposites can also be obtained from the polymerization of aniline in an aqueous GO solution, followed by the reduction of the GO to achieve graphene–PANI nanocomposites.[29]

Graphene–PANI nanocomposites are very recent materials, and although some procedures for synthesizing them have been successfully developed during the last 4 years, the processability of these materials in the post-synthesis steps remains a serious barrier to the development of devices based on these materials. For example, graphene–PANI nanocomposites must be deposited as films over ordinary substrates for several applications. If these nanocomposites are to be used in opto-electronic devices, these films need to be transparent. However, except for some electrochemical-based routes, for which several parameters are difficult to control and which are restricted to conductive substrates, the most commonly reported routes for the preparation of the graphene–PANI nanocomposites yielded insoluble powders that are difficult to disperse. The material therefore cannot be deposited as a homogeneous and/or transparent film. In this chapter we propose a simple and versatile route to solve this problem, based on interfacial polymerization. Interfacial polymerization is described in the following section.

9.4 Graphene–Polyaniline Nanocomposites Through Interfacial Polymerization

Interfacial polymerization is a versatile method to chemically synthesize structurally and morphologically homogeneous conjugated polymers. The basic principle of interfacial polymerization is to maintain a separation between the monomer and oxidizing agent (responsible for initiating the polymerization) in two liquid–liquid immiscible phases, which means that the polymerization can occur only at the liquid–liquid interface. The interface does not participate in the reaction but restricts the environment in which the polymerization reaction occurs. As a general route, interfacial polymerization can be used to produce highly organized and nanostructured polymers, copolymers, membranes and other materials.[30–33]

The interfacial polymerization can be performed under static or dynamic (*i.e.* stirring) conditions, which greatly affects the characteristics of the resultant material.[34–35] Michaelson *et al.* first proposed an interfacial reaction of aniline using a surfactant and a biphasic system (water–chloroform).[36] The material was obtained as a free-standing film located at the aqueous–organic interface. According to the authors, the fibrous material produced under static conditions was obtained at the interface as a result of enriched aniline micelles that migrated from the aqueous phase to the organic phase. Similar results were further extended to aniline derivatives.[37] The same static conditions were employed by Huang *et al.* to obtain PANI nanofibers,[38–39] for which the liquid–liquid interface was crucial for the nanofiber morphology that was formed.

A similar synthetic procedure is the self-stabilized dispersion polymerization, which also uses a biphasic liquid interface.[40] The polymerization reaction occurs under dynamic conditions in which both immiscible liquids are stirred. The aniline–anilinium hydrochloride acts as a stabilizer for the emulsion, and the oxidant is added dropwise to the dispersion. The authors assumed that the reaction would occur in the aqueous phase, similar to the traditional synthesis; however, the liquid interface only separated the fresh polymer chains from the reactive ends that still remained in the aqueous phase. This procedure produced highly crystalline polymers that have the highest conductivity value reported in the literature for PANI (1300 S cm^{-1}).[41] In both polymerization procedures, the liquid interface separates either the reagents or the products.

The utilization of interfacial polymerization for the synthesis of PANI-based nanocomposites was first described by our research group for the synthesis of silver nanoparticle–PANI materials.[42–43] In that work, pre-synthesized silver nanoparticles (NPs) with an average diameter of ~ 3.4 nm were homogeneously dispersed in a toluene solution of aniline, and then the mixture was added to an aqueous solution of the oxidizing agent. The dispersed nanoparticles acted as nucleation centres. As the polymerization proceeded at the interface of the organic (containing the aniline and the silver NPs) and aqueous (containing the oxidizing agent) phases, the polymer initially grew around the silver NPs, which resulted in silver–PANI nanocomposites with morphologies that could be controlled according to the reaction time.

Recently, we extended this route to CNT and graphene-based PANI nanocomposites.[34,44–45] Transparent films were obtained at the interface, which could easily be transferred to ordinary substrates. First, specific conditions for a PANI interfacial polymerization were developed based on stirring and a low concentration of monomer. Under those conditions, PANI was obtained as a self-standing film, which was similar to the films reported by Michaelson *et al.*;[36] however, surfactants were not present. The same procedure that was employed to produce the silver NPs–PANI nanocomposites was used to introduce the insoluble materials (*i.e.* CNTs or graphene) over the course of the polymerization, which resulted in the formation of a nanocomposite film at the interface.

Graphene–PANI nanocomposite films can be obtained at several graphene–PANI ratios (which are determined by the initial amounts of monomer and graphene at the beginning of the reaction), and maintain high optical quality, transparency and homogeneity. A schematic representation of the synthetic procedure is shown in Figure 9.2. The desired amount of graphene that was obtained by the oxidation and reduction of graphite flakes is dispersed in an organic solvent (*e.g.* toluene, hexane, chloroform or benzene) under sonication.[45] Subsequently, the appropriate amount of aniline is mixed into the dispersion, and the system is maintained under ultrasound for an additional period of time. The resulting mixture is transferred to a round-bottomed flask containing an aqueous solution of HCl and a suitable amount of pre-dissolved ammonium persulfate (oxidizing agent). Vigorous magnetic stirring results in an emulsion that enhances the contact between the two liquid phases due to the formation of small drops of organic solvent that contain both the graphene and aniline, which disperse in the aqueous solution. Gradually, the graphene and aniline dispersed in the organic droplets migrate to the organic–water interface, and the polymerization begins at the interface directly over the graphene sheets. The growing graphene–PANI nanocomposites at the interface stabilize the emulsion throughout the reaction. After a period of time, the system becomes green, as expected for the conducting form of PANI (emeraldine salt). When the magnetic stirring is interrupted, the dispersed organic drops agglomerate to reform the organic phase. Additionally, all of the graphene–PANI located at the interfaces of each organic droplet and the water also agglomerate in a continuous and homogeneous network located at the liquid–liquid interface, which results in free-standing nanocomposite films. If the same synthetic procedure is performed without aniline or graphene, high-

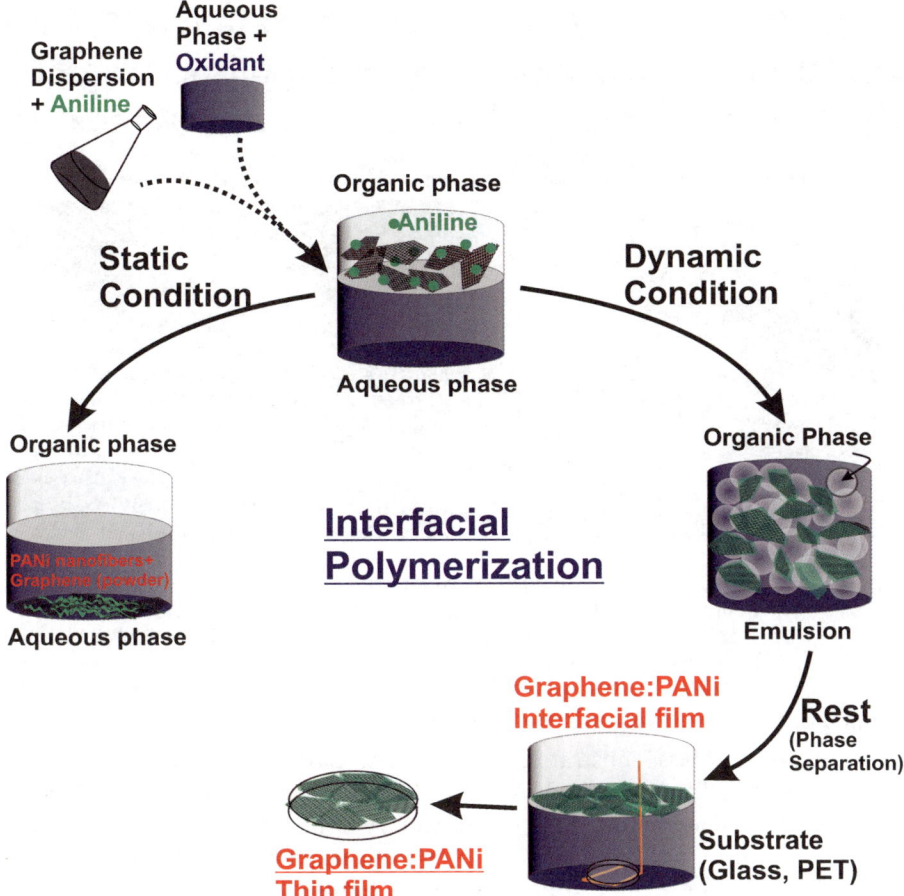

Figure 9.2 Scheme of the interfacial synthesis of graphene–PANI nanocomposites.

quality films of neat graphene and neat PANI can be obtained, respectively. However, the same experiment performed without stirring (under static conditions) produces a powder that is a polydisperse material, which precipitates from the aqueous phase.

The driving force for the formation of the film is the minimization of the interfacial energy caused by the location of the film at the interface, as described in previous reports, and is called a Pickering emulsion.[46–47] To deposit the nanocomposite films, cleaned substrates (*e.g.* glass slides, quartz slides, silicon wafers or ordinary plastics) are placed into a beaker into which the entire water–film–oil system is subsequently transferred. The substrates are simply lifted on to the film, which results in homogeneous coverage of the substrate by the film.

The top part of Figure 9.3 shows a neat graphene (left), a neat PANI (right) and two different graphene–PANI films (from the left to the right, weight/

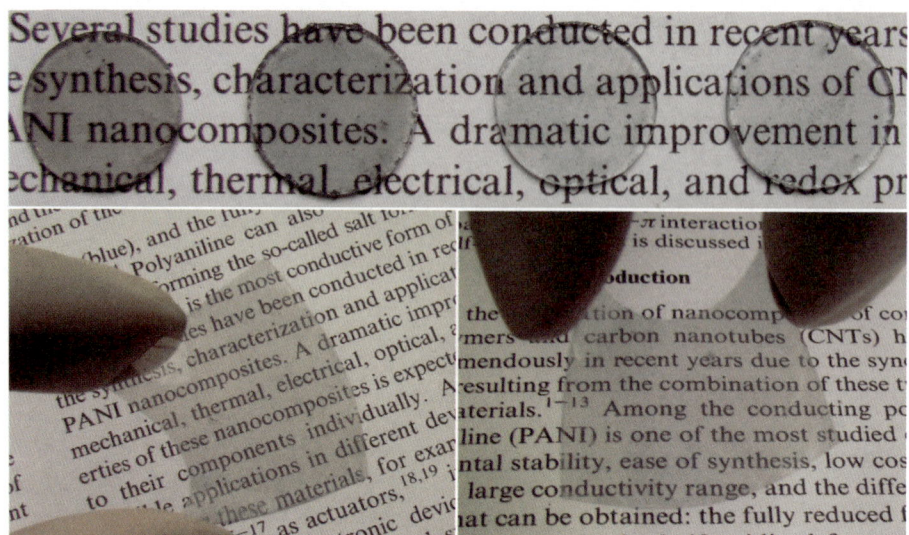

Figure 9.3   (Top) Thin films deposited over round 1 cm diameter glass substrates. From left to right: neat graphene, graphene–PANI-1:4, graphene–PANI-1:32 and neat PANI-emeraldine salt. (Bottom) Graphene–PANI-1:4 film deposited over plastic substrate.

volume ratio graphene:aniline 1:4 and 1:32, respectively) deposited over round 1 cm diameter glass substrates. The bottom part of Figure 9.3 shows a graphene–PANI film deposited over a plastic substrate, and the flexibility of the material is demonstrated. The film can be bent and flexed several times with no loss in its performance. The high optical quality and the mechanical integrity of the films can also be observed in Figure 9.3.

An interesting feature of the interfacial polymerization method is the possibility to obtain tailored materials through the control of experimental conditions, such as the concentration of graphene, the graphene–aniline and graphene–oxidant ratio, the presence or absence of stirring and the temperature. The aniline–graphene ratio is particularly important because the polymerization of aniline is very fast in the presence of graphene due to the heterogeneous nucleation (*i.e.* the graphene acts as a heterogeneous nuclei for the formation of PANI). The control of these synthetic variables allows some important characteristics of the films to be controlled including conductivity, thickness, transparency and roughness, which may be critical for future applications. Small changes in the synthetic parameters can sufficiently modify several characteristics of the films, which makes the characterization of the obtained materials critical for fully understanding their properties.

Figure 9.4 shows the normalized absorption spectra of films of GO, graphene (rGO), graphene–PANI-1:4 nanocomposite, graphene–PANI-1:32 nanocomposite and neat PANI in the conducting form (emeraldine salt, PANI-ES). For the GO sample, a sharp absorption peak was observed at

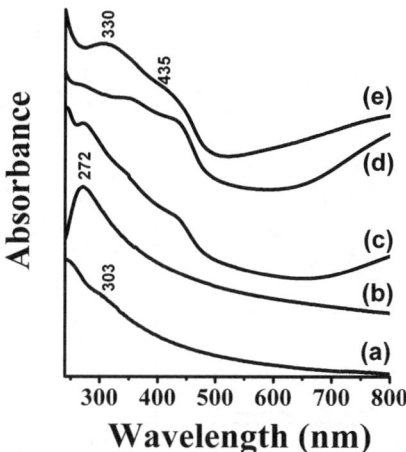

Figure 9.4   UV-vis absorption spectra of the films: (a) GO, (b) graphene (rGO), (c) graphene–PANI-1:4, (d) graphene–PANI-1:32, (e) PANI.

approximately 235 nm (not shown) and a broad shoulder was observed at 303 nm, which were attributed to non-ligand states caused by the presence of epoxy and peroxide species.[48] These bands were absent in the absorption spectrum of graphene, for which a strong absorption at 272 nm was observed. This peak was attributed to a π–π* transition of the C=C bonds.[48] The spectrum of PANI-ES showed three main features at approximately 350, 440 and 800 nm. The first feature was attributed to the bandgap transition, while the other two bands were due to interband transitions from the valence band to polaronic states created inside the gap.[49] The last band (800 nm) was also related to delocalized charge carriers (polarons/bipolarons), which are responsible for the electrical transport in PANI.[50–51] Similar bands were observed in the spectra for both of the nanocomposites, which were due to both the polymer and the graphene. The graphene band at 272 nm was not shifted in the nanocomposite, which indicates that the polymerization does not affect the conjugation of the graphene.[52]

The vibrational structures of the nanocomposites were investigated by Raman spectroscopy. The main advantage of using this technique is the possibility to selectively enhance vibrational modes of the conjugated polymers by choosing an appropriate excitation line (resonant Raman).[53] Additionally, Raman spectroscopy is intensely scattered by forms of sp^2 carbons (*e.g.* graphite, CNTs and graphene).[54] The Raman spectra of the graphene–PANI nanocomposites, as well as the neat GO, rGO and PANI-ES, collected using a laser line of 632.8 nm, are shown in Figure 9.5. Two prominent bands at 1330 and 1585 cm^{-1} were observed in the GO and rGO spectra, which correspond respectively to the D and G bands of graphitic materials.[55]

The PANI-ES spectrum in the high frequency region (1000–2000 cm^{-1}, located to the right in Figure 9.5) presents the typical bands of emeraldine salt:[56–57] 1168 and 1185 cm^{-1} (C–H bending for bipolaronic and polaronic

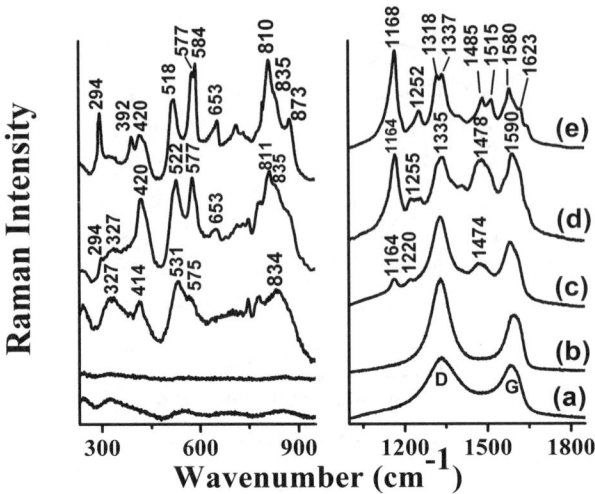

Figure 9.5 Raman spectra of the films: (a) GO, (b) graphene (rGO), (c) graphene–PANI-1:4, (d) graphene–PANI-1:32, (e) PANI.

lattices, respectively), 1252 cm^{-1} (C–N stretching in benzene diamine units), 1318 and 1337 cm^{-1} (C–N$^+$ stretching for polaron cation radical), 1485 cm^{-1} (C=N stretching of quinoid rings in non-protonated diimine units), 1515 cm^{-1} (C=N stretching in quinoid rings in protonated diimine units), 1580 cm^{-1} (C–C stretching in quinoid rings), 1596 cm^{-1} (C=C stretching in quinoid rings) and 1623 cm^{-1} (C–C stretching in benzenoid rings). Due to resonance effects, all of the vibrational modes relating to the oxidized fraction of the polymer were observed to be strongly enhanced in the PANI spectrum that was collected with the 632.8 nm laser. Analysis of the Raman spectra of the nanocomposites enabled the detection of the bands of both components (PANI and graphene), but several important spectral changes were observed as the proportion of graphene in the nanocomposites varied. For example, in both nanocomposites, the band observed at 1168 cm^{-1} in PANI-ES shifted to 1164 cm^{-1} and the quinoid modes (centred at 1478 cm^{-1}) intensified, which indicates that the polymer became more bipolaronic in the presence of graphene (*i.e.* the graphene induces bipolaronic stabilization).[58] Another observation is that the two polaronic bands at 1318 and 1337 cm^{-1} in PANI-ES merged into one single band centred at 1335 cm^{-1} in the nanocomposites spectra, which indicates that a more uniform polymer formed due to the presence of graphene.[59] These spectral observations may be related to the strong interaction of PANI-ES with the graphene. The polymer chains were expected to grow from the extended honeycomb structure of the sp^2 carbons, which would induce close contact and strong aromatic interactions.

The low-frequency region of the Raman spectrum (200–1000 cm^{-1}, located on the left in Figure 9.5) shows spectral information regarding the conformation and periodicity of polymer chains. The presence of graphene caused the

PANI bands to broaden, which indicates that the conformation of the polymer chains was affected by the graphene. The spectrum of PANI-ES shows a well-defined band at 294 cm^{-1}, which is related to the pseudo-orthorhombic structure of PANI.[60] The intensity of this band decreased as the amount of graphene increased and was not observed in the graphene–PANI-1:4 sample, which indicates a lack in the periodicity of the polymer as the polymer grows over the graphene sheets. The bands centred at 420 and 518 cm^{-1} were related to out-of-plane C–C deformation in aromatic rings; the presence of graphene shifted these bands to 414 and 531 cm^{-1}, respectively. This result indicates that a complex interaction between graphene with the conjugated chains of PANI exists. Spectroscopic evidence for the strong interaction between the components was the absence of the bands at 584, 711 and 810 cm^{-1} for the graphene–PANI-1:4, which were observed in the spectrum of neat PANI-ES. These bands are associated with deformations of the amine groups and out-of-plane deformations of the aromatic rings.[61] The absence of the bands in the graphene–PANI nanocomposite indicates that intimate contact existed between the phases in a planar motif. Finally, the modes associated to quinoid rings (bipolarons) (*e.g.* the band at 834 cm^{-1}) were intensified, which confirms that the graphene sheets stabilized the bipolarons. Interpretation of the Raman data indicates that the interaction between the graphene and PANI-ES occurs through the oxidizing segments of the polymer, which induces a bipolaronic structure and reduces the planar structure of the PANI.

Figure 9.6 shows the transmission electron microscopy (TEM) images of a graphene film and the graphene–PANI-1:4 nanocomposite film. The film of neat graphene is clearly formed by an agglomeration of interconnected graphene sheets. In the nanocomposite image, the polymer can be observed to have homogeneously grown over the sheets, which indicates the high interaction between the two components.

An interesting application for these materials is as transparent electrodes, which would be substitutes for indium tin oxide ITO-based materials. The sheet resistivity of the films was evaluated by the four probe methods. Very low sheet resistance was observed for the graphene film ($\sim$30 $\Omega/\square$) with low transparency (20% transmittance at 550 nm), which is not sufficient to enable the use of the film in applications as a transparent electrode. A significantly

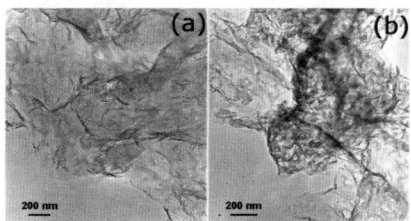

Figure 9.6   TEM images of (a) a graphene film obtained by interfacial polymerization; (b) the graphene-PANI-1:4 nanocomposite film.

lower resistivity was determined for the graphene–PANI-1:4 nanocomposite film (~ 4–$5\ \Omega/\square$); however, for practical applications, a transparency of 70% is not adequate. An excellent compromise between sheet resistivity ($60\ \Omega/\square$) and high transparency ($\sim 90\%$) was determined for the graphene–PANI-1:32 sample. These values are similar to the values observed for ITO-based electrodes and are among the best values reported for this type of application.[62] An explanation for the higher conductivity value determined for the graphene–PANI-1:4 film is that PANI can minimize the contact resistance between individual graphene sheets. This observation was also reported for a composite of carbon nanotube films with a very low ratio of a PANI derivative, which was prepared and measured under similar conditions as the graphene–PANI-1:4 film.[63] As the proportion of PANI increases from the 1:32 ratio, the resistance of the films should naturally rise. The improvement in the transparency resulted from the polymer-mediated increase in separation of the graphene sheets.

The electrochemical characterization of the films was performed by cyclic voltammetry, and the results are shown in Figure 9.7A. The voltammogram shows two main oxidation processes characteristic of PANI (indicated by A and D), which indicates that the polymer remains electroactive in the presence

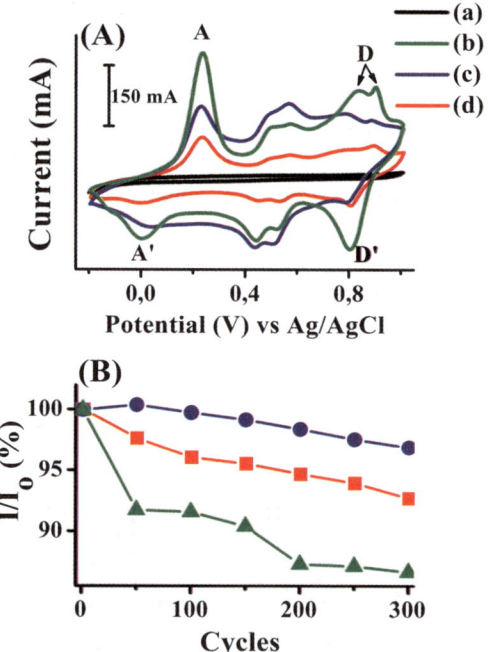

Figure 9.7   (A) Cyclic voltammograms of the films: (a) graphene, (b) PANI, (c) graphene–PANI-1:32,(d) graphene–PANI-1:4. (B) The relative intensity of the density of current of the anodic peak A after several voltammetric cycles.

of graphene. The redox pair marked 'A' corresponds to the oxidation of leucoemeraldine to emeraldine salt, while the redox pair marked 'D' corresponds to the oxidation of emeraldine to pernigraniline.[59] These two redox reactions are accompanied by colour changes (Figure 9.1) observed in all of the samples, which indicates that the materials display electrochromism even in the presence of high amounts of graphene.

The electrochemical stability of the films was evaluated by performing 300 consecutive cycles over a potential range of −0.2 to 0.6 V. After 300 cycles, the current density of the anodic peak A of the film of neat PANI decreased by 14% due to the electrodegradation of the polymer, while the current density of the graphene–PANI-1:32 and graphene–PANI-1:4 nanocomposites decreased only 3% and 8%, respectively (Figure 9.7B). These results clearly show that the presence of graphene also increased the electrochemical stability of the polymer. This effect on the electrochemical stability was also observed for PANI–CNT nanocomposites. This effect is related to the strong interaction between the polymer and the carbonaceous nanostructure.[34] During the voltammetric cycles, counter-ion exchange resulted in volume changes in the polymer, which were responsible for the loss in electrochemical response. However, the polymers suffer less damage when supported by carbonaceous nanostructured materials. The presence of carbon nanomaterial also positively affects the conductivity of the material and the diffusion length of ions.[64] The improved stability of the graphene–PANI-1:32 film revealed that graphene influences the polymerization of aniline in a complex manner, because the high graphene proportion (graphene–PANI-1:4) did not yield the best results.

9.5 Conclusion and Final Remarks

The synergistic effects due to the chemical or physical interactions between the polymer and graphene in conducting polymer–graphene-like nanocomposites produced novel and improved mechanical, thermal, optical, chemical, catalytic and electrical properties, which causes these materials to be very attractive for use in various devices. Due to the complexity of the redox structures of the polymers, as well as the strong dependence on the properties of the structures, the discovery of chemical routes to produce this novel class of nanostructured materials in a controlled method is imperative.

The preparation of polymers by interfacial polymerization has grown substantially in recent years due to the possibility of controlling important characteristics of the polymer (*e.g.* particle size, morphology, crystallinity and structure). This study demonstrated that interfacial polymerization can be extended to the preparation of nanocomposite free-standing films, in which both the synthesis and processing of the nanocomposites are performed in a single step. This route was successfully applied with various combinations of conducting polymers and nanocarbon materials and likely can be extended to be a general route for the fabrication of nanocomposite materials as a whole.

References

1. D. W. Hatchett, M. Josowicz, *Chem. Rev.*, 2008, **108**, 746.
2. M. Baibarac, I. Baltog, S. Lefrant, J. Y. Mevellec, O. Chauvet, *Chem. Mater.*, 2003, **15**, 4149.
3. M. K. Ram, H. Gomez, F. Alvi, E. Stefanakos, Y. Goswami, A. Kumar, *J. Phys. Chem. B*, 2011, **115**, 21987.
4. W. R. Small, F. Masdarolomoor, G. G. Wallace, M. i. h. Panhuis, *J. Mater. Chem.*, 2007, **17**, 4359.
5. L. Nyholm, G. Nyström, A. Mihranyan, M. Strømme, *Adv. Mater.*, 2011, **23**, 3751.
6. P. M. Beaujuge, C. M. Amb, J. R. Reynolds, *Acc. Chem. Res.*, 2010, **43**, 1396.
7. R. L. Caygill, C. S. Hodges, J. L. Holmes, S. P. J. Higson, G. E. Blair, P. A. Millner, *Biosens. Bioelectron.*, 2012, **32**, 104.
8. A. Roigé, M. Campoy-Quiles, J. O. Ossó, M. I. Alonso, L. F. Vega, M. Garriga, *Synth. Met.*, 2012, **161**, 2570.
9. E. Giorgetti, G. Margheri, S. Sottini, M. Muniz-Miranda, *Synth. Met.*, 2003, **139**, 929.
10. Y.-J. Cheng, S.-H. Yang, C.-S. Hsu, *Chem. Rev.*, 2009, **109**, 5868.
11. G. Daoust, M. Leclerc, *Macromolecules*, 1991, **24**, 455.
12. P. M. Beaujuge, J. M. J. Fréchet, *J. Am. Chem. Soc.*, 2011, **133**, 20009.
13. K.-Y. Shen, C.-W. Hu, L.-C. Chang, K.-C. Ho, *Sol. Energy Mater. Sol. Cells*, 2012, **98**, 294.
14. J. Stejskal, R. G. Gilbert, *Pure Appl. Chem.*, 2002, **74**, 857.
15. A. Varela-Alvarez, J. A. Sordo, *J. Chem. Phys.*, 2008, **128**, 174706.
16. H. Qiu, M. Wan, B. Matthews, L. Dai, *Macromolecules*, 2001, **34**, 675.
17. X. Wang, J. Liu, X. Huang, L. Men, M. Guo, D. Sun, *Polym. Bull.*, 2008, **60**, 1.
18. Q. Wu, Y. Xu, Z. Yao, A. Liu, G. Shi, *ACS Nano*, 2010, **4**, 1963.
19. Y. Bo, H. Yang, Y. Hu, T. Yao, S. Huang, *Electrochim. Acta*, 2011, **56**, 2676.
20. Y. Du, S. Z. Shen, W. Yang, R. Donelson, K. Cai, P. S. Casey, *Synth. Met.*, 2012, **161**, 2688.
21. G. Wang, W. Xing, S. Zhuo, *Electrochim. Acta*, 2012, **66**, 151.
22. H. Bai, Y. Xu, L. Zhao, C. Li, G. Shi, *Chem. Commun.*, 2009, 1667.
23. D.-W. Wang, F. Li, J. Zhao, W. Ren, Z.-G. Chen, J. Tan, Z.-S. Wu, I. Gentle, G. Q. Lu, H.-M. Cheng, *ACS Nano*, 2009, **3**, 1745.
24. H. Bai, C. Li, G. Shi, *Adv. Mater.*, 2011, **23**, 1089.
25. K. Zhang, L. L. Zhang, X. S. Zhao, J. Wu, *Chem. Mater.*, 2010, **22**, 1392.
26. V. C. Tung, M. J. Allen, Y. Yang, R. B. Kaner, *Nat. Nanotechnol.*, 2009, **4**, 25.
27. A. V. Murugan, T. Muraliganth, A. Manthiram, *Chem. Mater.*, 2009, **21**, 5004.
28. J. Shen, Y. Hu, M. Shi, X. Lu, C. Qin, C. Li, M. Ye, *Chem. Mater.*, 2009, **21**, 3514.
29. C. Valleés, P. Jimeénez, E. Munñoz, A. M. Benito, W. K. Maser, *J. Phys. Chem. B*, 2011, **115**, 10468.
30. C. Scott, D. Wu, C.-C. Ho, C. C. Co, *J. Am. Chem. Soc.*, 2005, **127**, 4160.

31. K. Bouchemal, F. Couenne, S. Briançon, H. Fessi, M. Tayakout, *AIChE J.*, 2006, **52**, 2161.
32. N. Nuraje, K. Su, N.-l. Yang, H. Matsui, *ACS Nano*, 2008, **2**, 502.
33. X.-G. Li, J. Li, Q.-K. Meng, M.-R. Huang, *J. Phys. Chem. B*, 2009, **113**, 9718.
34. R. V. Salvatierra, M. M. Oliveira, A. J. G. Zarbin, *Chem. Mater.*, 2010, **22**, 5222.
35. V. V. Yashin, A. C. Balazs, *J. Chem. Phys.*, 2004, **121**, 11440.
36. J. C. Michaelson, A. J. McEvoy, *J. Chem. Soc., Chem. Commun.*, 1994, 79-80.
37. A. Falcou, A. Longeau, D. Marsacq, P. Hourquebie, A. Duchêne, *Synth. Met.*, 1999, **101**, 647.
38. J. X. Huang, R. B. Kaner, *J. Am. Chem. Soc.*, 2004, **126**, 851.
39. J. Huang, S. Virji, B. H. Weiller, R. B. Kaner, *J. Am. Chem. Soc.*, 2002, **125**, 314.
40. S. H. Lee, D. H. Lee, K. Lee, C. W. Lee, *Adv. Funct. Mater.*, 2005, **15**, 1495.
41. K. Lee, S. Cho, S. Heum Park, A. J. Heeger, C.-W. Lee, S.-H. Lee, *Nature*, 2006, **441**, 65.
42. M. M. Oliveira, E. G. Castro, C. D. Canestraro, D. Zanchet, D. Ugarte, L. S. Roman, A. J. G. Zarbin, *J. Phys. Chem. B*, 2006, **110**, 17063.
43. M. M. Oliveira, D. Zanchet, D. Ugarte, A. J. G. Zarbin in *Progress in Colloid and Polymer Science,* ed. F. Galembeck, Springer Berlin, Heidelberg, 2004, p. 126.
44. R. V. Salvatierra, L. G. Moura, M. M. Oliveira, M. A. Pimenta, A. J. G. Zarbin, *J. Raman Spectrosc.*, 2012, DOI 10.1002/jrs.3144..
45. S. H. Domingues, R. V. Salvatierra, M. M. Oliveira, A. J. G. Zarbin, *Chem. Commun.*, 2011, **47**, 2592.
46. E. Vignati, R. Piazza, T. P. Lockhart, *Langmuir*, 2003, **19**, 6650.
47. J. Zhou, X. Qiao, B. P. Binks, K. Sun, M. Bai, Y. Li, Y. Liu, *Langmuir*, 2011, **27**, 3308.
48. S. Saxena, T. A. Tyson, S. Shukla, E. Negusse, H. Chen, J. Bai, *Appl. Phys. Lett.*, 2011, **99**, 013104.
49. W. S. Huang, A. G. MacDiarmid, *Polymer*, 1993, **34**, 1833.
50. R. Colle, P. Parruccini, A. Benassi, C. Cavazzoni, *J. Phys. Chem. B*, 2007, **111**, 2800.
51. D. Chinn, J. DuBow, J. Li, J. Janata, M. Josowicz, *Chem. Mater.*, 1995, **7**, 1510.
52. D. Li, M. B. Muller, S. Gilje, R. B. Kaner, G. G. Wallace, *Nat. Nanotechnol.*, 2008, **3**, 101.
53. G. Louarn, M. Lapkowski, S. Quillard, A. Pron, J. P. Buisson, S. Lefrant, *J. Phys. Chem.*, 1996, **100**, 6998.
54. L. M. Malard, M. A. Pimenta, G. Dresselhaus, M. S. Dresselhaus, *Phys. Rep.*, 2009, **473**, 51.

55. M. A. Pimenta, G. Dresselhaus, M. S. Dresselhaus, L. G. Cancado, A. Jorio, R. Saito, *Phys. Chem. Chem. Phys.*, 2007, **9**, 1276.
56. M. C. Bernard, A. Hugot-Le Goff, *Electrochim. Acta*, 2006, **52**, 595.
57. R. Mazeikiene, V. Tomkute, Z. Kuodis, G. Niaura, A. Malinauskas, *Vib. Spectrosc.*, 2007, **44**, 201.
58. J. E. P. Silva, D. L. A. Faria, S. I. C. Torresi, M. L. A. Temperini, *Macromolecules*, 2000, **33**, 3077.
59. M. Lapkowski, K. Berrada, S. Quillard, G. Louarn, S. Lefrant, A. Pron, *Macromolecules*, 1995, **28**, 1233.
60. P. Colomban, S. Folch, A. Gruger, *Macromolecules*, 1999, **32**, 3080.
61. M. Cochet, G. Louarn, S. Quillard, J. P. Buisson, S. Lefrant, *J. Raman Spectrosc.*, 2000, **31**, 1041.
62. D. S. Hecht, L. Hu, G. Irvin, *Adv. Mater.*, 2011, **23**, 1482.
63. Y. Ma, W. Cheung, D. Wei, A. Bogozi, P. L. Chiu, L. Wang, F. Pontoriero, R. Mendelsohn, H. He, *ACS Nano*, 2008, **2**, 1197.
64. H. Zhang, G. Cao, Z. Wang, Y. Yang, Z. Shi, Z. Gu, *Electrochem. Commun.*, 2008, **10**, 1056.

CHAPTER 10

Crystallization Properties of Isotactic Polypropylene–Graphene Nanocomposites

JIA-ZHUANG XU[1], ZHONG-MING LI*[1] AND
BENJAMIN S. HSIAO*[2]

[1] College of Polymer Science and Engineering and State Key Laboratory of Polymer Materials Engineering, Sichuan University, Chengdu 610065, China; [2] Department of Chemistry, Stony Brook University, Stony Brook, NY 11794-3400, USA
*E-mail: zmli@scu.edu.cn or bhsiao@notes.cc.sunysb.edu

10.1 Introduction

One of the most promising applications of graphene is in polymer nanocomposites, which have inspired a growing scientific interest by reason of their widely potential applications in high-technology areas.[1] Polymer–graphene nanocomposites with a very low filler loading exhibit significant improvements in mechanical, thermal, gas barrier, electrical and flame retardant properties compared to the virgin polymers.[2–3] Therefore, the discovery of graphene is considered to open a new strategy for the fabrication of functional nanocomposites due to its extraordinary properties and plentiful resource of natural graphite.[4] Much effort is devoted to improving or maximizing the practical performances of polymer nanocomposites. The properties of nanocomposites are primarily based on those of the polymer matrix, which are found to be strongly affected by its microstructures,

RSC Nanoscience & Nanotechnology No. 26
Polymer–Graphene Nanocomposites
Edited by Vikas Mittal
© The Royal Society of Chemistry 2012
Published by the Royal Society of Chemistry, www.rsc.org

especially their crystalline structures. As two-thirds of industrial polymers are crystallizable and nanofillers are generally their effective nucleating agents, nanofiller-induced polymer crystallization is of profound importance for achieving high-performance nanocomposites.

Generally, nanofiller-driven crystallization of polymers is attributed to their high specific surface area, which reduces the nucleation barrier, thereby exerting positive effects on crystallization.[5] However, this non-procedural conclusion cannot give a convincing explanation to clarify the dynamic process of nanofiller-induced polymer crystallization. The assembly kinetics of polymer chains, especially their early connection and epitaxial form on the surface of nanofillers, are still in the initial stage of exploration.

Although various theoretical models have been proposed to depict the panorama of polymer crystallization (*i.e.* Lauritzen–Hoffman theory,[6] orientation fluctuation,[7] density fluctuation[8] and the multistage growth model[9]), it is commonly agreed that conformational ordering of the molecular chains is needed to provoke crystallization. The ordering of polymer chains consists of intrachain conformational ordering, as well as interchain positional and orientational ordering. It seems that following the trajectory of intrachain conformational ordering is the first important step in comprehending graphene-induced polymer crystallization. The template effect of graphene nanosheets on the landscaping of poly(L-lactic acid) (PLLA) chains was reported in our previous study, and their strong nucleating ability was explained in terms of surface-induced conformational order.[10,11] Unfortunately, little information on intrachain conformation adjustment could be extracted because of the complex and strong interchain interactions of PLLA.[12] In another scenario, the effects of different flow fields have great influence on the final crystallinity and crystalline morphology during typical polymer processing operations (*e.g.* extrusion, injection, and blow moulding).[13–15] The subject of flow-induced crystallization has attracted a great deal of attention in the community.[16–19] It is well established that shear flow significantly enhances the crystallization kinetics of polymers. The accelerating effect can be attributed to the formation of oriented molecular chains, which nucleate and form a crystal structure. The lifetime of the shear-induced precursors is closely related to the shear intensity and crystallization conditions.[18,20] For the development of polymer–graphene nanocomposites, the combined effects of shear fields and graphene on the crystallization behaviour of polymers are of particular interest.[21] Here two kinds of nucleation origins can coexist: (1) homogeneous nucleating sites where shear-induced row nuclei (shish) may occur; (2) heterogeneous nucleating sites which are facilitated by the presence of nanoparticles.[22] The local velocity gradient may increase by orders of magnitude in the vicinity of nanoparticles, where a synergistic effect on the crystallization kinetics of polymer can be obtained. However, the presence of nanoparticles can also limit the motion of molecular chains, resulting in suppression of perfection and crystallinity of polymer crystals.[23–25] Therefore, graphene-induced polymer crystallization under shear flow is a much more complex process compared with the pure polymer.

This chapter reviews our recent work on the intrachain ordering and shear-induced crystallization of polymer–graphene nanocomposites. Isotactic polypropylene (iPP) was chosen as the model polymer. Most absorption bands of iPP in the mid-infrared region belong to the regularity bands and are related to the intrachain vibration within a single chain, which is helpful in tracing its intrachain conformation ordering and crystallization evolution. Also, the shear-induced crystallization and polymorphism of iPP have been extensively investigated, which enables us to better understand the combined effects of graphene nanosheets (GNSs) and shear fields on the crystallization behaviour, such as crystallization kinetics and crystalline structure.

10.2 Intrachain Conformational Ordering of iPP–Graphene Nanocomposites

To give a better idea of the intrachain conformational ordering of iPP–GNSs, geometrical structures of GNSs are shown in Figure 10.1.[26] This typical atomic force microscopy (AFM) image shows that the synthesized nanocomposites are only 1–2 nm thick, and their average calculated length is 0.87 μm, with a large distribution. Not all GNSs are single-atom layers, but the length of the GNSs is nearly 3 orders of magnitude greater than their thickness, so they can be regarded as (2D) nanosheets. Solution coagulation was used to prepare iPP–GNSs. The nanocomposites with 0.05 wt% and 0.1 wt% GNSs are called PPG05 and PPG10, respectively, while neat iPP is described as PPG0.

Fourier-transform infrared spectroscopy (FTIR) is highly sensitive to conformation changes and packing density of molecular chains. By tracing the intensity and shape changes of the characteristic bands or peaks online, we

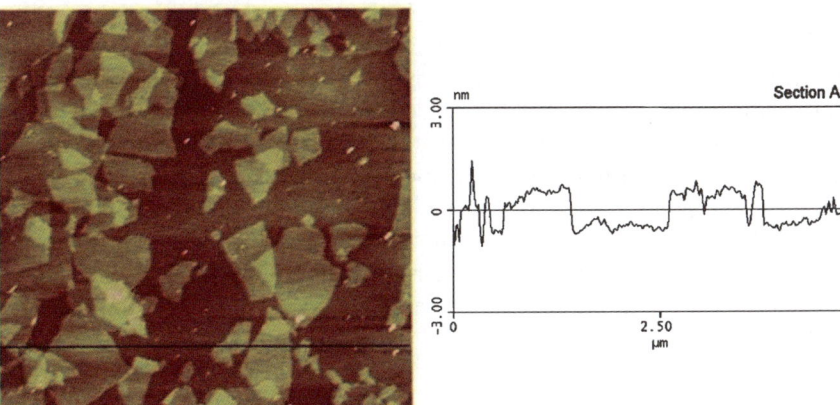

Figure 10.1 AFM image of GNSs.[26]

reported the intrachain ordering of iPP, which is suggested to be the first choice for getting a deep insight to GNS-driven iPP crystallization.[27]

Figure 10.2 shows the time-resolved FTIR spectra of PPG0, PPG05 and PPG10 isothermally crystallizing at 145 °C. Intensity changes and band shifts of the crystalline-sensitive bands apparently occur as time increases, indicating that FTIR spectra are highly sensitive to structural changes during crystallization. The IR bands at 940, 1220, 1303, 1167, 841, 998, 900 and 973 cm^{-1} are nominated as regularity bands and successively correspond to 3_1 helical structures with degree of order from high to low.[28] The gain in conformational ordering bands is directly attributable to the augmentation of the helical population. The regularity bands representing long helices are used to evaluate conformation evolution of iPP chains. The bands for short helices are not adopted because short helical structures already exist in iPP melt, with the observation of a strong peak at 973 cm^{-1} (helical length with 3–4 monomers). These short helices must experience some transformation in order to construct longer helices. Propagation and incorporation have been proposed, but no suitable method has so far been able to detect how these growth modes compete,[29] so the fluctuation of the short helices cannot be satisfactorily elucidated. According to Doi–Edwards dynamics theory, the critical persistence length of iPP for isotropic-to-nematic transition is 11 monomers in 3_1 helical conformation.[28,30–31] The conformational band at 998 cm^{-1} corresponds to 10 monomer units, which is suggested to be more sensitive to stable-to-unstable transition when crystallization is triggered. It is reasonable to choose 998 cm^{-1} for statistical analysis of the conformation evolution during crystallization, and 1303 cm^{-1}, corresponding to helical length with 13 monomers, is taken as crystalline signal.[32] Following the changing progress of those two bands may help to understand the crystallization process, especially unveiling how GNSs accelerate iPP crystallization in the early stages.

Figure 10.3a illustrates the normalized intensity of the crystalline band (I_{1303}) as a function of crystallization time for iPP and its GNSs. The crystallization of iPP is strongly accelerated in the presence of GNSs, indicating the effective nucleating ability of GNSs, even at a very low concentration. As the GNS content increases from 0.05 to 0.1 wt%, the half-crystallization time ($t_{1/2}$) of iPP–GNSs reduces from 27.5 to 15.0 min. More nucleation sites provided by GNSs would further elevate the crystallization kinetics of iPP. Interestingly, the corresponding variations of their conformational ordering band follow the same changing tendency as presented in Figure 10.3b. GNSs also speed conformational ordering of iPP associated with accelerating the crystallization. By establishing relationship with GNS-induced conformation ordering, we got a deeper insight into GNS-driven polymer crystallization.

2D correlation spectroscopy is a very powerful tool for studying the structural changes during polymer crystallization.[12] Deriving from the analysis of the asynchronous spectra, the order of structure changes is emphasized under an external field (*i.e.* temperature, time). Figure 10.4 presents the

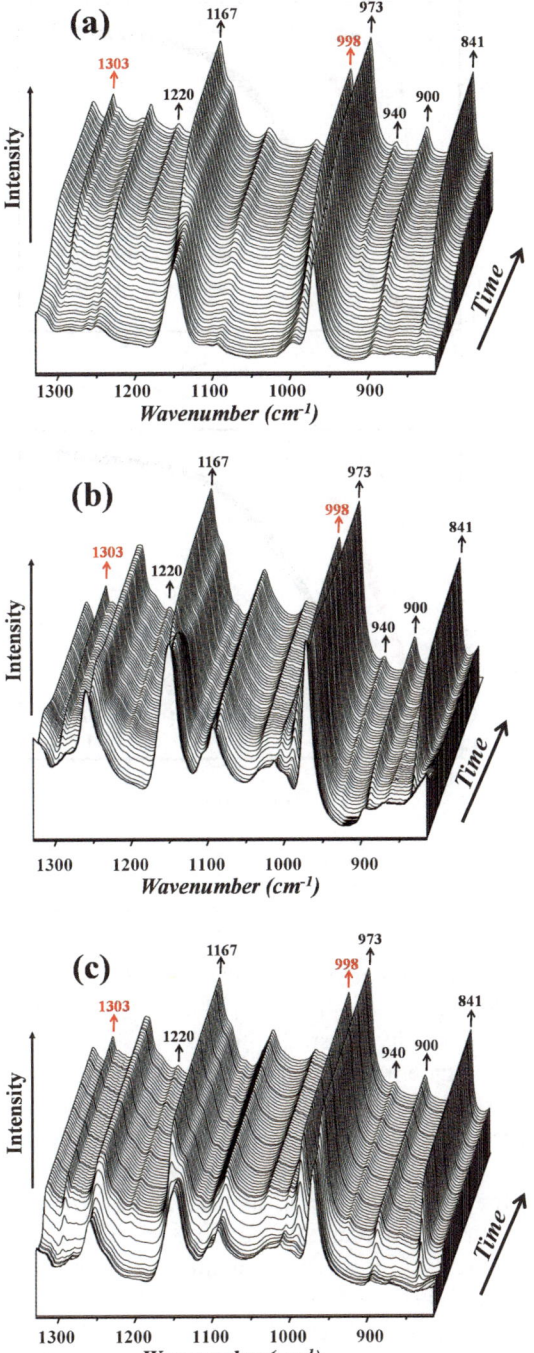

Figure 10.2 Time-resolved spectra in the range 1330–820 cm^{-1} of (a) PPG0, (b) PPG05 and (c) PPG10 isothermally crystallizing at 145 °C.

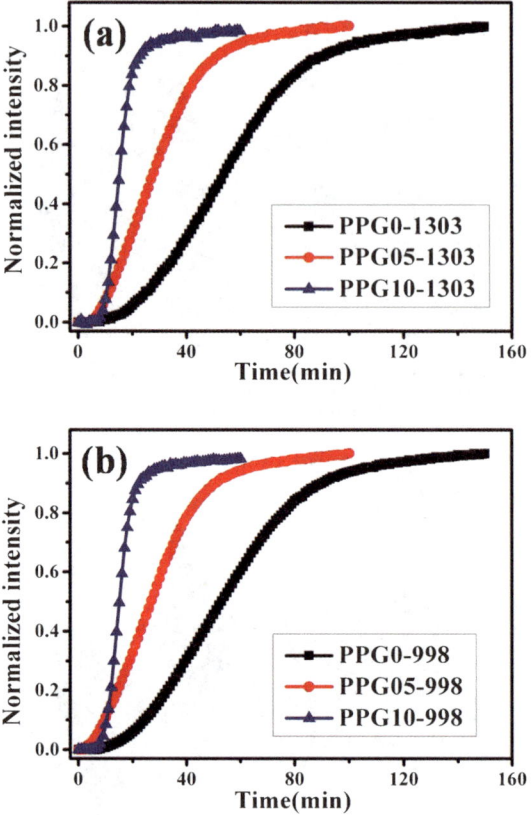

Figure 10.3 Normalized intensity of crystalline band at 1303 cm^{-1} (a) and conformational ordering band at 998 cm^{-1} (b) as a function of time for PPG10, PPG05 and PPG0 isothermally crystallizing at 145 °C.

synchronous $\Phi(v_1, v_2)$ and asynchronous $\Psi(v_1, v_2)$ 2D correlation spectra in the range of 1300–1000 cm^{-1} with 1100–900 cm^{-1} of iPP and its GNS nanocomposites isothermally crystallizing at 145 °C. The intensity in $\Phi(v_1, v_2)$ represents the coincidental changes of spectral variations measured at v_1 and v_2 during crystallization of iPP, while the intensity in $\Psi(v_1, v_2)$ reflects sequential or successive changes of specific structures. According to Noda's rules,[33–34] the signs of the cross-peak (v_1, v_2) are the same (both positive or both negative) in the synchronous and asynchronous spectra if the band at v_1 varies prior to that at v_2, whereas the sequence is reversed when the signs are different. From Figure 10.4, a common phenomenon could be observed in all three systems, $\Phi(1303, 998) > 0$ and $\Psi(1303, 998) < 0$, indicating that the intensity change occurs in the 998 cm^{-1} band before the 1303 cm^{-1} band. That is, the conformational ordering of long ordering structures acted as a vanguard preceding the crystallization of iPP.

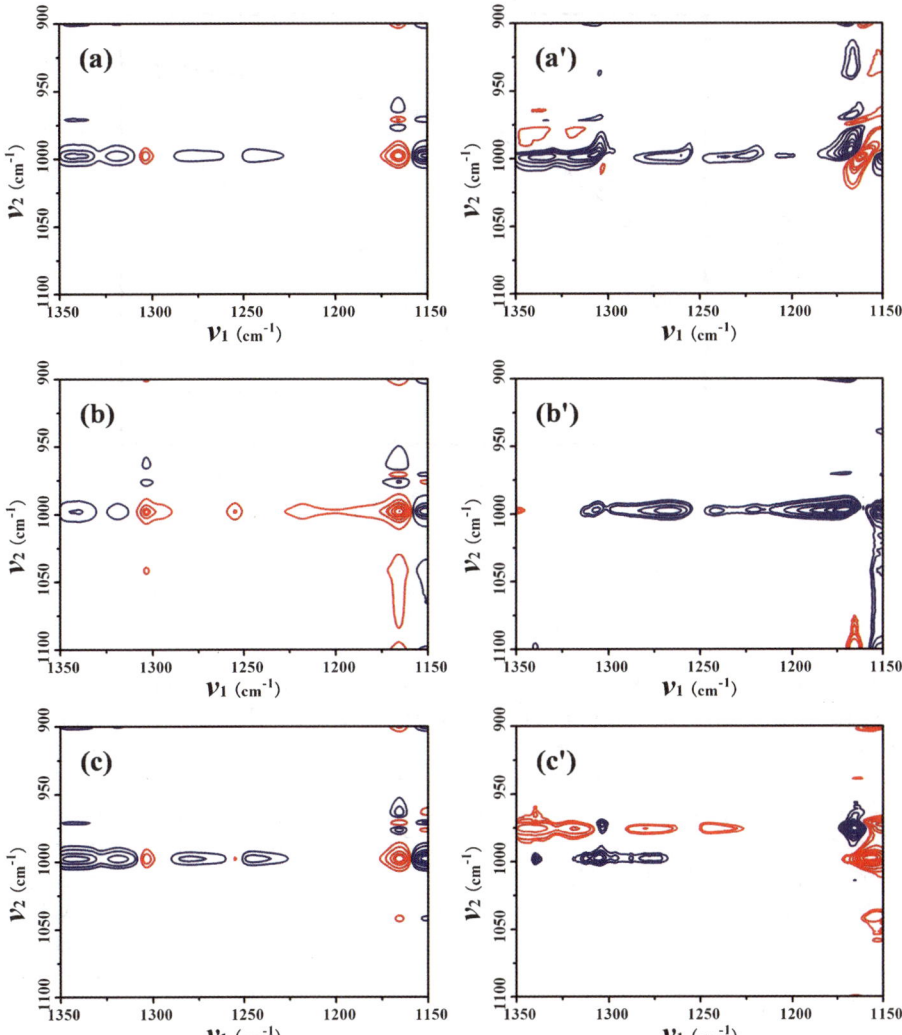

Figure 10.4   Synchronous and asynchronous correlation spectra of PPG0 (a, a'), PPG05 (b, b'), and PPG10 (c, c') in the region of 1350–1150 cm^{-1} (v_1) and 1100–900 cm^{-1} (v_2) calculated from the time-resolved spectra obtained during crystallization at 145 °C.

Here, we proposed a synchronous reference method to evaluate the changing extent of conformational ordering and crystallization. The semi-logarithmic curves of I_{998} with the development of I_{1303} for iPP and its nanocomposites are shown in Figure 10.5. The line K is representative of synchronous reference frame, *i.e.* $I_{998} = I_{1303}$. It can be seen that the curves relating to PPG0, PPG05, and PPG10 are over the K line to start with and subsequently coincide with it. I_{998} diverges positively from I_{1303} at initial stage of crystallization. It is

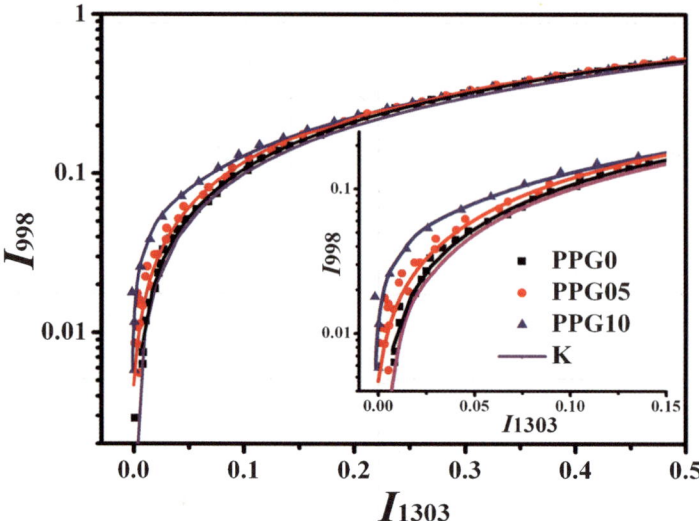

Figure 10.5   Normalized intensity of 998 cm^{-1} as the evolution of normalized intensity of 1303 cm^{-1} for PPG0, PPG05, and PPG10 isothermally crystallizing at 145 °C. Line K is represents $I_{998} = I_{1303}$.

suggested that conformational ordering occurs prior to crystal growth, in accordance with the former 2D correction results. Then the continuous packing of long helices into the crystal lattice or lamellar layer leads to the synchronization of conformational ordering and crystal growth. A slight deviation is observed for PPG0. This implies that the spontaneous formation of ordered segments in the bulk melt is finite, which is consistent with the report by Li *et al.* They observed that a large quantity of conformational ordering segments existed at the boundary of growing spherulites.[32] When GNSs are introduced, a salient increase of I_{998} occurs compared to the development of I_{1303} at the early period of crystallization. A great deal of long helical structure is easily formed in the presence of GNSs.

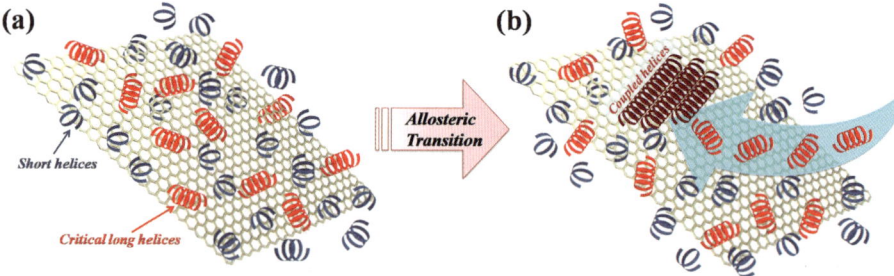

Figure 10.6   Schematic diagrams of intrachain conformation ordering iPP in the proximity of GNSs: (a) pileup of long ordering helices; (b) early stage of crystallization.

These interesting results allow us to conceive a perspective of the physical origin for GNS-induced iPP crystallization at an early stage, which is schematically portrayed in Figure 10.6. Intrachain conformational ordering of iPP chains is prone to happen probably because of the interactions between GNSs and iPP chains, such as interaction between protruding methyl groups of iPP and graphene layer of sp^2-bonded carbon.[35] So at the initial stage of crystallization a large number of long helices are stacked up near the surface of GNSs, as shown in Figure 10.6a. This may be attributed to the surface-induced conformational ordering or surface-induced polymer crystallization.[26,36] The recent investigation by Li *et al.* also confirmed that the molecular chains of polyethylene (C-axis of the crystal) are parallel to the basal plane of the reduced graphene sheets.[36] When the stable system is sluiced (nucleation), the coupled helices or crystal lamellae appear because isolated single long helices are not sufficient to trigger crystallization.[29,37] Preexisting long helices could then pack rapidly into the formed coupled helices (Figure 10.6b). The crystallization rate is thereby significantly speeded up compared to neat iPP. Essentially, there is a competition between the formation and consumption of long helices as crystal growth proceeds. This competition may be maintained until the completion of crystallization. Our results reveal that the presence of GNSs facilitates the formation of long helical segments, leading to a considerable enhancement of crystallization kinetics of iPP, though the epitaxial mode is not yet clear.[27]

10.3   Crystallization Kinetics of iPP–Graphene Nanocomposites Under Shear Flow

10.3.1   Crystallization Under Quiescent Conditions

The nucleating ability of GNSs was first evaluated under quiescent conditions before a shear field was applied. Synchrotron wide-angle X-ray diffraction (WAXD) measurements were carried out at the Advanced Polymers Beamline (X27C, = 0.1371 nm) in the National Synchrotron Light Source (NSLS), Brookhaven National Laboratory (BNL). A 2D MAR CCD X-ray detector (MAR-USA) with a resolution of 1024 × 1024 pixels (pixel size = 158.44 μm) was used to acquire 2D-WAXD images at 30 s intervals. Figure 10.7 shows selected 2D WAXD patterns of neat iPP and its GNSs isothermally crystallized at 145 °C under quiescent condition. A diffuse scattering ring is found in the first pattern ($t = 0$ min) of PPG0, indicating that no crystals is formed in the iPP melt, which is also the case in PPG05 and PPG10. This confirms that the previous crystal structure of iPP is completely erased when cooled to the T_c. After an induction period, several isotropic crystal reflection peaks can be seen in all three samples, indicating the process of crystallization under quiescent conditions. The intensities of these isotropic diffraction peaks become stronger with time until the completion of crystallization. The time evolution of integrated 1D WAXD curves of PPG0, PPG05, and PPG10 is

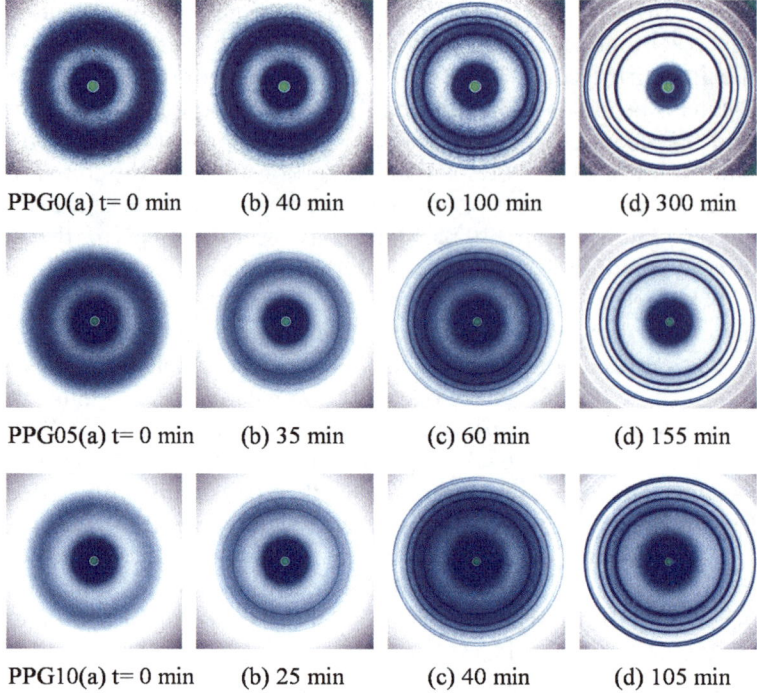

PPG0(a) t= 0 min   (b) 40 min   (c) 100 min   (d) 300 min

PPG05(a) t= 0 min   (b) 35 min   (c) 60 min   (d) 155 min

PPG10(a) t= 0 min   (b) 25 min   (c) 40 min   (d) 105 min

Figure 10.7   Selected 2D wide-angle X-ray diffraction (WAXD) patterns of neat iPP and its GNSs isothermally crystallized at 145 °C under quiescent conditions.

illustrated in Figure 10.8. It can be seen that only α-crystal reflection peaks, corresponding to (110), (040), and (130) reflections, appear in all three samples. A qualitative comparison of their time axes shows that the crystallization rates of iPP containing GNSs are evidently accelerated.

Normalized crystallinity of PPG0, PPG05 and PPG10 as a function of time is plotted in Figure 10.9. It can be concluded that the addition of GNSs significantly enhances the crystallization rate of iPP. Table 10.1 illustrates the $t_{1/2}$ of the three samples, where $t_{1/2}$ for PPG05 and PPG10 is seen to be reduced by more than 50% compared to that of PPG0, indicating the highly effective nucleation ability of GNSs. As the GNS content increases from 0.05 to 0.1 wt%, $t_{1/2}$ of the nanocomposites further decreases from 70.5 to 46.8 min. This proves that with the increasing nanofiller content, more nucleation sites provided by GNSs can further shorten the induction period and accelerate the crystallization rate.

10.3.2   Crystallization Under Shear Flow

A step shear at a rate of 20 s^{-1} for 5 s is applied to investigate the effects of shear flow on the crystallization of iPP–GNSs (see Figure 10.10a). Selected 2D

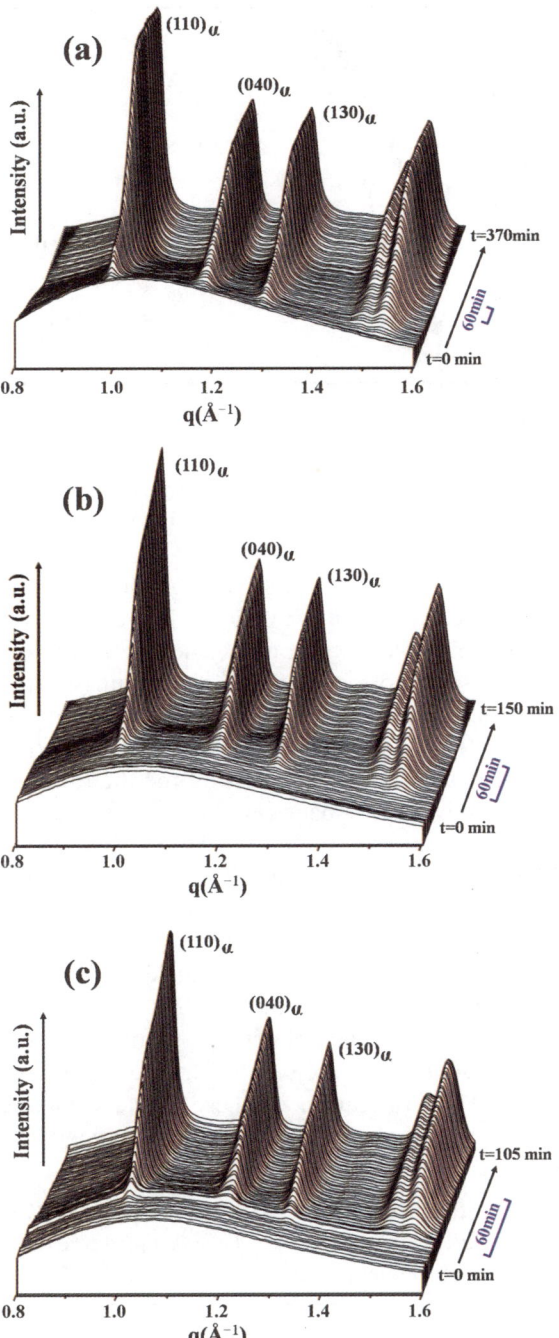

Figure 10.8   Time development of integrated 1D WAXD profiles of (a) PPG0, (b) PPG05 and (c) PPG10 isothermally crystallized at 145 °C under quiescent conditions.

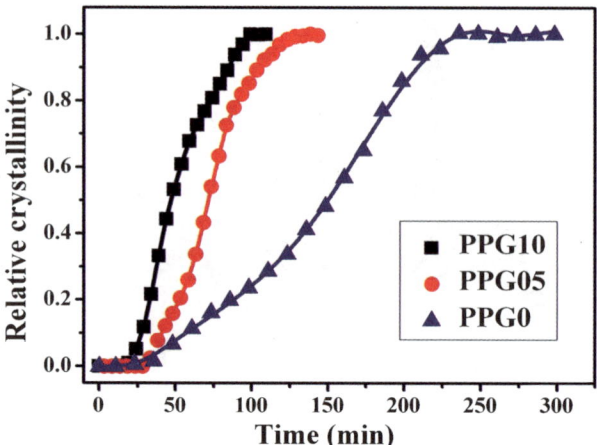

Figure 10.9 Normalized crystallinity of PPG0, PPG05 and PPG10 at 145 °C under quiescent conditions as a function of time.

WAXD patterns of neat iPP and its GNS nanocomposites isothermally crystallized at 145 °C after the cessation of shear are illustrated in Figure 10.11. All three samples exhibit much shorter time to initiate crystalline reflections compared with those crystallized under quiescent conditions. This confirms that the application of shear plays an important role in promoting the crystallization of iPP. However, for PPG0, almost no anisotropic diffraction rings can be observed. This can be understood for the following reasons. The applied shear is relatively weak (shear rate 20 s^{-1} for 5 s), and the degree of supercooling is not too great ($\Delta T = 20$ °C), as the normal melting temperature of α-iPP is 165 °C). This results in highly flexible polymer chains in a weakly oriented melt. The rapid relaxation of these chains leads to the formation of unoriented crystals. This is consistent with the stage of shear-induced oriented precursors breaking down into point-like nuclei at high T_c, resulting in isotropic growth of iPP crystals.[38] In contrast, 2D WAXD patterns of PPG05 and PPG10 exhibit strong evidence of crystal orientation (see Figures 10.11b and 10.11c). Clear reflection arcs are seen in the meridian, indicating oriented crystals are formed. The orientation of these crystals is consistent with the epitaxial growth from the row-nuclei, *i.e.* it is perpendicular to the flow

Table 10.1 Half-crystallization time ($t_{1/2}$) of PPG0, PPG05 and PPG10 isothermally crystallized at 145 °C under quiescent conditions and after step shear.

	$t_{1/2}$ *under quiescent condition (min)*	$t_{1/2}$ *under shear condition (min)*
PPG0	150.3	29.5
PPG05	70.5	3.7
PPG10	46.8	2.6

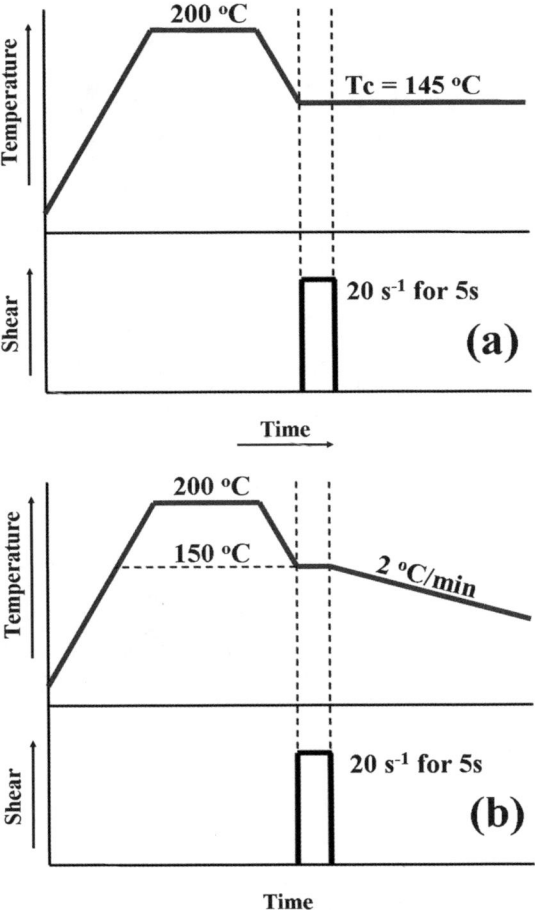

Figure 10.10 Schematic of shear conditions and thermal history as a function of time for WAXD and SAXS experiments: (a) isothermal crystallization and (b) non-isothermal crystallization.

direction. The growth of oriented crystals in PPG05 and PPG10 can be attributed to the combined effects of shear and GNSs. This phenomenon has also been observed in another system, where the stability of flow-induced precursors was found to improve in the presence of nanoparticles.[39] Due to the polymer chain–nanoparticle interaction, precursor formation and chain alignment can be enhanced, especially in the case of anisotropic nanoparticles. Under the shear field, the sheets of GNSs with 2D flaky geometry are likely to align parallel along the flow direction and form an oriented network, where the motion of extended iPP chains is restricted. In other words, GNSs exhibit a synergistic effect with shear in promoting the crystallization and orientation of iPP crystals.

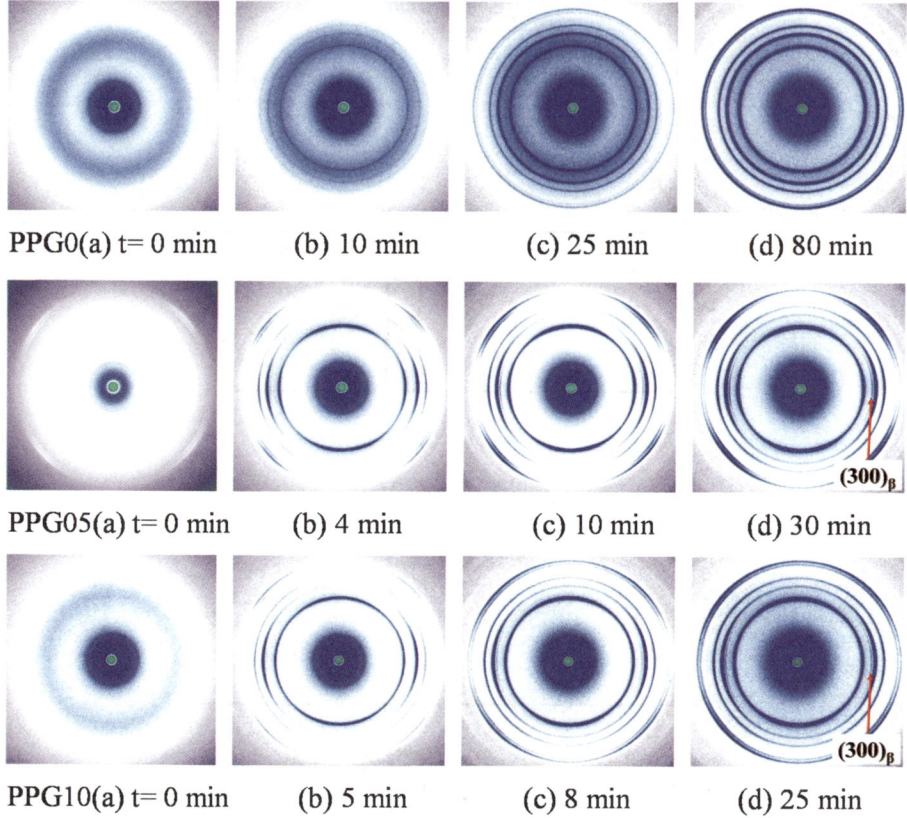

PPG0(a) t= 0 min   (b) 10 min   (c) 25 min   (d) 80 min

PPG05(a) t= 0 min   (b) 4 min   (c) 10 min   (d) 30 min

PPG10(a) t= 0 min   (b) 5 min   (c) 8 min   (d) 25 min

Figure 10.11   Selected 2D WAXD patterns of neat iPP and its GNSs isothermally crystallized at 145 °C after cessation of shear (at a shear rate of 20 s^{-1} for 5 s). The shear direction is vertical.

Figure 10.12 shows the time evolution of integrated 1D WAXD profiles of neat iPP and its GNS nanocomposites isothermally crystallized at 145 °C after shear. For PPG0 (Figure 10.12a), only α-crystals appear, even though the shear flow often promotes the formation of β-crystals. As seen in Figure 10.11, the rapid relaxation of the row-nuclei may occur in a low degree of supercooling; thus, the relaxed row-nuclei would lose their β-nucleating ability.[40] It is interesting to note that a small fraction of β-crystals can be observed in the PPG05 and PPG10 samples. We attribute this behaviour to the restrictive effect of GNSs, which preserves the self-orientation nuclei (shish) that can induce the growth of β-crystals.

Figure 10.13 shows the normalized crystallinity as a function of time of PPG0, PPG05 and PPG10 after shear. Clear enhancement of the crystallization kinetics of iPP is obtained both in neat iPP and in its GNS nanocomposites. As illustrated in Table 10.1, $t_{1/2}$ for PPG0 (29.5 min) after shear is reduced to about 20% of the value for PPG0 under quiescent

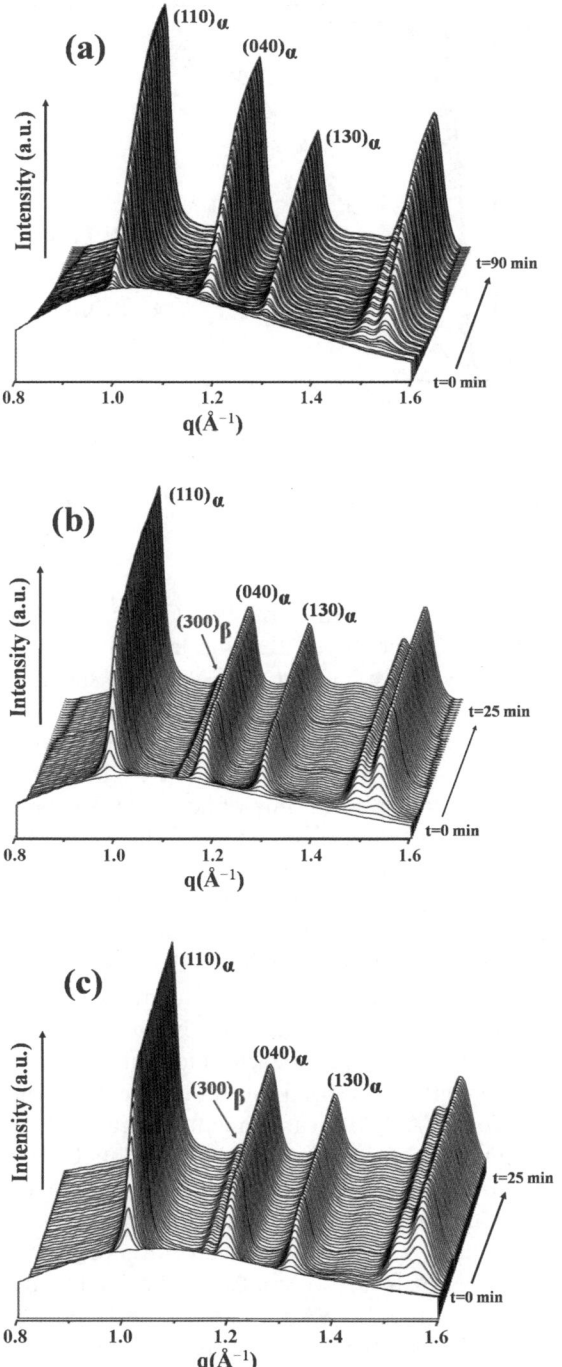

Figure 10.12 Time evolution of 1D WAXD profiles of (a) PPG0, (b) PPG05 and (c) PPG10 isothermally crystallized at 145 °C after the cessation of shear.

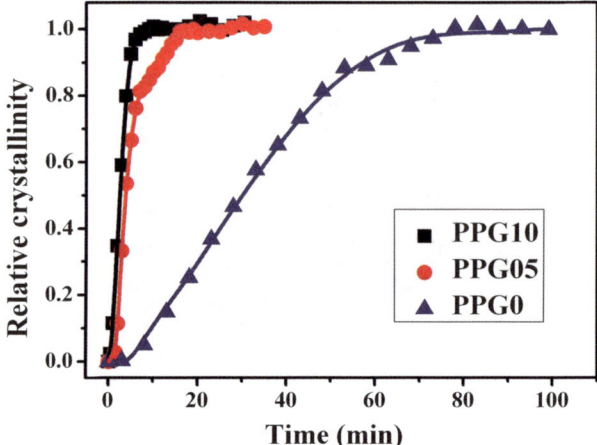

Figure 10.13   Normalized crystallinity of PPG0, PPG05 and PPG10 as a function of
time, which were isothermally crystallized at 145 °C after the cessation
of shear.

conditions. The crystallization rate of iPP–GNSs after shear, however,
increases by at least 18 times over that under quiescent conditions. This
verifies that there is a synergistic effect of GNSs and shear in promoting the
crystallization of iPP. Generally, the addition of nanoparticles can increase the
viscosity of the system and prolong the relaxation time of molecular
chains.[39,41] Thus, it is logical to consider that the enhanced nucleating sites
for growth of iPP crystals consist not only of heterogeneous agents (GNSs) but
also self-nucleation (rod-like nuclei or shish) induced by shear. These self-
formation nuclei (homogeneous nuclei) further accelerate the crystallization
rate of PPG05 and PPG10 under shear field compared to those crystallized
under quiescent conditions (Table 10.1). Under shear, the $t_{1/2}$ value of PPG05
and PPG10 falls from 3.7 min to 2.6 min when the GNS content changes from
0.05 wt% to 0.1 wt%. This indicates that the content effect of GNS on the
crystallization rate under shear is not very pronounced.

To better understand the structure development in iPP–GNSs during shear,
in situ SAXS measurement was carried out under the same thermomechanical
conditions as *in situ* WAXD measurement (Figure 10.2a). Figure 10.14
illustrates selected 2D SAXS patterns of neat iPP and its GNS nanocomposites
isothermally crystallized at 145 °C after shear. In all three systems, a pair of
weak scattering maxima appears immediately in the meridian and becomes
stronger with time. There is no sign of equatorial streak, which is probably
because the size of the shish is too small and below the detection limit of
SAXS.[16,42–43] For PPG0, the 2D SAXS patterns indicate obvious structural
anisotropy, rather different from the corresponding 2D WAXD results
(Figure 10.11). The reason for these differences may be partially due to the
different detecting scales of WAXD and SAXS measurements.[16,43–44]
However, it is also conceivable that although the crystal structure appears to

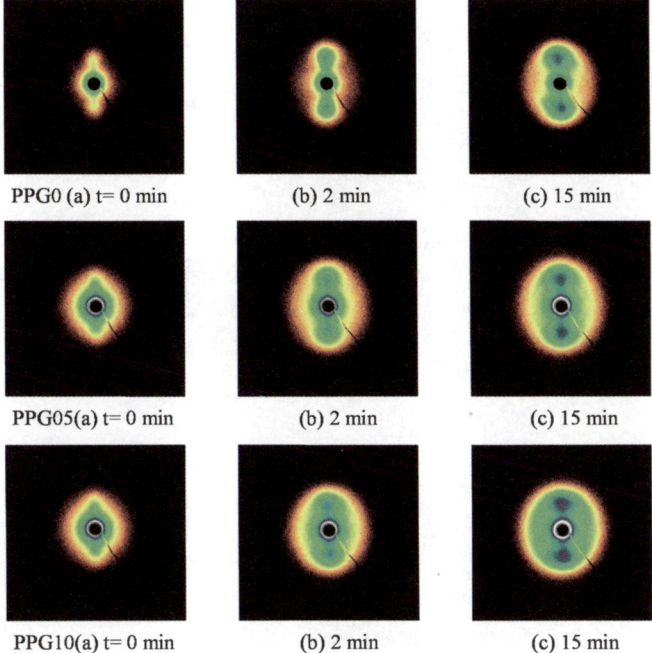

PPG0 (a) t= 0 min   (b) 2 min   (c) 15 min

PPG05(a) t= 0 min   (b) 2 min   (c) 15 min

PPG10(a) t= 0 min   (b) 2 min   (c) 15 min

Figure 10.14 Selected 2D SAXS patterns of PPG0, PPG05 and PPG10 isothermally crystallized at 145 °C after the cessation of shear (at a rate of 20 s^{-1} for 5 s).

be random, the large arrangement of lamellae possess some preferred orientation. The vertical (meridional) scattering streak indicates that the shear-induced oriented structures are preserved at high T_c even after subsequent crystallization.

Time-resolved integrated 1D SAXS profiles of neat iPP and its nanocomposites isothermally crystallized at 145 °C after shear are illustrated in Figure 10.15. It is shown that the scattered intensity gradually increases with time for all three systems, and it reaches a plateau value indicating the completion of crystallization. All profiles in Figures 10.15a–c are plotted on the same time scale, so we can conclude that GNSs show a strong ability to enhance the crystallization rate of iPP, which is consistent with the WAXD results.

Figure 10.16 illustrates the changes of average spacing between the adjacent lamellae (*i.e.* long period, L_B) in PPG0, PPG05 and PPG10. In all three systems, L_B is found to decrease rapidly at the early stage of crystallization and then reaches a plateau value at the late stage (*e.g.* L_B of PPG05 decreases from 36.9 nm to 22.4 nm). It is further noted that L_B for PPG05 (22.4 nm) and PPG10 (23.0 nm) is slightly smaller than that for PPG0 (24.6 nm). This observation indicates that higher nucleation density (due to the presence of GNSs) leads to smaller long period.[42] The difference between the plateau

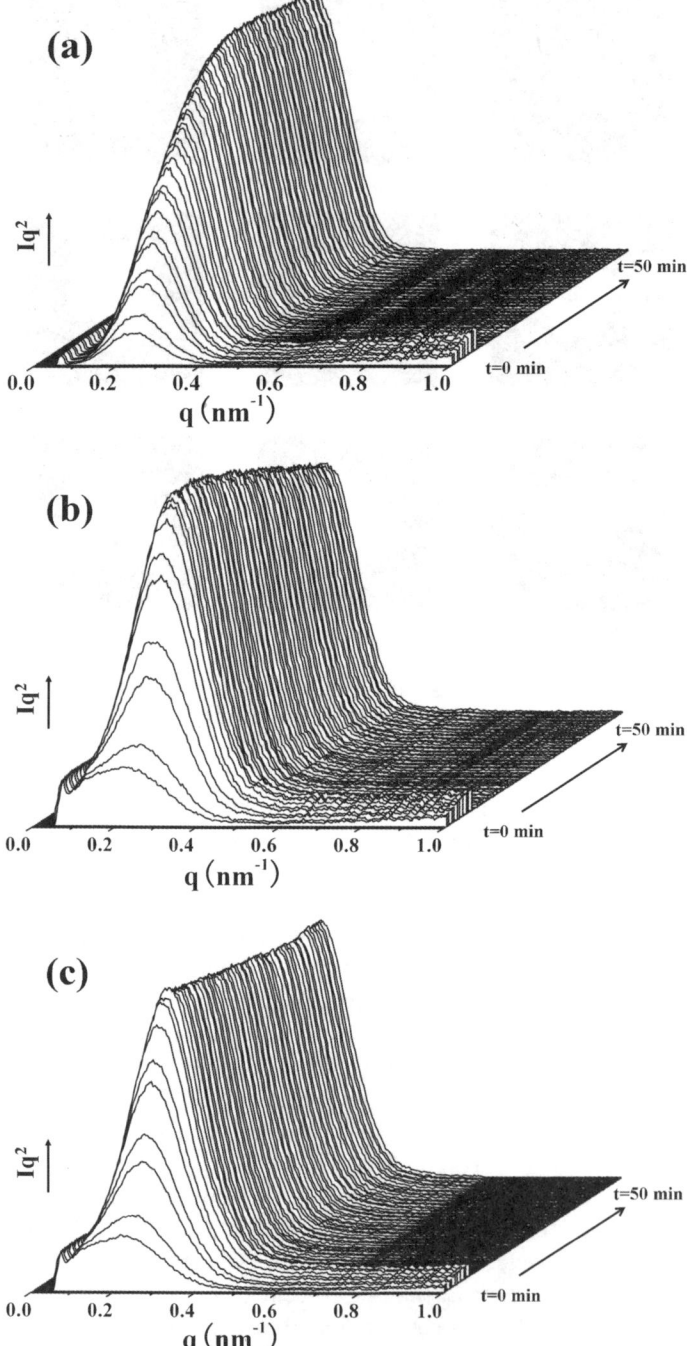

Figure 10.15 Time-resolved integrated 1D SAXS profiles of (a) PPG0, (b) PPG5 and (c) PPG10 isothermally crystallizing at 145 °C after shear.

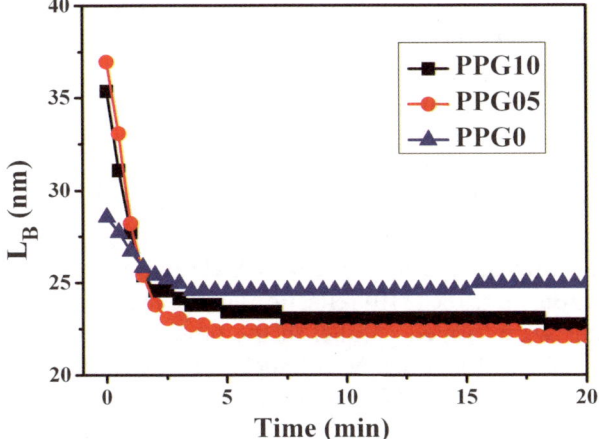

Figure 10.16 Plots of long period as a function of crystallization time for PPG0, PPG05 and PPG10 after the cessation of shear.

values of L_B for PPG0, PPG05 and PPG10 confirms that GNSs are good nucleating agents for iPP crystallization even under shear.

10.3.3 Combined Effect of GNSs and Shear Flow on Crystallization of iPP

The results described above provided a microscopic insight into the crystallization behaviour of iPP–GNSs under shear. As shown in Figure 10.1, it is confirmed that GNSs are 2D nanosheets with a very large specific surface area. With such geometry, the molecular chains are prone to interact with the surface of GNSs and reduce the nucleation barrier of iPP. The surface of GNSs can attract nearby iPP chains, allowing them to form a conformational ordered structure that would trigger the crystallization process.[27] As a result, both the induction period and the crystallization rate of iPP can be significantly affected.

When a relatively weak shear field (at a rate of 20 s^{-1} for 5 s) is applied, the crystallization kinetics of all three systems is accelerated (see Table 10.1). For PPG0 crystallized after the cessation of shear, isotropic diffraction rings (instead of anisotropic scattering features) are seen (Figure 10.11). This may be explained as follows. Generally, only polymer molecules above the critical orientation molecular weight (M^*) could remain oriented after shear at a given shear rate ($\dot{r}$), following the relationship $M^* \propto \dot{r}^{-\alpha}$ where α is a positive exponent.[45] The memory of the flow field can be erased if the sheared sample is kept quiescent at a high temperature for a sufficiently long time. This is because the flow-induced precursors are intrinsically unstable at high temperatures.[18] Considering the case of sheared PPG0, the sample was crystallized at a low degree of supercooling ($\Delta T = 20\ °C$), where the mobility of iPP chains was high. This would result in an isotropic crystal structure due to two possibilities: (1) the relaxation of shear-induced row-nuclei leads to a random crystal structure (this has also been found in the studies of poly(butylene terephthalate) and isotactic polystyrene);[17,46] (2) the size and concentration of oriented crystals is too small to be detected by WAXD. However, the lamellar structure could remain with a certain degree of orientation in the flow direction, which is observable by SAXS (Figure 10.14).

For PPG05 and PPG10 crystallized after shear, it was found that anisotropic diffraction rings appeared clearly in 2D WAXD patterns (Figure 10.11). The arc-like diffractions (*e.g.* the (110) reflection) in the meridian indicated that the oriented crystals were formed parallel to the flow direction. This observation confirms that GNSs dramatically enhance the crystallization kinetics and also affect the morphology. García-Gutiérrez *et al.* also reported that the orientation of poly(butylene terephthalate) crystals under shear changed dramatically in the presence of carbon nanotubes (CNTs).[24] Both poly(butylene terephthalate) (PBT) chains and CNTs were found to align along the flow direction. The epitaxial crystallization of PBT was able to occur on the surface of CNTs, resulting in the NHSK structure. Compared with CNTs, GNSs exhibit 2D flaky geometry. The direct formation of the NHSK structure seems unreasonable because the length of GNSs (about 0.87 μm) is too large for kebabs to vegetate, but the epitaxial growth of iPP on the surface of GNSs is evident due to the significant increase of crystallization rate of iPP under quiescent conditions. In addition, it is well known that a network structure can be formed by the interactions of nanoparticles in the polymer matrix, where this network also restricts the motion of molecular chains.[25,47–49] With the application of shear, the isotropic network structure (due to the nanoparticle interactions) can transform into an antisotropic one, where some GNS sheets become parallel to the flow direction, allowing some iPP chains to also align accordingly. Additionally, the relaxation time of the oriented polymer chains is prolonged due to the restriction effect of GNSs, where the row-nuclei can survive longer and evolve into the shish entity.

There should be two kinds of nucleating sites in sheared iPP–GNSs: (1) the heterogeneous nucleating sites provided by GNSs, which are inclined to align along the flow field; (2) the self-nucleating sites (or homogeneous nucleating sites, in the form of row-nuclei) due to the effect of shear. By the cooperative interactions of these two types of nucleating sites, faster crystallization kinetics can be induced in sheared PPG05 and PPG10 when compared to that in sheared PPG0. The WAXD results strongly suggest that there is a synergistic effect of GNSs and shear flow on the crystallization behaviour of iPP. Meanwhile, the relatively small dependence of crystallization rate of iPP on the concentration of GNSs may be due to the limited amount of preserved row-nuclei formed in the GNS network (Figure 10.13 and Table 10.1).

From Figure 10.11, another phenomenon was noted: the appearance of β-crystals in both iPP–GNS samples (however, no β-crystals formed in pure iPP). To confirm the origin of these β-crystals, we first discuss why no β-crystals were found in the sheared PPG0. Figure 10.17 illustrates the azimuthal profiles of 2D WAXD patterns of PPG0 for the $(110)_\alpha$ reflection at the initial stage of isothermal crystallization after shear. As stated earlier, row-nuclei can relax into point-like nuclei in sheared PPG0 at high temperature, causing the isotropic crystal growth (Figure 10.11), whereas large oriented lamellar structures are also present on the large scale (Figure 10.14). A similar phenomenon has been reported previously. Varga *et al.* reported that by pulling glass fibre in iPP melt, α row-nuclei could be developed, and this surface could induce the growth of β-crystals leading to the formation of cylindrite.[40] However, the capability of α row-nuclei to induce β-crystals was lost by repeated crystallization and melting. The relaxation of row-nuclei may also decrease the ability to induce β-crystals. Therefore, only the α-crystals appeared in PPG0 isothermally crystallized after shear, while in PPG05 and PPG10 relatively strong anisotropy could be observed at the initial stage of crystallization, which also suggests that oriented nuclei play an important role in the formation of β-crystals.

A non-isothermal crystallization experiment, which was expected to preserve row-nuclei by subsequent cooling after shear, was then carried out as shown in Figure 10.10b. Figure 10.18a shows the initial azimuthal distribution of PPG0 ($t = 4$ min) and PPG10 ($t = 3$ min), where clear crystal orientation is observed in PPG0, verifying that shear-induced row-nuclei were retained and further developed into oriented crystals. The final integrated 1D WAXD profiles of PPG0 and PPG10 at the end of the non-isothermal study (80 °C) are shown in Figure 10.18b. It can be seen that a fair amount of β-crystals emerges in PPG0. This evidence definitively confirms that row-nuclei survived by the cooling

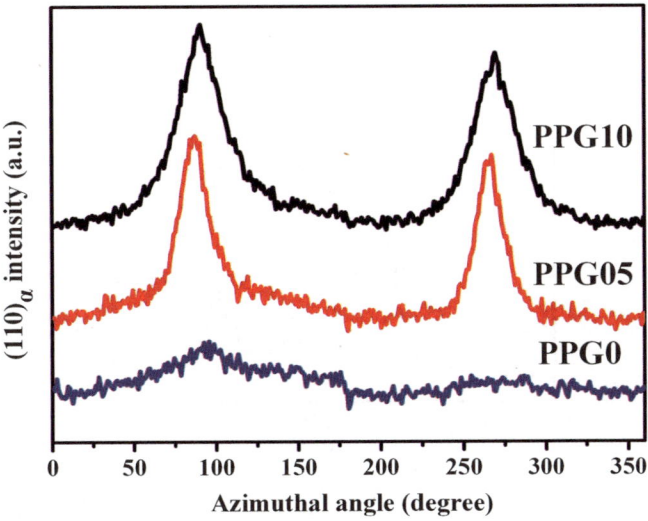

Figure 10.17 Azimuthal profiles of 2D WAXD patterns at the $(110)_\alpha$ reflection for PPG0, PPG05 and PPG10 in the initial stage of crystallization.

treatment, where the presence of row-nuclei induced β-crystals. However, although anisotropy in crystal structure is observed at the initial period of crystallization in PPG10, no detectable β-crystals can be observed. This means that the influence of GNSs on the formation of α-crystals overwhelms that of β-crystals, which can be induced by the survival of row-nuclei. It is conceivable that the heterogeneous GNS nucleating sites do not favour the formation of β-crystals.

According to the results above, the following conclusions can be drawn as follows. GNSs have a strong ability to induce the crystallization of α-crystals, while the application of shear can induce row-nuclei. However, the stability of the row-nuclei is enhanced by the confinement of the GNS network, thus holding the ability to induce β-crystals. There is direct competition between the two nucleating sites (GNSs and row-nuclei of iPP). GNSs manifest more potential on the nucleation ability for iPP than row-nuclei induced by shear.

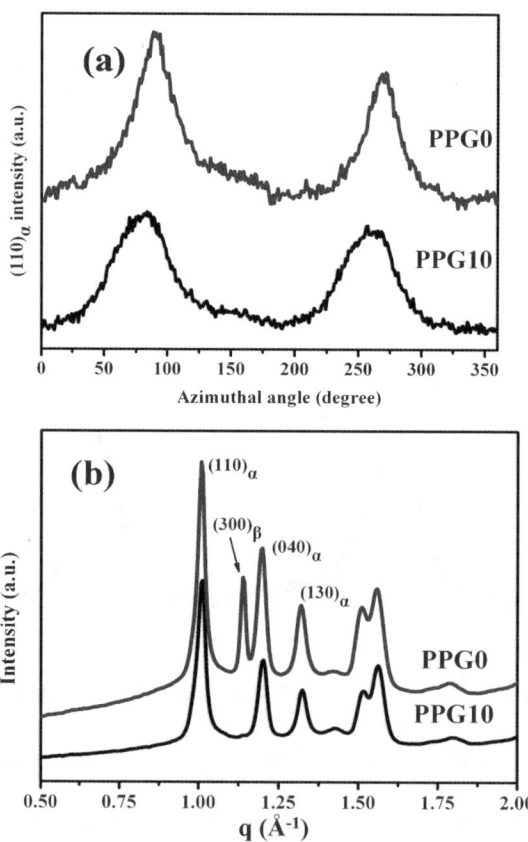

Figure 10.18   Initial azimuthal distribution for (a) $(110)_\alpha$ reflection and (b) final integrated 1D WAXD profiles of PPG0 and PPG10 during non-isothermal crystallization process after the cessation of shear (20 s^{-1} for 5 s at 150 °C).

As a result, when shear is applied to iPP–GNSs, the dominant formation of α-crystals can be attributed to the nucleation by GNSs. The appearance of a small amount of β-crystals is due to the row-nuclei surviving through the restricting effect of the GNS network on molecular chains.[50]

10.4   Summary

The crystallization properties of iPP–GNS nanocomposites have been discussed at the molecular level and under shear condition. GNSs exhibited strong nucleating ability in accelerating the crystallization rate of iPP. The half-crystallization time of nanocomposites only containing 0.05 wt% GNSs was reduced by at least 50% compared to that of neat iPP. Conformational ordering was demonstrated to play a vanguard role in iPP crystallization, especially at the very early stage. The presence of GNSs facilitated the formation of long helical structures, which was suggested to be the reason for GNS-induced iPP crystallization. As relatively weak shear (at a rate of 20 s^{-1} for 5s at T_c = 145 °C) was applied to the iPP melt, obvious evidence of crystal orientation for iPP–GNS nanocomposites appeared compared to neat iPP. GNSs presented an amplification effect of shear on the shear-induced row-nuclei and orientation of iPP crystals, where two kinds of nucleating origins coexisted in sheared nanocomposites: heterogeneous nucleating sites (GNSs); self-nucleating sites (row-nuclei) induced by shear. GNSs and shear flow exhibited a synergistic effect on promoting the crystallization kinetics of iPP. The results from this study offered beneficial help for understanding graphene-induced iPP crystallization and controlling the crystalline structure of iPP–GNSs.

Acknowledgement

The authors are grateful for financial support from the National Outstanding Youth Foundation of China (Grant No. 50925311) and the National Natural Science of China (Grant No. 51120135002, 51121001).

References

1. J. R. Potts, D. R. Dreyer, C. W. Bielawski, R. S. Ruoff, *Polymer*, 2011, **52**, 5.
2. H. Bai, C. Li, G. Q. Shi, *Adv. Mater.*, 2011, **23**, 1089.
3. H. Kim, A. A. Abdala, C. W. Macosko, *Macromolecules*, 2010, **43**, 6515.
4. T. Kuilla, S. Bhadra, D. H. Yao, N. H. Kim, S. Bose, J. H. Lee, *Prog. Polym. Sci.*, 2010, **35**, 1350.
5. B. P. Grady, F. Pompeo, R. L. Shambaugh, D. E. Resasco, *J. Phys. Chem. B*, 2002, **106**, 5852.
6. S. Z. D. Cheng, B. Lotz, *Polymer*, 2005, **46**, 8662.
7. M. Imai, K. Kaji, T. Kanaya, *Phys. Rev. Lett.*, 1993, **71**, 4162.

8. J. Baert, P. Van Puyvelde, *Macromol. Mater. Eng.*, 2008, **293**, 255.
9. G. Strobl, *Prog. Polym. Sci.*, 2006, **31**, 398.
10. X. Hu, H. N. An, Z. M. Li, Y. Geng, L. B. Li, C. L. Yang, *Macromolecules*, 2009, **42**, 3215.
11. H. S. Xu, X. J. Dai, P. R. Lamb, Z. M. Li, *J. Polym. Sci., Part B: Polym. Phys.*, 2009, **47**, 2341.
12. J. M. Zhang, H. Tsuji, I. Noda, Y. Ozaki, *J. Phys. Chem. B*, 2004, **108**, 11514.
13. Á. Kmetty, T. Bárány, J. Karger-Kocsis, *Prog. Polym. Sci.*, 2010, **35**, 1288.
14. G. Kalay, M. J. Bevis, *J. Polym. Sci., Part B: Polym. Phys.*, 1997, **35**, 241.
15. G. J. Zhong, L. B. Li, E. Mendes, D. Byelov, Q. Fu, Z. M. Li, *Macromolecules*, 2006, **39**, 6771.
16. R. H. Somani, L. Yang, B. S. Hsiao, P. K. Agarwal, H. A. Fruitwala, A. H. Tsou, *Macromolecules*, 2002, **35**, 9096.
17. L. B. Li, W. H. de Jeu, *Macromolecules*, 2004, **37**, 5646.
18. F. Azzurri, G. C. Alfonso, *Macromolecules*, 2005, **38**, 1723.
19. T. Hashimoto, H. Murase, Y. Ohta, *Macromolecules*, 2010, **43**, 6542.
20. G. Kumaraswamy, A. M. Issaian, J. A. Kornfield, *Macromolecules*, 1999, **32**, 7537.
21. J. Jancar, J. F. Douglas, F. W. Starr, S. K. Kumar, P. Cassagnau, A. J. Lesser, S. S. Sternstein, M. J. Buehler, *Polymer*, 2010, **51**, 3321.
22. A. Rozanski, B. Monasse, E. Szkudlarek, A. Pawlak, E. Piorkowska, A. Galeski, J. M. Haudin, *Eur. Polym. J.*, 2009, **45**, 88.
23. W. R. Hwang, G. W. M. Peters, M. A. Hulsen, H. E. H. Meijer, *Macromolecules*, 2006, **39**, 8389.
24. M. C. Garcia-Gutierrez, J. J. Hernandez, A. Nogales, P. Pantine, D. R. Rueda, T. A. Ezquerra, *Macromolecules*, 2008, **41**, 844.
25. N. Patil, L. Balzano, G. Portale, S. Rastogi, *Carbon*, 2010, **48**, 4116.
26. J. Z. Xu, T. Chen, C. L. Yang, Z. M. Li, Y. M. Mao, B. Q. Zeng, B. S. Hsiao, *Macromolecules*, 2010, **43**, 5000.
27. J. Z. Xu, Y. Y. Liang, G. J. Zhong, H. L. Li, C. Chen, L. B. Li, Z. M. Li, *J. Phys. Chem. Lett.*, 2012, **3**, 530.
28. X. Y. Zhu, D. Y. Yan, Y. P. Fang, *J. Phys. Chem. B*, 2001, **105**, 12461.
29. H. N. An, B. J. Zhao, Z. Ma, C. G. Shao, X. Wang, Y. P. Fang, L. B. Li, Z. M. Li, *Macromolecules*, 2007, **40**, 4740.
30. T. Shimada, M. Doi, K. Okano, *J. Chem. Phys.*, 1988, **88**, 7181.
31. Y. Geng, G. L. Wang, Y. H. Cong, L. G. Bai, L. B. Li, C. L. Yang, *Macromolecules*, 2009, **42**, 4751.
32. Y. H. Cong, Z. H. Hong, Z. M. Qi, W. M. Zhou, H. L. Li, H. Liu, W. Chen, X. Wang, L. B. Li, *Macromolecules*, 2010, **43**, 9859.
33. I. Noda, *Appl. Spectrosc.*, 1993, **47**, 1329.
34. I. Noda, A. E. Dowrey, C. Marcott, G. M. Story, Y. Ozaki, *Appl. Spectrosc.*, 2000, **54**, 236A.
35. K. B. Lu, N. Grossiord, C. E. Koning, H. E. Miltner, B. van Mele, J. Loos, *Macromolecules*, 2008, **41**, 8081.

36. S. Cheng, X. Chen, Y. G. Hsuan, C. Y. Li, *Macromolecules*, 2011, **45**, 993.
37. H. N. An, X. Y. Li, Y. Geng, Y. L. Wang, X. Wang, L. B. Li, Z. M. Li, C. L. Yang, *J. Phys. Chem. B*, 2008, **112**, 12256.
38. A. Nogales, G. R. R. Mitchell, A. S. Vaughan, *Macromolecules*, 2003, **36**, 4898.
39. N. Patil, L. Balzano, G. Portale, S. Rastogi, *Macromolecules*, 2010, **43**, 6749.
40. J. Varga, J. KargerKocsis, *J. Polym. Sci., Part B: Polym. Phys.*, 1996, **34**, 657.
41. K. Wang, Y. Xiao, B. Na, H. Tan, Q. Zhang, Q. Fu, *Polymer*, 2005, **46**, 9022.
42. L. Yang, R. H. Somani, I. Sics, B. S. Hsiao, R. Kolb, H. Fruitwala, C. Ong, *Macromolecules*, 2004, **37**, 4845.
43. R. H. Somani, L. Yang, B. S. Hsiao, T. Sun, N. V. Pogodina, A. Lustiger, *Macromolecules*, 2005, **38**, 1244.
44. Z. G. Wang, B. S. Hsiao, E. B. Sirota, P. Agarwal, S. Srinivas, *Macromolecules*, 2000, **33**, 978.
45. R. H. Somani, B. S. Hsiao, A. Nogales, S. Srinivas, A. H. Tsou, I. Sics, F. J. Balta-Calleja, T. A. Ezquerra, *Macromolecules*, 2000, **33**, 9385.
46. M. C. G. Gutierrez, G. C. Alfonso, C. Riekel, F. Azzurri, *Macromolecules*, 2004, **37**, 478.
47. J. T. Xu, Y. Q. Zhao, Q. Wang, Z. Q. Fan, *Macromol. Rapid Commun.*, 2005, **26**, 620.
48. M. D'Haese, P. Van Puyvelde, F. Langouche, *Macromolecules*, 2010, **43**, 2933.
49. L. M. C. Dykes, J. M. Torkelson, W. R. Burghardt, R. Krishnamoorti, *Polymer*, 2010, **51**, 4916.
50. J. Z. Xu, C. Chen, Y. Wang, H. Tang, Z. M. Li, B. S. Hsiao, *Macromolecules*, 2011, **44**, 2808.

Subject Index